Synchrotron Radiation Techniques in Industrial, Chemical, and Materials Science

Library of Congress Cataloging-in-Publication Data

Synchrotron radiation techniques in industrial, chemical, and
materials science / edited by Kevin L. D'Amico, Louis J. Terminello,
and David K. Shuh.
     p.   cm.
  "Proceedings of the combined symposia on applications of
synchrotron research to materials science, held August 21-23, 1994,
in Washington, D.C., and applications of synchrotron radiation in
chemistry and related fields, held August 27-September 1, 1995, in
Chicago, Illinois"--T.p. verso.
  Includes bibliographical references and index.
  ISBN 978-0-306-45389-2   ISBN 978-1-4615-5837-8 (eBook)

  DOI 10.1007/978-1-4615-5837-8
  1. Synchrotron radiation--Industrial applications--Congresses.
I. D'Amico, Kevin L. II. Terminello, Louis J. III. Shuh, David K.
TP249.S975   1996
660--dc20   96-41921
                                                       CIP

Proceedings of the combined symposia on Applications of Synchrotron Research to Materials
Science, held August 21 – 23, 1994, in Washington, D.C.; and Applications of Synchrotron
Radiation in Chemistry and Related Fields, held August 27 – September 1, 1995, in Chicago,
Illinois

ISBN 978-0-306-45389-2

© 1996 Springer Science+Business Media New York
Originally published by Plenum Press,New York in 1996

Synchrotron Radiation Techniques in Industrial, Chemical, and Materials Science

Edited by

Kevin L. D'Amico
X-Ray Analytics, Ltd.
Hinsdale, Illinois

Louis J. Terminello
Lawrence Livermore National Laboratory
Livermore, California

and

David K. Shuh
Lawrence Berkeley National Laboratory
Berkeley, California

Springer Science+Business Media, LLC

PREFACE

The individual papers that comprise this monograph are derived from two American Chemical Society (ACS) Fall National Meetings that focused on the current uses of synchrotron radiation (SR) research techniques. The first Symposium was held in Washington, DC, in August 1994, and the second convened in Chicago, IL, in August 1995. The intent of these symposia was to present a broad overview of several current topics in industrial, chemical, and materials-based SR research to a chemically inclined audience. The SR techniques covered were divided roughly into the three general fields of industrial, chemical, and materials science for this purpose. Included within these four categories are environmental, geologic, atomic/molecular, analytical, solid state physics, surface science, and biological applications of SR. There is little doubt that structural biology and environmental science are the largest growth areas in SR research as this monograph goes to press.

The spirit of these symposia was to bring together the expert synchrotron radiation user with new and potential users of SR techniques. There are now a preponderance of particle storage rings, located throughout the world, devoted exclusively to the production of SR. There have been great improvements in the particle accelerators and storage rings from which SR emanates. These newest third generation SR sources are the result of the successful collaboration between SR users and accelerator physicists which has made a reality out of experiments never before possible. Several thorough and detailed overviews of synchrotron radiation properties and its uses have been written by H. Winnick, G. Magaritondo, and others. The little orange X-Ray Data Booklet published by the Center for X-ray Optics at the Lawrence Berkeley National Laboratory (LBNL) has been a trusted reference companion to users of SR for many years. Most of the SR sources have World Wide Web (WWW) home pages that describe the particular attributes of the respective facilities and can be easily accessed via the X-ray WWW Server at http://xray.uu.se/.

The Editors would like to express their thanks to the authors for their contributions to this book, as well as for their excellent presentations at both ACS Meetings. The Editors would also like to thank the ACS, Divisions of Industrial and Nuclear Chemistry for sponsoring the session in 1994, and in particular Drs. Dale. L. Perry and Ralph Gatrone. We would also like to thank Dr. Patricia Baisden and the ACS Division of Nuclear Chemistry for their sponsorship of the Symposium in 1995.

Besides the presenting authors, several other individuals have contributed to this volume. We would like to thank Karen Sitzberger for her help with the organization of the symposia and the production of this book. Several colleagues helped with reviewing the contributions to this monograph: Prof. David Templeton, Dr. Mike Soltis, and Prof. G. Dan Waddill. We are also grateful to Dr. Neville Smith of the Advanced Light Source (LBNL) and Dr. Joe Dehmer from the Advanced Photon Source (Argonne National Laboratory) for overview talks on the future of SR research at their third generation facilities at the Meeting in 1995.

K. L. D'Amico
L. J. Terminello
D. K. Shuh

CONTENTS

INDUSTRIAL PROTEIN X-RAY CRYSTALLOGRAPHY:
AN OVERVIEW

Joel D. Oliver

The Procter & Gamble Company
Miami Valley Laboratories
P. O. Box 538707
Cincinnati, Ohio 45253-8707

INTRODUCTION

This presentation provides a change of direction in several respects relative to the other presentations in this symposium. First, rather than dealing with small chemical molecules, I will be considering complex biological macromolecules. Second, whereas the research described in the other presentations has typically been based on limited amounts of data, protein crystal structure determinations require vast amounts of data (from 15,000 to 25,000 independent measurements for a 30 kDa protein structure at 2.0 Å resolution). Third, the compounds reported in the other presentations are relatively stable both chemically and with respect to ionizing radiation. But proteins are highly sensitive to both chemical and radiation-induced degradation. And finally, this presentation will describe a general strategy to use synchrotron X-radiation to solve industrially important protein structures, whereas the other presentations have provided detailed case studies of how synchrotron X-ray data have been used to address industrially important problems.

As a final introductory comment, this presentation is truly an overview of industrial protein X-ray crystallography. It does not represent an exhaustive survey of the field and any oversights are unintentional.

BACKGROUND

The first question one might well ask is, "Why should industry be interested in pursuing protein X-ray crystallography?" Here are some of the reasons for that interest.

First, protein engineering efforts to modify proteins to provide improved properties are greatly benefited if the relation between protein structure and protein function is well understood. Protein crystallography is ideally suited to provide the required structural knowledge. Once the initial structure of the subject protein has been determined, the determination of subsequent structures of protein variants, produced by either site-directed or random mutagenesis, is normally rather straightforward and can provide results quickly enough to be useful to the research team.

Second, protein crystallography can play a vital role in the discovery of new drugs. Most of the drugs currently on the market were developed from lead molecules discovered by a synthesize-and-test procedure. That procedure is very expensive and slow to converge on a marketable product. Most pharmaceutical companies currently prefer to isolate the biological receptor for the disease state and develop synthetic compounds to interact with that receptor to mediate the disease progression. If the receptor and its complexes with drug candidates can be crystallized, then protein crystallography can provide the detailed knowledge to design subsequent compounds and in principle converge more quickly on the commercial product. This detailed knowledge regarding the interaction between the drug candidate and its receptor should result in drugs with better specificity and reduced side effects. This modern design procedure, sometimes referred to as structure-based drug design, has already been used successfully by pharmaceutical companies and the resultant drugs should appear on the market soon. The same design procedure can also be, and is being, used in the development of novel agricultural chemicals.

Third, protein crystallography can also lead to the understanding of the mechanism of action of enzymes and receptors. This knowledge may then suggest new industrial applications of those enzymes or of compounds that mediate receptor interactions.

Fourth, the structures of proteins can also guide efforts to resolve development and production problems, for instance to enhance production levels or eliminate undesirable post-translational modifications of proteins.

And finally, in an era of increasing government regulation, protein structures from crystallography may enhance the likelihood of governmental approval of a new commercial product that is based on or interacts with a biological macromolecule.

Other factors that have stimulated industrial interest in protein crystallography are the rapid increase (cf. Figure 1) in the number of known crystal structures in the public domain, as maintained in the Brookhaven Protein Data Bank (PDB),[1] and the diversity of biological functions dependent upon those proteins. The rate of growth of new PDB entries in 1993 was 1273 new protein structures. This spike in the growth rate resulted in a doubling of the number of publicly available protein structures and was the result of staffing increases and the conscientious efforts of the PDB staff to reduce the backlog of deposited structures. Obviously, the PDB staff did a marvelous job that year. Currently, the PDB is handling approximately 90 new protein structure depositions per month. Many of those structures have direct interest to industry and are now being vigorously pursued. And there are many other structures of macromolecules determined at academic and industrial laboratories that have not been made publicly available due to their commercial relevance.

This growth in the number of known protein structures parallels the historical development of the discipline as described in the preface to Max Perutz's book on protein structure.[2] In 1952, Dr. Perutz showed that the phase problem for protein structures could be solved via the technique of isomorphous replacement. For the successful application of that technique, one must first crystallize the protein and then produce at least two heavy-atom labeled isomorphous derivatives of the protein crystal. The locations of the heavy atoms in the two derivative crystals can then be determined from the diffraction data from the crystals and used to determine the phases of the native structure.

However, the application of this method to determine protein structures proceeded very slowly. By 1959, only two protein structures were known. By 1965, a third structure had been determined. By 1970, a total of eleven protein structures had been solved. Through 1977 new protein structures were determined at a rate of about six per year. By 1982 and

1985 , the rate was up to 10 and 17 new protein X-ray structures/year, respectively. By 1990 new protein structures were being produced from crystallography at the rate of more than 100 per year.

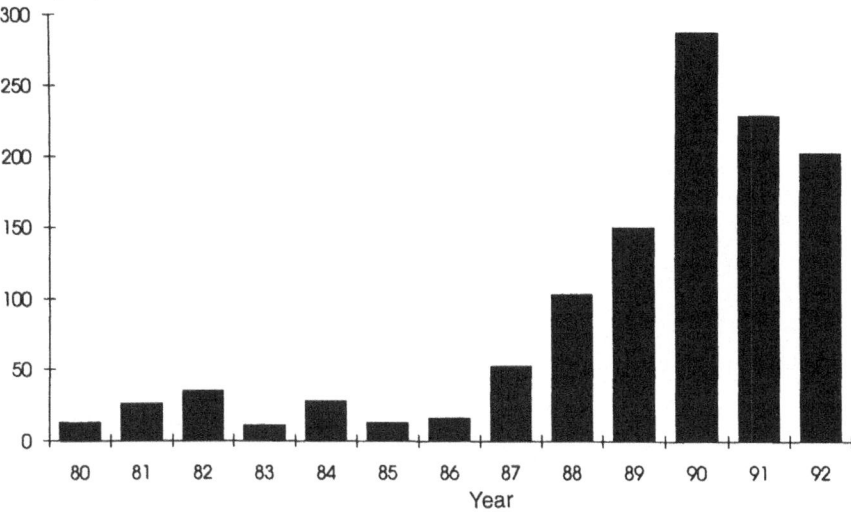

Figure 1. Graphical representation of the annual growth rate in the number of new entries in the Brookhaven Protein Data Bank.

As the total number of known protein structures and the rate of determining new protein structures both grew, industrial application became likely. Finally, in 1982 Merck hired the first industrial protein crystallographer, Dr. Manuel Navia. Very quickly other companies began to implement in-house protein crystallography and the numbers of industrial protein crystallographers and companies employing these scientist grew very rapidly as shown in Figure 2. Currently, world-wide there are approximately 190 protein crystallographers employed by about 60 companies.

This very rapid growth in protein crystallography during the last twelve years was the result of a number of factors. These factors include the advent of recombinant DNA technology. As a result of this technology large quantities of many new interesting proteins became available for investigation. Also, since these proteins are from recombinant sources, the purification schemes employed for purification are generally simpler and less damaging to the protein. Thus the homogeneity and reproducibility of the protein samples are much better than for samples of those same proteins isolated from their natural sources. Consequently, the probability for developing suitable crystallization conditions is much higher.

Advances in protein chemistry, such as HPLC, FPLC and affinity chromatography, have improved the purity and homogeneity of the protein samples available for crystallization experiments. Concomitantly, the prospects of obtaining X-ray quality crystals have improved.

The availability of synchrotron X-ray beamlines that are configured for convenient data collection for protein crystals has resulted in improved data quality and improved structural results. Since the data are collected so quickly at a synchrotron X-ray beamline, the protein crystals experience much less radiation-induced decomposition than those same crystals experience with even the best in-laboratory X-ray source. As a result, the data quality is better and usually the crystals will scatter to higher resolution, both of which improve the precision of the resultant crystal structure.

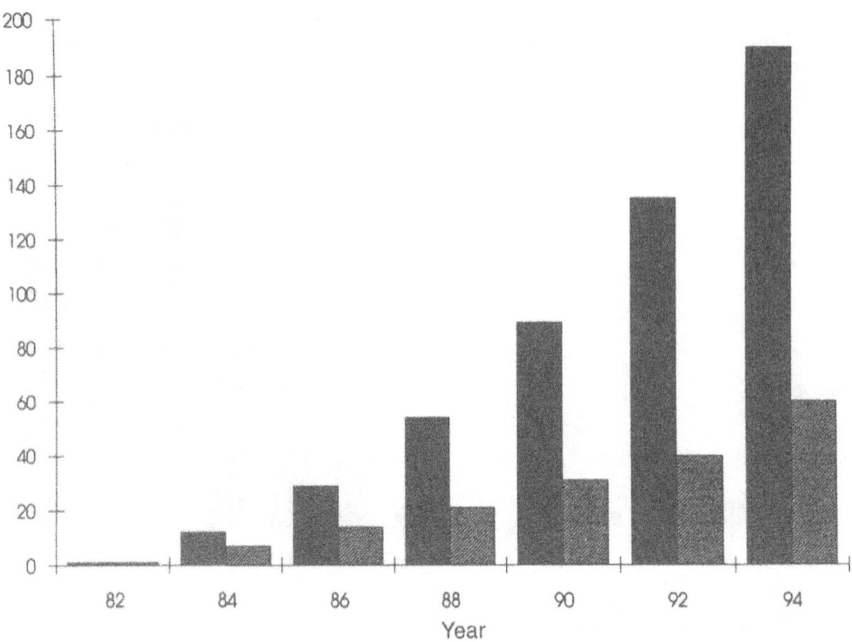

Figure 2. Growth rate of industrial protein crystallography. The solid bar gives the numbers of crystallographers and the slashed bar represents the number of companies.

Likewise, the development of reliable 2-D X-ray area detectors, crossed-wire proportional counters, image plates, and charge-coupled device detectors has also provided major improvements in X-ray data quality for protein crystals. These improvements are equally true for in-house, as well as synchrotron, X-ray sources and result from the increased speed of data collection and the high redundancy of the measured diffraction data.

The improvements in the performance:price ratio for computers and graphics workstations (by several orders of magnitude during the recent past) have virtually eliminated the cost barriers to performing the extensive calculations that are required to determine and refine a protein structure at high resolution. Calculations that required months of computer time and tens of thousands of dollars only ten years ago can now be done in only a few days in one's own laboratory with a modern graphics workstation.

In industry, the high consensus value of structure-based design of commercially important molecules has resulted in a rapidly growing commitment to protein crystallography. Once an initial structure of the target protein has been determined and if its complexes with compounds of interest can be crystallized (frequently using the same conditions as for the native protein crystals), 3-D structures can often be determined within a few weeks to a month after the compounds and protein samples are given to the crystallographer. This time frame is appropriate for the structures to provide the experimental basis for molecular modeling experiments which should converge on subsequent generations of molecules with enhanced properties and quickly identify a suitable commercial candidate molecule.

REPRESENTATIVE PROTEIN STRUCTURES DETERMINED FROM INDUSTRIAL MACROMOLECULAR X-RAY CRYSTALLOGRAPHY

An impressive number of protein structures have already been investigated in industrial X-ray laboratories. Representative examples are given in Table 1.

Table 1. Protein Crystal Structures Determined in Industrial Laboratories

- Detergent Enzymes
    - Subtilisin Proteases (Ciba-Geigy,[3] Genencor,[4] Genex,[5] Procter & Gamble[6])
- Acid Proteases
    - Human Renin (SmithKline,[7] Upjohn[8])
    - Porcine Pepsin (Abbott[9])
- Human Immunodeficiency Virus [HIV] Enzymes
    - HIV-1 Protease (Abbott,[10] Agouron,[11] Lilly,[12] Merck[13], SmithKline[14])
    - HIV-2 Protease (Upjohn[15])
    - HIV-1 Reverse Transcriptase, Ribonuclease H Domain (Agouron,[16] Upjohn[17])
- Simian Immunodeficiency Virus Protease (SmithKline[18])
- Phospholipase A-2 (Lilly,[19] Upjohn[20])
- Matrix Metalloproteases
    - Human Fibroblast Collagenase (Glaxo,[21] Roche,[22] Sterling Winthrop[23])
- Interleukins (Glaxo,[24] Hoffman LaRoche,[25] Upjohn,[26-28] Synergen[29])
- Rhinovirus Coat Protein (Sterling Winthrop[30])
- Rhinovirus 3C Protease (Agouron[31])
- Mammalian Growth Hormone (Genentech,[32] Lilly,[33] Monsanto[34])
- Purine Nucleoside Phosphorylase (BioCryst[35])
- Alkaline Phosphatase (Abbott[36])
- Thymidylate synthase (Agouron[37])
- Bifunctional Dihydrofolate Reductase-Thymidylate Synthase (Agouron[38])
- Thrombin/Inhibitor Complexes (Bristol-Myers Squibb,[39] Ciba-Geigy[40], DuPont-Merck,[41] Lilly[42])
- Cyclophilin/cyclosporin A (Sandoz[43])
- Transforming Growth Factor β2 (Ciba-Geigy[44])
- 5-Enol-pyruvylshikimate-3-phosphate Synthase (Monsanto[45])
- Glycinamide Ribonucleotide Transformylase (Agouron[46])
- DNA Polymerase β (Agouron[47])

These molecules include detergent enzymes in support of efforts to improve their properties through genetic engineering , the viral protease involved in AIDS, the enzyme involved early in the arachadonic acid cascade which produces inflammation, matrix metalloproteases which are involved in arthritic conditions, immune-active proteins, the virus responsible for the common cold (which will be discussed later in this presentation), a hormone to treat human dwarfism, a nucleoside-scavenging enzyme and the enzyme responsible for blood clotting and some severe cardiovascular circulatory disorders. For several of these proteins tens, and even hundreds, of crystal structures have been determined in attempts to tailor the properties of the enzymes or to develop molecules which effectively moderate the biological activities of those proteins. Most of these structures are highly proprietary and will not be publicly disclosed for many years, if ever.

THE VALUE OF SYNCHROTRON X-RADIATION

The extent of industrial involvement in protein X-ray crystallography and the contributions that have resulted are certainly impressive. However, since this presentation is part of a symposium on the industrial applications of synchrotron radiation, why is synchrotron X-radiation important for protein X-ray crystallographic work being conducted in industry? There are a number of benefits from using synchrotron X-radiation in the determination of protein crystal structures that result from the intrinsic properties of this unique radiation.

First, there are benefits that result from the extreme brilliance of synchrotron X-ray sources. As a consequence of this brilliance, smaller crystals can be examined. This is very important because it is quite common for one to initially obtain small crystals and much time and effort is then spent in growing crystals large enough for X-ray diffraction studies. Being able to use smaller crystals reduces the time required to determine the protein's structure.

Also, this greater brilliance generally increases the resolution limit to which protein crystals diffract. The precision of the structures that result from these data is better than would be obtained from the use of data collected with an in-laboratory X-ray source. This higher precision allows the fine details of structure/function relationships to be determined that could not be realized otherwise. Better understanding of the structure/function relationships may then lead to competitively superior products.

Another direct benefit of greater brilliance is a reduction in the extent of radiation damage that occurs during X-ray data collection. Since the rate of crystal decomposition is usually more strongly dependent on exposure time than on source brilliance, more high-quality data can be collected from a crystal before it decomposes to the point that its diffraction data are no longer useful. This effect reduces the systematic errors in the data that arise from decomposition corrections and the scaling of the intensity data collected from several crystals.

A reduction in the elapsed time to complete the diffraction data collection is also a result of the greater brilliance of synchrotron X-ray sources. If wide bandpasss Laue diffraction conditions are used, the time for data collection can be reduced to only a few seconds and time-resolved structural studies[48-50] may be possible for many, if not all, proteins of industrial importance. These time-resolved structure studies can provide atomic-scale understanding of how that protein produces its biological action. This detailed understanding of the protein's mechanism of action may then suggest novel ideas for mediating the biological action of that protein, which in turn may provide leads to potential commercial products.

The tunability of X-radiation from a synchrotron can enable the use of a more efficient structure determination procedure commonly known as multi-wavelength anomalous diffraction, or MAD phasing.[51] For MAD phasing to be possible, the protein must contain one or more heavy atoms (e.g. Se atoms from selenomethionyl residues or intrinsic transition metals) which give rise to an X-ray absorption edge within the spectral range of the synchrotron X-ray source. Diffraction data sets are then collected at three or four wavelengths near that absorption edge. Algebraic formulae are then used to extract the native phases from these multiple data sets. In principle MAD phasing is simpler and should provide structures more quickly than does the traditional multiple isomorphous replacement technique.

The pulsed nature of a synchrotron X-ray may be useful in performing time-resolved structural studies on very short (ms to ps) time frames. By using Laue diffraction conditions for data collection with a very brilliant source, such as a third-generation synchrotron X-ray source, one pulse, or at most a few pulses, may be sufficient to stimulate 55 to 75% of all the possible reflections from a protein crystal. Structural studies in these time frames will revolutionize our understanding of the mechanistic details which underlie the biological properties of many proteins. Hosts of new commercial applications may arise.

The extremely high natural collimation of X-radiation emitted from a synchrotron source can be used advantageously to investigate the structures of biological macromolecules which produce crystals with very large unit cells (200 to 1000 Å). Avoidance of overlap of adjacent reflections from these crystals is virtually impossible with even the best in-laboratory X-ray source, but is straightforward with a synchrotron X-ray source.

LOCATIONS OF APPROPRIATE SYNCHROTRON X-RADIATION SOURCES

Given then that synchrotron X-ray sources play an essential role in modern protein X-ray crystallography and that the geographical locations of these sources determine their accessibilities, where are the synchrotron X-ray sources that are equipped to collect diffraction data from protein crystals? A world map that illustrates these locations is shown in Figure 3.

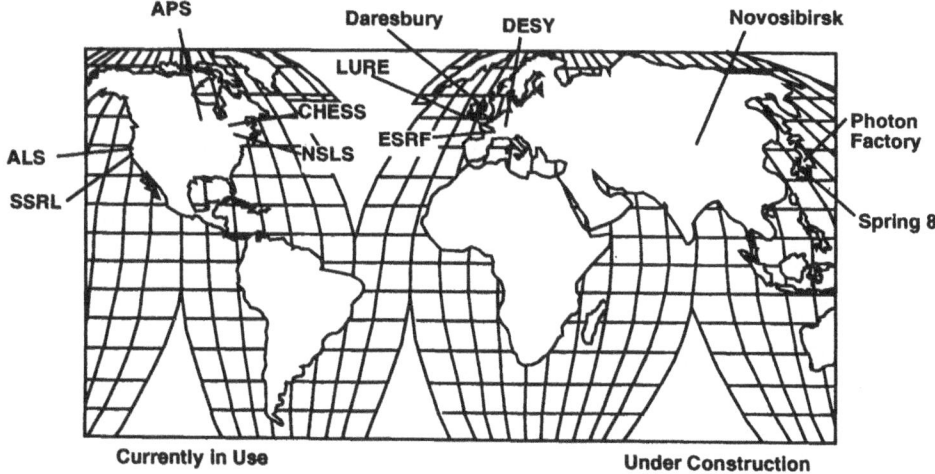

Figure 3. A world map showing the locations of all the synchrotron facilities with beamlines equipped for protein crystallography (the Spring 8 facility and the APS facility are currently under construction).

The current protein X-ray data collection facilities are located at the Photon Factory in Japan, at the synchrotron in Novosibirsk in the former U.S.S.R., at DESY in Germany, at LURE and the European Synchrotron Radiation Facility in France, at the Daresbury facility in the U.K., at the Cornell High Energy Synchrotron Source and the National Synchrotron Light Source in New York, and at the Advanced Light Source and the Stanford Synchrotron Radiation Laboratory in California. Protein X-ray data collection beamlines are also planned for the Spring 8 facility in Japan and the Advanced Photon Source (APS) in Chicago, both of which are now under construction.

THE INDUSTRIAL MACROMOLECULAR CRYSTALLOGRAPHY ASSOCIA-TION (IMCA)

Several years ago an ambitious plan was developed to build a third-generation synchrotron, the APS, on the campus of Argonne National Laboratories. This synchrotron source was designed to be a dedicated source of X-radiation. Some of its design parameters are ring energy of 7 GeV, beam current of 100 ma, ring circumference of 1110 meters, 35 bending magnet sources, 34 insertion devices, and a projected beam lifetime of 10 hours. The plan to locate the APS near Chicago where it would be easily accessed by industrial firms in the Midwest and on the East Coast of the U.S. prompted the formation of a consortium of companies now know as the IMCA. The purpose of the IMCA was to provide easy access by its member companies to protein X-ray diffraction data collection facilities at this third-generation X-ray synchrotron source. The complete rationale for forming the IMCA is given in Table 2.

Table 2. The Rationale for Forming the IMCA

- To gain rapid and frequent access to synchrotron X-ray data collection for both proprietary and nonproprietary research.
  - through participation in beamline X8C at NSLS;
  - through the development of user-friendly, highly-productive beamlines and onsite laboratory facilities at the APS.

- To promote competitiveness of industrial biotechnology and protein crystallography in the U.S.

- To provide a strong voice in the development of national user facilities for structural biology research.

IMCA MEMBER COMPANIES

The companies comprising the IMCA and the research focus of the association are described below.

IMCA Member Companies: Abbott Laboratories
Bristol-Myers Squibb
Glaxo, Inc.
Eli Lilly Company
Merck & Company
Miles, Inc.
Monsanto Company
Parke-Davis Pharmaceutical Research
The Procter & Gamble Distribution Company
SmithKline Beecham
Sterling Winthrop, Inc.
The Upjohn Company

Focus: Structural characterization of biological macromolecules to aid in the design of biologically active compounds for industrial use, understanding the mechanism of action of biological macromolecules and guiding protein engineering efforts to develop molecules with improved properties.

THE ADVANCED PHOTON SOURCE (APS)

The unique properties of the APS and the resultant possibilities for performing all the types of protein crystallographic research mentioned earlier in this presentation were the major driving forces for the formation of the IMCA. The APS site is shown in Figure 4.

Figure 4. Aerial photograph of the APS construction site showing the locations of components and facilities (notice the user residence facility which is located close to the synchrotron for the convenience of guest users).

The proximity of the APS to and its convenient access by the IMCA member companies was another important factor leading to the formation of the association. Figure 5 displays a map with the locations of the APS and the twelve member companies of the IMCA. Convenient access to the APS is provided either through commercial airlines or via personal automobiles for some of the more closely located companies.

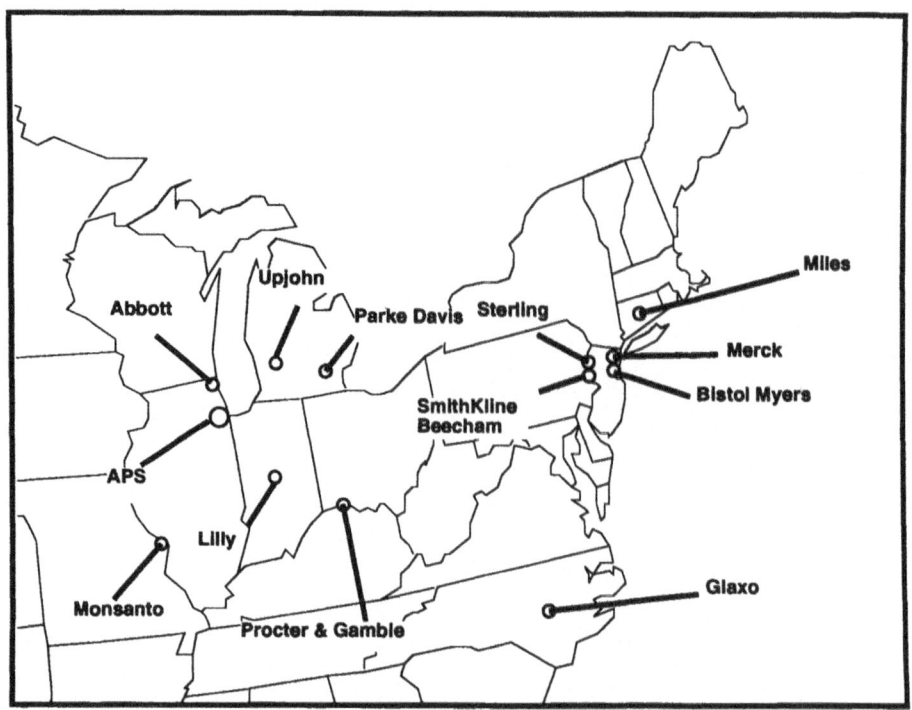

Figure 5. A map of the midwest and northeast portions of the U.S. with the locations of the twelve member IMCA companies and the APS.

The IMCA companies will develop their X-ray data collection facility, which is described in Table 3, as a Collaborating Access Team, or CAT, called IMCA CAT.

To build the IMCA's two protein X-ray data collection beamlines, the twelve member companies will collaborate with the Illinois Institute of Technology (IIT). The CAT Director for the construction phase is Professor Timothy I. Morrison of IIT. The $6.8M that will be required for the construction of the two beamlines will be provided by the IMCA member companies. The insertion device, a type B wiggler, and other "behind the wall" components will be provided by the APS. Consequently 25% of the beam time of the IMCA's facility will be available through open proposals. Each of the beamlines will be capable of providing X-radiation in the energy range 6.3 to 25 KeV and will be suitable for data collection in the monochromatic, Laue, or multi-wavelength mode.

Table 3. The IMCA CAT Beamlines at the APS

Collaboration:	IMCA Companies and Illinois Institute of Technology
CAT Director (construction):	Professor T. I. Morrison
Studies:	Macromolecular X-ray Diffraction
Estimated cost:	$6.8 Million through 1996
Funding Source:	IMCA Member Companies
Development Staff:	6.5 full-time equivalents (FTEs) at Peak
Operational Staff:	4.0 FTEs
Number of Sectors:	One
Beamlines:	Bending Magnet
	Wiggler B
Energy Range:	6.3 to 25.0 KeV
Types of Diffraction Studies:	Monochromatic, Multi-wavelength, Laue

In addition to the IMCA, three other consortia plan to construct biological X-ray beamlines at the APS. The names of those consortia, the institutions that will construct the beamlines and the types of X-ray studies to be done at the respective beamlines are given in Table 4.

Table 4. Biological X-ray Beamlines Approved at the Advanced Photon Source

Consortium	Builder	Purpose/Capabilities
Bio-CAT	IIT	Non-crystalline materials, SAXS, EXAFS, X-ray absorption
Bio-CARS	U. Chicago	Large-cell and real-time studies, monochromatic, multi-wavelength, Laue diffraction
Structural Biology Center	ANL	National user facility, monochromatic, multi-wavelength, Laue diffraction

SOME FUTURE DEVELOPMENTS THAT ARE NEEDED

The APS and the protein X-ray crystallographic facilities that will be available there will provide enormous capabilities. But to derive the maximum benefit from these capabilities, several future developments are needed.

One aspect where major developments are needed is the overall process of manipulating the crystals and collecting the X-ray diffraction data. The need for these developments can be illustrated by considering the consequences of the extreme brilliance of the third-generation synchrotron X-ray sources. For perspective, a protein X-ray data collection that requires 15 (X-ray exposure) days with a conventional in-laboratory X-ray diffractometer, requires only 1/2 to 1 1/2 days with current 2-D X-ray area detector systems. By also using a current, second-generation synchrotron X-ray source, those same unique X-ray data can be generated from only 10 to 40 minutes of X-ray exposure of the crystals. (However, the elapsed time in collecting these data is typically 1 to 2 hours due to the delays in transferring the data frames to the host computer.) Since the APS will be about 10^3 times more brilliant than current second-generation synchrotron X-ray sources, these same data will require from 130 to 500 ms of X-ray exposure! Clearly then, speeding up all the aspects of the protein crystallographic research at the APS not involving crystal manipulations and X-ray exposure procedures is a major challenge.

For instance, reliable technologies to increase the throughput of synchrotron X-ray data collection (e.g. via crystal-handling robots and multiple experimental stations per beamline) will be very important. Also, new 2-D X-ray detectors that are capable of operating at the high count rates and the short cycle times of the APS need to be developed. New data processing software that is capable of dealing quickly with huge volumes of X-ray diffraction data must be written. And highly integrated software systems for protein structure determination/refinement/analysis/-display must be available so the structural biologists at the APS can extract the biological significance of their resultant crystals structures. Synthetic molecules (e.g. caged-inhibitors) must be synthesized to instantaneously release the active inhibitors if short time-frame time-resolved crystallography is to be successfully exploited at the APS.

AN INDUSTRIAL MACROMOLECULAR CRYSTALLOGRAPHY EXAMPLE

Up to this point this presentation has outlined the strategy for using protein X-ray crystallography and synchrotron X-ray sources to benefit industry. Next I would like to discuss an actual example of how these techniques have been used to benefit an industrial research project. The target of that research is a class of proteins called the human rhinoviruses, which are the major cause of the common cold in humans.[52] There are at least 100 serotypically distinct rhinoviruses.[53] These viruses are picornaviruses and thus are non-enveloped, small, + sense single-stranded RNA viruses. The surface of the viral particle consists of a coat of viral proteins called a capsid. Each viral capsid has icosahedral symmetry and is comprised of 60 copies of each of four viral polypeptides, VP1-VP4.

The structure of the first human rhinovirus, HRV14, was determined in the laboratory of Dr. Michael Rossmann at Purdue University.[54] The crystal structure determination required the use of data collected from a synchrotron X-ray source because the HRV14 crystals are extremely sensitive to X-radiation damage. In fact the crystals cease to diffract to high resolution in less time than is required to collect a complete diffraction data set with a typical laboratory rotating anode X-ray source. Each of the three viral polypeptides, VP1-VP3 (Figure 6), contains an eight-stranded antiparallel β-barrel. The smallest of the four viral capsid polypeptides, VP4, was found to reside on the interior surface of the capsid, adjacent to the viral RNA.

Shortly after the structure of HRV14 was determined, a collaboration was formed between the Rossmann group and a research group at Sterling Winthrop which had a long-standing interest in a series of capsid-binding compounds known to inhibit the replication of many picornaviruses.[55] These antiviral compounds block replication at the uncoating step and, in some serotypes, also block attachment.[56-57]

Structure-based drug design (SBDD) for this capsid-binding class of antiviral compounds started with the discovery that these compounds bind in a hydrophobic pocket in one of the viral capsid polypeptides, VP1.[58] This binding site is within the VP1 β-barrel and is separated from the exterior of the capsid by a region of structure called the GH loop. Subsequent structure determinations of numerous compounds have increased the data base used for drug design.[59] Compounds from Sterling Winthrop, as well as other compounds, have been examined in this SBDD program. Due to the serotypic diversity of rhinoviruses, it is clear that structures of more than a single serotype would be required for the design of broad spectrum antiviral agents. Towards this end, the structures of a number of rhinovirus serotypes have now been determined (HRV14s, HRV1a, HRV16, and HRV50). Each of these structure determinations has required the use of a synchrotron X-ray source for collection of the diffraction data.[64-67]

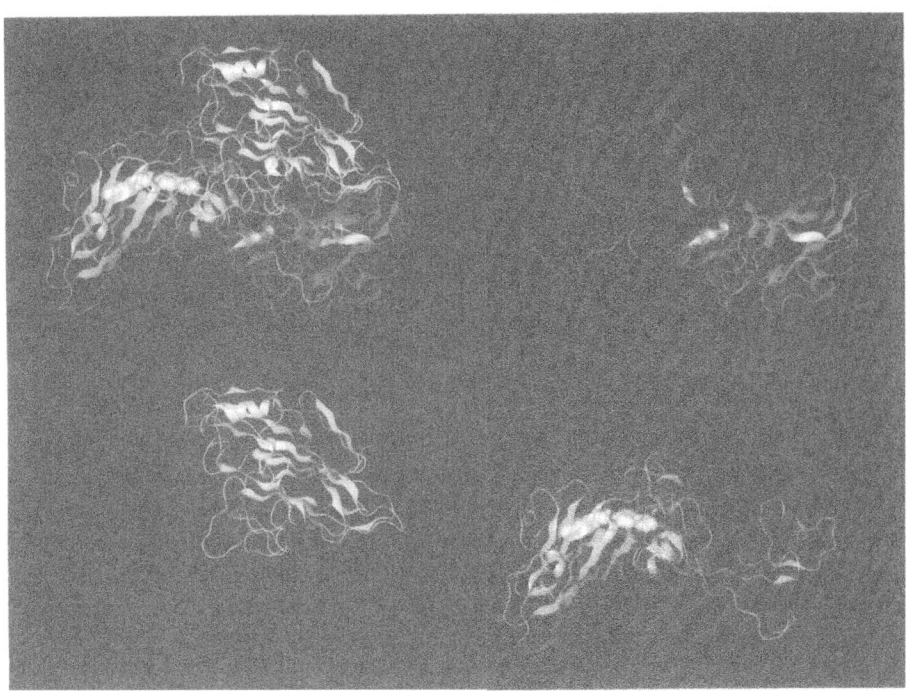

Figure 6. A four-panel illustration of the viral capsid. The upper left panel shows VP1 and the antiviral compound (as a ball-and-stick representation) which occupies the hydrophobic pocket. VP2 is shown in the upper right panel. VP3 is illustrated in the lower left panel. The lower right panel shows the entire assembly of molecules, VP1-VP4 and the antiviral compound. (VP4 lies behind the VP1-VP3 assembly and can hardly be seen.)

Synchrotron X-radiation diffraction data also made possible a study which provided interesting information regarding the molecular mechanism of action of the capsid-binding compounds. Rhinoviruses are acid labile, and do not survive when exposed to pH less than 5. When exposed to solutions of this pH region, rhinoviruses form non-infectious particles whose structure is similar or identical to the particles observed during the infectious process. It is also known that rhinoviruses require exposure to the low pH of the endosomal compartment of the host's cells for productive uncoating (and subsequent viral replication) to occur.[68] Further, when complexed with Sterling Winthrop antiviral (WIN) compounds, rhinoviruses are less susceptible to acid-induced inactivation than are the native viral particles.[55,69] All these data suggest that the knowledge of the structure of an acidified rhinovirus might provide important clues to the pH-dependent mechanism of capsid uncoating.

The effort to determine the structure of such an acidified rhinovirus proved to be somewhat problematic. The crystals of HRV14 were too acid labile and dissolved when directly exposed to a low-pH environment. The eventual solution to this problem required the use of the intense F1 beamline station at the Cornell High Energy Synchrotron Source (CHESS). For this study, the crystals were sealed in X-ray capillaries along with an aliquot of a volatile acid. The acid solution was separated from the crystal by a vapor phase. During the course of the diffraction experiment the acid diffused through the vapor phase and eventually destroyed the crystal, usually within 10-20 minutes. The high brilliance of the F1 station allowed the X-ray data to be collected as a series of 15 to 30 second, 0.3° oscillation photographs while the crystal was still intact and able to diffract the X-radiation.

Once they were processed, these data clearly showed three regions of the HRV14 capsid disordered upon acidification: (1) the putative Ca^{2+} ion which lies on the icosahedral five-fold axis, (2) the N-terminal region of VP3 and VP4 which surround the icosahedral five-fold axis on the interior of the capsid, and (3) the GH loop adjacent to the hydrophobic binding pocket within VP1.[70]

The disorder of the GH loop is especially intriguing because this loop is surrounded by a number of mutation sites which convey acid stability to the virus particle. An attractive theory which accommodates these data is that the GH loop undergoes an acid-induced conformational change *in vivo* which is similar to the one observed in the *in vitro* X-ray study. This conformational change increases the likelihood of productive uncoating of the viral capsid. This structural change can be inhibited either by binding an antiviral compound to the region that is adjacent to the GH loop in the VP1 pocket, or alternatively by mutation of the primary structure of VP1 in this region, thereby inhibiting productive infection of the virus.

These experiments have provided valuable insight into the mechanism of infection of human rhinoviruses and would not have been possible without the use of synchrotron X-radiation sources to provide the diffraction data. In particular, the acidification experiment would have been much more difficult, if not impossible, to conduct if an X-ray source less brilliant than the CHESS F1 beamline station were used. One should anticipate that the scope of possible X-ray experiments of biological macromolecules will significantly expand as the brilliance of synchrotron X-ray sources continues to increase.

SUMMARY

Synchrotron X-radiation (SR) has already made an enormous impact on the field of macromolecular X-ray crystallography. SR has provided useful X-ray data for smaller crystals than can be used with in-laboratory X-ray sources, has enabled the initial forays into time-resolved crystallography for biological macromolecules and has resulted in a new, more straightforward technique (MAD phasing) for solving protein crystal structures. The beneficial access to the third-generation SR sources, such as the APS, should lead to even more rapid understanding and manipulation of the biological processes mediated by biological macromolecules. The IMCA is positioned to generate the necessary X-ray diffraction data that the member companies will use to leverage the molecular structures that result to aid the development of new, commercially successful products.

ACKNOWLEDGMENTS

The author sincerely appreciates the assistance of several individuals who provided information used in this presentation. Dr. Keith D. Watenpaugh of the Upjohn Company and Dr. Noel D. Jones formerly of Eli Lilly and Company and currently with Molecular Structure Corporation, who are the current chairman and former chairman, respectively, of the IMCA, provided slides and many useful points regarding the history of industrial protein X-ray crystallography and the strategy that led to the formation of the IMCA. Dr. Vincent L. Giranda of Sterling Winthrop, Inc. kindly supplied the technical details for the HRV example which illustrated how an industrial R&D project can benefit from protein structural information developed using X-ray data collected at a synchrotron X-ray source. Finally, I would like to thank the organizers of this Symposium, Drs. Louis J. Terminello and Kevin L. D'Amico, for inviting me to participate on the program.

REFERENCES

1. F. C. Bernstein, T. F. Koetzle, G. J. B. Williams, E. F. Meyer, Jr., M. D. Brice, J. R. Rodgers, O. Kennard, T. Shimanouchi, and M. Tasumi, The protein data bank: a computer based archival file for macromolecular structures, *J. Mol. Biol.*, 112:535 (1977).

2. M. Perutz. "Protein Structure New Approaches to Disease and Therapy," W. H. Freeman and Company, New York (1992).

3. D. W. Heinz, J. P. Priestle, J. Rahuel, K. S. Wilson, and M. G. Grütter, Refined crystal structures of subtilisin novo in complex with wild-type and two mutant eglins comparison with other serine protease inhibitor complexes, *J. Mol. Biol.*, 217:353 (1991).

4. R. Bott, M. Ultsch, A. Kossiakoff, T. Graycar, B. Katz, and S. Power, The three-dimensional structure of *Bacillus amyloliquefaciens* subtilisin at 1.8 Å and an analysis of the structural consequences of peroxide inactivation, *J. Biol. Chem.*, 263:7895 (1988).

5. P. N. Bryan, M. L. Rollence, M. W. Pantoliano, J. Wood, B. C. Finzel, G. L. Gilliland, A. J. Howard, and T. L. Poulos, Proteins of enhanced thermostability: characterization of a thermostable variant of subtilisin, *Proteins Struc. Funct. and Genet.*, 1:326 (1986).

6. C. R. Erwin, B. L. Barnett, J. D. Oliver, and J. F. Sullivan, Effects of salt bridges on the stability of subtilisin BPN', *Protein Eng.*, 4:87 (1990).

7. L. W. Lim, R. A. Stegeman, N. K. Leimgruber, J. K. Gierse, and S. S. Abdel-Meguid, Preliminary crystallographic study of glycosylated recombinant human renin, *J. Mol. Biol.*, 210:239 (1989).

8. K. D. Watenpaugh, H. M. Einspahr, B. C. Finzel, L. L. Clancy, A. M. Mulichak, D. R. Holland, S. W. Muchmore, R. A. Poorman, J. O. Hui, R. L. Heinrikson, K. Murakami, A. Shoda, L. L. Maggiora, and T. K. Sawyer, Crystallographic studies of a renin-renin inhibitor complex and comparison with other aspartyl proteinase-inhibitor complexes, Twelfth American Peptide Symposium, Cambridge. MA, June 16-21 (1991).

9. L. Chen, J. W. Erickson, T. J. Rydel, C. H. Park, D. Neidhart, J. Luly, and C. Abad-Zapatero, Structure of a pepsin/renin inhibitor complex reveals a novel crystal packing induced by minor chemical alterations in the inhibitor, *Acta Crystallogr.*, B48:476 (1992).

10. J. Erickson, D. J. Neidhart, J. VanDrie, D. J. Kempf, X. C. Wang, D. W. Norbeck, J. J. Plattner, J. W. Rittenhouse, M. Turon, N. Wideburg, W. E. Kohlbrenner, R. Simmer, R Helfrich, D. A. Paul, and M. Knigge, Design, activity, and 2.8 Å crystal structure of a C_2 symmetric inhibitor complexed to HIV-1 protease, *Science*, 249:527 (1990).

11. M. D. Varney, K. Appelt, V. Kalish, M. R. Reddy, J. Tatlock, C. L. Palmer, W. H. Romines, B.-W. Wu, and L. Musick, Crystal-structure-based design and synthesis of novel C-terminal inhibitors of HIV protease, *J. Med. Chem.*, 37:2274 (1994).

12. S. W. Kaldor, M. Hammond, B. A. Dressman, J. E. Fritz, T. A. Crowell, R. W. Schevitz, J.-P. Wery, D. K. Clawson, and K. Appelt, New dipeptide isosteres useful for the inhibition of HIV-1 protease, Abstracts of Papers of the 206th Meeting of the American Chemical Society, 144-Medi., August 22 (1993).

13. M. A. Navia, P. M. D. Fitzgerald, B. M. McKeever, C.-T. Leu, J. C. Heimbach, W. K. Herber, I. S. Sigal, P. L. Darke, and J. P. Springer, Three-dimensional structure of aspartyl protease from human immunodeficiency virus HIV-1, *Nature (London)*, 337:615 (1989).

14. S. S. Abdel-Meguid, B. Zhao, K. H. M. Murthy, E. Winborne, J.-K. Choi, R. L. DesJarlais, M. D. Minnich, J. S. Culp, C. Debouck, T. A. Tomaszek, Jr., T. D. Meek, and

G. B. Dreyer, Inhibition of human immunodeficiency virus-1 protease by a C_2-symmetric phosphinate. synthesis and crystallographic analysis, *Biochemistry*, 32:7972 (1993).

15. A. M. Mulichak, J. O. Hui, A. G. Tomasselli, R. L. Heinrikson, K. A. Curry, C.-S. Tomich, S. Thaisrivongs, T. K. Sawyer, and K. D. Watenpaugh, The crystallographic structure of the protease from human immunodeficiency virus type 2 with two synthetic peptidic transition state analog inhibitors, *J. Biol. Chem.*, 268:13103 (1993).

16. J. F. Davies, II, Z. Hostomska, Z. Hostomsky, S. R. Jordan, and D. A. Matthews, Crystal structure of the ribonuclease H domain of HIV-1 reverse transcriptase, *Science*, 252:88 (1991).

17. D. Chattopadhay, B. C. Finzel, S. H. Munson, D. B. Evans, S. K. Sharma, N. A. Strakalaitis, D. P. Brunner, F. M. Eckenrode, Z. Dauter, Ch. Betzel, and H. M. Einspahr, Crystallographic analyses of an active HIV-1 ribonuclease H domain show structural features that distinguish it from the inactive form, *Acta. Cryst.*, D49:423 (1993).

18.B. Zhao, E. Winborne, M. D. Minnich, J. S. Culp, C. Debouck, and S. S. Abdel-Meguid, Three-dimensional structure of a simian immunodeficiency virus protease/inhibitor complex. Implications for the design of human immunodeficiency virus type 1 and 2 protease inhibitors, *Biochemistry*, 32:13054 (1993).

19. J.-P. Wery, R. W. Schevitz, D. K. Clawson, J. L. Bobbitt, E. R. Dow, G. Gamboa, T. Goodson, Jr., R. B. Hermann, R. M. Kramer, D. B. McClure, E. D. Mihelich, J. E. Putnam, J. D. Sharp, D. H. Stark, C. Teater, M. W. Warrick, and N. D. Jones, Structure of recombinant human rheumatoid arthritic synovial fluid phospholipase A_2 at 2.2 Å resolution, *Nature (London)*, 352:79 (1991).

20. D. R. Holland, L. L Clancy, S. W. Muchmore, T. J. Rydel, H. M. Einspahr, B. C. Finzel, R. L. Heinrikson, and K. D. Watenpaugh, The crystal structure of a lysine 49 phospholipase A_2 from the venom of the cottonmouth snake at 2.0-Å resolution, *J. Biol. Chem.*, 265:17649 (1990).

21. B. Lovejoy, A. Cleasby, A. M. Hassell, K. Longley, M. A. Luther, D. Weigl, G. McGeehan, A. B. McElroy, D. Drewry, M. H. Lambert, and S. R. Jordan, Structure of the catalytic domain of fibroblast collagenase complexed with an inhibitor, *Science*, 263:375 (1994).

22. N. Borkakoti, F. K. Winkler, D. H. Williams, A. D'Arcy, M. J. Broadhurst, P. A. Brown, W. H. Johnson, and E. J. Murray, Structure of the catalytic domain of human fibroblast collagenase complexed with an inhibitor, *Nat. Struct. Biol.*, 1:106 (1994).

23. J. C. Spurlino, A. M. Smalllwood, D. D. Carlton, T. M. Banks, K. J. Vavra, J. S. Johnson, E. R. Cook, J. Falvo, R. C. Wahl, T. A. Pulvino, J. J. Wendoloski, and D. L. Smith, 1.56 Å Structure of mature truncated human fibroblast collagenase, *Proteins Struc. Funct. and Genet.*, 19:98 (1994).

24. M. V. Milburn, A. M. Hassell, M. H. Lambert, S. R. Jordan, A. E. I. Proudfoot, P. Graber, and T. N. C. Wells, A novel dimer configuration revealed by the crystal structure at 2.4 Å resolution of human interleukin-5, *Nature (London)*, 363:172 (1993).

25. B. J. Graves, M. H. Hatada, W. A. Hendrickson, J. K. Miller, V. S. Madison, and Y. Satow, Structure of interleukin 1α at 2.7-Å resolution, *Biochemistry*, 29:2679 (1990).

26. H. M. Einspahr, L. L. Clancy, D. R. Holland, S. W. Muchmore, K. D. Watenpaugh, and B. C. Finzel, The crystal structure of human interleukin-1α, in: "Current Research in Protein Chemistry," J. J. Villafranca, ed., Academic Press, San Diego (1990).

27. B. C. Finzel, L. L. Clancy, D. R. Holland, S. W. Muchmore, K. D. Watenpaugh, and H. M. Einspahr, Crystal structure of recombinant human interleukin-1β at 2.0 Å Resolution, *J. Mol. Biol.*, 209:779 (1989).

28. L. L. Clancy, B. C. Finzel, A. W. Yem, M. R. Deibel, Jr., N. A. Strakalaitis, and D. P. Brunner, Initial crystallographic analysis of a recombinant human interleukin-1 receptor antagonist protein, *Acta Cryst.*, D50:197 (1994).

29. G. P. A. Vigers, P. Caffes, R. J. Evans, R. C. Thompson, S. P. Eisenberg, and B. J. Brandhuber, X-ray structure of Interleukin-1 Receptor Antagonist at 2.0 Å Resolution, *J. Biol. Chem.*, 269:12874 (1994).

30. V. L. Giranda, Structure-based drug design of antirhinoviral compounds, *Structure*, 2:695 (1994).

31. D. A. Matthews, W. W. Smith, R. A. Ferre, B. Condon, G. Budahazi, W. Sisson, J. E. Villafranca, C. A. Janson, H. E. McElroy, C. L. Gribskov, and S. Worland, Structure of human rhinovirus 3C protease reveals a trypsin-like polypeptide fold, RNA-binding site, and means for cleaving precursor polyprotein, *Cell*, 77:761 (1994).

32. A. M. de Vos, M. Ultsch, and A. A. Kossiakoff, Human growth hormone and extracellular domain of its receptor: crystal structure of the complex, *Science*, 255:306 (1992).

33. N. D. Jones, J. DeHoniesto, P. M. Tackitt, and G. W. Becker, Crystallization of authentic recombinant human growth hormone, *Biotechnology*, 5:499 (1987).

34. S. S. Abdel-Meguid, H.-S. Shieh, W. W. Smith, H. E. Dayringer, B. N. Voiland, and L. A. Bentle, Three-dimensional structure of a genetically engineered variant of porcine growth hormone, *Proc. Natl., Acad. Sci. USA*, 84:6434 (1987).

35. S. E. Ealick, Y. S. Babu, C. E. Bugg, M. D. Erion, W. C. Guida, J. A. Montgomery, and J. A. Secrist, Application of crystallographic and modeling methods in the design of purine nucleoside phosphorylase inhibitors, *Proc. Natl. Acad. Sci. USA*, 88:11540 (1991).

36. L. Chen, D. Neidhart, W. K. Kohlbrenner, W. Mandecki, S. Bell, J. Sowadski, and C. Abad-Zapatero, 3-D structure of a mutant (Asp101-Ser) of *E. coli* alkaline phosphatase with higher catalytic activity, *Protein Eng.*, 5:605 (1992).

37. M. D. Varney, G. P. Marzoni, C. L. Palmer, J. G. Deal, S. Webber, K. M. Welsh, R. G. Bacquet, C. A. Bartlett, C. A. Morse, C. L. J. Booth, S. M Herrmann, E. F. Howland, R. W. Ward, and J. White, Crystal-structure-based design and synthesis of benz[cd]indole-containing inhibitors of thymidylate synthase, *J. Med. Chem.*, 35:663 (1992).

38. D. R. Knighton, C.-C. Kan, E. Howland, C. A. Janson, Z. Hostomska, K. M. Welsh, and D. A. Matthews, Structure of and kinetic chanelling in bifunctional dihydrofolate reductase-thymidylate synthase, *Nature Struc. Biol.*, 1:186 (1994).

39. L. T. Taberno, C. Y. Chang, S. L. Ohringer, W. F. Lau, E. J. Iwanowicz, W. Han, T. C. Wang, S. M. Seiler, D. G. M. Roberts, and J. S. Sack, "Structure of a retro-binding peptide inhibitor complexed with human α-thrombin," submitted for publication.

40. M. G. Grütter, J. P. Priestle, J. Rahuel, H. Grossenbacher, W. Bode, J. Hofsteenge, and S. R. Stone, Crystal structure of the thrombin-hirudin complex: a novel mode of serine protease inhibition, *EMBO Journal*, 9:2361 (1990).

41. V. L. Nienaber and P. C. Weber, Crystal structure of Boc-(D)-Phe-Pro-Arg thrombin at 2.0Å resolution, American Crystallographic Association Meeting, Atlanta, GA, June 25-July 1 (1994).

42. N. Y. Chirgadze, D. K. Clawson, P. D. Gesellchen, R. B. Hermann, R. E. Kaiser, Jr., J. L. Olkowski, D. J. Sall, R. W. Schevitz, G. F. Smith, R. T. Shuman, J.-P. Wery, and N. D. Jones, The x-ray structure at 2.2 Å resolution of a ternary complex containing human alpha-thrombin, a hirudin peptide (54-65) and an active site inhibitor, Joint Meeting of the American Crystallographic Association and the Pittsburgh Diffraction Conference, Pittsburgh, PA, August 9-14 (1992).

43. J. Kallen, C. Spitzfaden, M. G. M. Zurini, G. Wider, H. Widmer, K. Wüthrich, and M. D. Walkinshaw, Structure of human cyclophilin and its binding site for cyclosporin A determined by X-ray crystallography and NMR spectroscopy, *Nature (London)*, 353:276 (1991).

44. M. P. Schlunegger and M. G. Grütter, An unusual feature revealed by the crystal structure at 2.2 Å resolution of human transforming growth factor-β2, *Nature (London)*, 358:430 (1992).

45. W. C. Stallings, S. S. Abdel-Meguid, L. W. Kim, H.-S. Shieh, H. E. Dayringer, N. K. Leimgruber, R. A. Stegeman, K. S. Anderson, J. A. Sikorski, S. R. Padgette, and G. M. Kishore, Structure and topological symmetry of the glyphosphate target 5-*enol*-pyruvylshikimate-3-phosphate synthase: a distinctive protein fold, *Proc. Natl. Acad. Sci. USA*, 88:5046 (1991).

46. R. J. Almassy, C. A. Janson, C.-C. Kan, and Z. Hostomska, Structures of apo and complexed *Escherichia coli* glycinamide ribonucleotide transformylase, *Proc. Natl. Acad. Sci. USA*, 89:6114 (1992).

47. J. F. Davies, II, R. J. Almassy, Z. Hostomska, R. A. Ferre, and Z. Hostomsky, 2.3 Å Crystal structure of the catalytic domain of DNA polymerase, *Cell*, 76:1123 (1994).

48. J. Hadju, P. A. Machin, J. W. Campbell, T. J. Greenhough, I. J. Clifton, S. Zurek, S. Gover, L. N. Johnson, and M. Elder, Millisecond X-ray diffraction and the first electron density map from Laue photographs of a protein crystal, *Nature (London)*, 329:178 (1987).

49. I. Schlichting, S. C. Almo, G. Rapp, K. Wilson, K. Petratos, A. Lentfer, A. Wittinghofer, W. Kabsch, E. F. Pai, G. A. Petsko, and R. S. Goody, Time-resolved X-ray crystallographic study of the conformational change in Ha-Ras p21 protein on GTP hydrolysis, *Nature (London)*, 345:309 (1990).

50. P. T. Singer, A. Smalås, R. P. Carty, W. F. Mangel, and R. M. Sweet, The hydrolytic water molecule in trypsin, revealed by time-resolved Laue crystallography, *Science*, 259:669 (1993).

51. H. M. K. Murthy, W. A. Hendrickson, W. H. Orme-Johnson, E. A. Merritt, and R. P. Phizackerley, Crystal structure of *Clostridium acidi-urici* ferrodoxin at 5-Å resolution based on measurements of anomalous X-ray scattering at multiple wavelengths, *J. Biol. Chem.*, 263:18430 (1988).

52. R. R. Rueckert, Picornaviridae and their replication, *in*: "Virology," B. N. Fields and D. M. Knipe, eds., Raven Press, New York (1990).

53. C. R. Uncapher, C. M. DeWitt,and R. J. Colonno, The major and minor group receptor families contain all by one human rhinovirus serotype, *Virology*, 180:814 (1991).

54. M. G. Rossmann, E. Arnold, J. W. Erickson, E. A. Frankenberger, J. P. Griffith, H.-J. Hecht, J. E. Johnson, G. Kamer, M. Luo, A. G. Mosser, R. R. Rueckert, B. Sherry, and G. Vriend, Structure of a human common cold virus and functional relationship to other picornaviruses, *Nature (London)*, 317:145 (1985).

55. M. P. Fox, M. J. Otto, and M. A McKinlay, Prevention of rhinovirus and poliovirus uncoating by WIN 51711, an new antiviral drug, *Antimicrob. Agents Chemother.*, 30:110 (1986).

56. D. C. Pevear, M. J. Fancher, P. J. Felock,M. G. Rossmann, M. S. Miller, G. Diana, A. M. Treasurywala, M. A. McKinlay, and F. J. Dutko, Conformational change in the floor of the human rhinovirus canyon blocks adsorption to HeLa cell receptors, *J. Virol.*, 63:2002 (1989).

57. D. A. Shepard, B. A. Heinz, and R. R. Rueckert, WIN 52035-2 inhibits both attachment and eclipse of human rhinovirus 14, *J. Virol.*, 67:2245 (1993).

58. T. J. Smith, M. J. Kremer, M. Luo, G. Vriend, E. Arnold, G. Kamer, M. G. Rossmann, M. A. McKinlay, G. D. Diana, and M. J. Otto, The site of attachment in human rhinovirus 14 for antiviral agents that inhibit uncoating, *Science*, 233:1286 (1986).

59. A. Zhang, R. G. Nanni, T. Li, G. F. Arnold, D. A. Oren, A. Jacobo-Molina, R. L. Williams, G. Kamer, D. A. Rubenstein, Y. Li, E. Rozhon, S. Cox, P. Buontempo, J. O'Connell, J. Schwartz, G. Miller, B. Bauer, R. Versace, P. Pinto, A. Ganguly, V.

Girijavallabhan, and E. Arnold, Structure determination of antiviral compound SCH 38057 complexed with human rhinovirus 14, *J. Mol. Biol.*, 230:857 (1993).

60. J. Badger, I. Minor, M. J. Kremer, M. A. Oliveira, T. J. Smith, J. P. Griffith, D. M. A. Guerin, S. Krishnaswamy, M. Luo, M. G. Rossmann, M. A. McKinlay, G. D. Diana, F. J. Dutko, M. Fancher, R. R. Rueckert, and B. A. Heinz, Structural analysis of a series of antiviral agents complexed with human rhinovirus 14, *Proc. Natl. Acad. Sci. USA*, 85:3304 (1988).

61. M. S. Chapman, I. Minor, M. G. Rossmann, G. D. Diana, and K. Andries, Human rhinovirus 14 complexed with antiviral compound R 61837, *J. Mol. Biol.*, 217:455 (1991).

62. V. L. Giranda, G. R. Russo, P. J. Felock, T. Draper, J. Guiles, F. J. Dutko, G. D. Diana, D. C. Pevear, and M. McMillan, The structures of four methyltetrazole-containing compounds in human rhinovirus serotype 14, *Acta Crystallogr. D*, in press.

63. K. H. Kim, P. Willingmann, Z. X. Gong, M. J. Kremer, M. S. Chapman, I. Minor, M. A. Oliveira, M. G. Rossmann, K. Andries, G. D. Diana, F. J. Dutko, M. A. McKinlay, and D. C. Pevear, A comparison of the anti-rhinoviral drug binding pocket in HRV14 and HRV1A, *J. Mol. Biol.*, 230:206 (1993).

64. E. Arnold and M. G. Rossmann, Analysis of the structure of a common cold virus, human rhinovirus 14, refined at a resolution of 3.0 Å, *J. Mol. Biol.*, 211:763 (1990).

65. S. Kim, T. J. Smith, M. S. Chapman, M. G. Rossmann, D. C. Pevear, F. J. Dutko, P. J. Felock, G. D. Diana, and M. A. McKinlay, Crystal structure of human rhinovirus serotype 1A (HRV1A), *J. Mol. Biol.*, 210:91 (1989).

66. M. A. Oliveira, R. Zhao, W.-M. Lee, M. J. Kremer, I. Minor, R. R. Rueckert, G. D. Diana, D. C. Pevear, F. J. Dutko, M. A. McKinlay, and M. G. Rossmann, The structure of human rhinovirus 16, *Structure*, 1:51 (1993).

67. V. L. Giranda, personal communication.

68. L. Perez and L. Carrasco, Entry of poliovirus into cells does not require a low-pH step, *J. Virol.*, 67:4543 (1993).

69. M. Gruenberger, D. Pevear, G. D. Diana, E. Kuechler, and D. Blaas, Stabilizaiton of human rhinovirus serotype 2 against pH-induced conformational change by antiviral compounds, *J. Gen. Virol.*, 72:431 (1993).

70. V. L. Giranda, B. A. Heinz, M. A. Oliveira, I. Minor, K. H. Kim, P. R. Kolatkar, M. G. Rossmann, and R. R. Rueckert, Acid-induced structural changes in human rhinovirus 14: possible role in uncoating, *Proc. Natl. Acad. Sci. USA*, 89:10213 (1992).

RECENT ADVANCES IN THE USE OF SYNCHROTRON RADIATION
FOR MACROMOLECULAR CRYSTALLOGRAPHY

Robert M. Sweet

Biology Department
Brookhaven National Laboratory
Upton, NY 11973

INTRODUCTION

The field of protein crystallography is in exponential growth. This expansion is owed principally to these factors: easy production of pure specimens by use of modern cloning and expression methods, improved methods for crystal growth, rapid advances in the use of cryogenics to prevent radiation damage to specimens, continually improving software for every step in the crystallographic process, better computer and detector hardware, and finally improved x-ray sources. There are roughly eight synchrotron sources and twenty-four individual facilities around the world that are used by macromolecular crystallographers. The National Synchrotron Light Source (NSLS) at Brookhaven National Laboratory represents one of these powerful x-ray sources. Brookhaven's Biology Department operates several facilities at the NSLS for measurement of diffraction data. I can survey our recent experiences and plans for the future to reveal some of the promise and achievements of these methods.

Although data measured at synchrotron x-ray sources still represent only a fraction of the total structures determined each year, many of the most difficult problems are solved at these facilities. There are several reasons for this. The high brightness of the beam allows one to measure data from small or weakly diffracting crystals. It's also this brightness (which really means *intensity* and *collimation*) that makes synchrotron radiation so useful for study of crystals of viruses. The synchrotron x-ray beam is polychromatic; this provides several possibilities. One is that we can use the full white beam for diffraction, employing the "Laue" method for measurement of data. Individual exposures (diffraction snapshots) can easily be in the millisecond time scale or shorter. This allows one to examine transient states that one can contrive to have appear during enzyme-catalyzed reactions or other dynamic events in a crystal. A second use for the polychromatic nature of the x-ray beam is to exploit the anomalous diffraction that occurs from atoms when the x-ray photon energy is near an absorption edge of that atom. In this case, the anomalous diffraction helps dramatically by providing nearly direct solutions of crystal structures.

Synchrotron Radiation Techniques in Industrial, Chemical, and Materials Science
Edited by D'Amico *et al.*, Plenum Press, New York, 1996

Another reason why facilities at synchrotron radiation sources are so effective in the solving of big crystal structures is simply that so much effort is put into providing the best possible apparatus and software. In the first case, robust and versatile data-collection apparatus permits users to explore difficult experiments, for example extreme cryogenics, without worrying about the performance of the equipment. In the second case, improvements in software allow inexperienced experimenters to take excellent data with little supervision. We can review each of these advances in turn.

SYNCHROTRON DATA-COLLECTION APPARATUS

Modern macromolecular crystallographic apparatus is often suitable for use either with a conventional x-ray source or with a synchrotron, and there is substantial overlap. There are three issues to be resolved -- the type of detector, the crystal-orienter, and the software. Frequently the choices are made based on the availability of a suitable commercial product. We can discuss the two most popular styles of detection: imaging storage phosphors, and electronic imagers.

Storage-phosphor Imaging Plates

The overwhelming majority of synchrotron facilities employ some sort of a detector based on a reusable x-ray-sensitive film or plate that depends on a solid-state x-ray-storage phosphor. Also, the great majority of the apparatus for the digitizing of the signal from these imaging plates is a commercial product, used both at synchrotron sources and with ordinary anode-emission type sources.

The storage phosphor is generally a barium halide, doped with a rare earth such as europium. When an x-ray photon is absorbed, usually by the BaX_2 matrix, the excitation energy is transferred eventually to the Eu atoms. These atoms are promoted to a metastable F-center state. From this state a visible-light (blue) photon will be emitted when the atom is illuminated with a lower-energy (red) photon. This photo-stimulated emission can be captured and measured to give an estimate of the original incident x-ray intensity. The residual image on the plate can be erased by further illumination. The measured values for the x-ray intensity are fairly accurate because the natural gain in the system is high (lots of light for few x-ray photons) and because the "dark current" of the system is relatively low. A major advantage of the systems as x-ray detectors for diffraction studies is that it's fairly inexpensive to make the detectors large (200 - 500 mm) and to produce fairly fine spatial resolution (0.1 mm). An important difficulty is that the fastest commercial readouts for the phosphors still take several minutes.

A Special-Purpose Diffractometer

Sometimes for a synchrotron source, elaborate arrangements are made to get the best possible data. At Brookhaven we have constructed a hybrid diffractometer, shown in Figure 1. Here an imaging-plate scanner, manufactured as part of a complete but simple one-axis camera system by MAR Research of Hamburg, Germany, has been mounted on the Θ arm of an Enraf-Nonius FAST diffractometer. This provides the flexibility of the Enraf orienter and Θ arm with the data-collection efficiency of the 300 mm-diameter MAR detector ($\pi \times 10^6$ usable pixels). The entire system is mounted on a motorized table that provides convenient alignment of the diffractometer to the x-ray beam. Both diffractometer and alignment apparatus are under control of a single computer that is part of a dedicated local computing network with three work stations and about 15 GB of disk storage.

Figure 1. The hybrid MAR Research/Enraf-Nonius diffractometer at beamline X12-C of the NSLS.

Advanced Software for Experimental Control

The data-collection process at a synchrotron beamline is slightly more complex than that employed with conventional sources and apparatus. Part of this complexity arises from the control that is required of the beamline itself -- one must be able to realign apparatus to follow slight motions of the beam coming from the synchrotron, and one has the possibility to adjust the x-ray wavelength.

To make control of the apparatus as simple as possible has great benefit both for the user and for the operator of the facility. The users' time is precious at a synchrotron, so he must work efficiently. On the other hand, the machine often runs around the clock, so workers often put in very long hours and try to function with very little sleep. We have found that an important component to the user interface is the software that drives the system. We have constructed an integrated system wherein a single console and a small number of carefully organized program windows suffice to give the user complete control over his experiment.

An example of one of the control windows from this software is shown in Figure 2. Here one can see data-collection in progress, with the simple beamline-control window at the

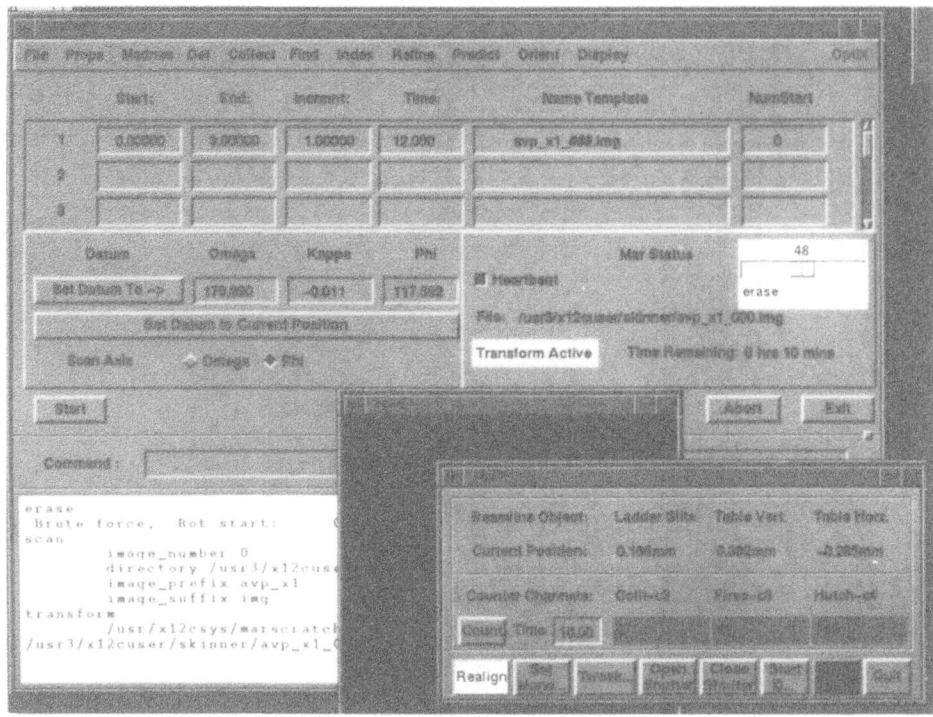

Figure 2. The data-collection window from MARMAD, the integrated data-collection program from beamline X12-C of the NSLS.

lower right. Although this monochrome figure can't show it, extensive use is made of color coding to indicate the operation in progress. An important function of the software is that it automatically documents the experiment -- files are written to record important operations and to specify every experimental parameter to which the software has access. The software performs a substantial amount of error checking -- this is an aspect of the program that requires continual development. We have found that a salary spent on software construction and maintenance more than repays itself in terms of reliability of the system, and ultimately in terms of the quality and quantity of data that can be measured.

Below will be found a discussion of a type of crystallography that involves the repetitive measurement of data at selected wavelengths: MAD phasing. Our data-collection software allows the user to program a sequence of data sweeps at a sequence of wavelengths. Not only is the action of the diffractometer coordinated with the function of the x-ray monochromator, but the software even measures the absorption spectrum from the heavy atoms in the specimen crystal and analyzes this spectrum to discriminate particular features. This allows data collection to proceed at precisely the right wavelength to give the strongest signal.

This software is written in a modular style that allows versatile incorporation of both new features and new hardware. It has been used twice already to test modern electronic imaging detectors based on charge-coupled devices in place of the MAR detector. The modifications required roughly a day's work in each case.

Charge-Coupled Device Detectors

Charge-coupled devices are useful for electronic imaging for many scientific purposes. They can provide fairly fine spatial resolution, a large number of picture elements (pixels),

Figure 3. The Brandeis modular detector based on a CCD chip and a fiber-optic taper.

reasonable precision (and accuracy) of intensity measurement, and fairly rapid readout. In addition, they are physically stable and not too expensive. They have been incorporated into a variety of area-sensitive detectors for x-radiation for several years.[1,2,3,4] A number of workers, including at least four commercial firms, have settled on a general design for a generation of detectors that are about to hit the streets.

The consensus design is exemplified by a detector constructed by the group of Walter Phillips of Brandeis University, shown in Figure 3. It is similar in most respects to detectors produced by the groups of Edwin Westbrook of Argonne National Laboratory and Sol Gruner of Princeton. It is comprised of a 25-mm square CCD chip bonded to the small end of a 4:1 fibre-optic taper. Each of the roughly 1000 x 1000 pixels of the CCD have an effective detection area on the front face of the taper of 100 μm square. An x-ray-sensitive phosphor that produces visible light when it is illuminated with x-rays is bonded to the front surface of the taper. Under typical conditions during synchrotron operation, when roughly 1-Å wavelength x-rays are used, the detector will measure diffraction patterns with quite high efficiency -- on the order of 75% of the photons will be measured. This is possible because sufficient visible-light photons are generated by the original "conversion" phosphor that, eventually, approximately ten photoelectrons appear in the CCD holding cell for each x-ray photon striding the front face of the detector. Readout time for individual images will be on the order of one second, commensurate with the length of individual exposures.

An important advantage of the compact form of these detectors is that they are small enough to be stacked together. Westbrook and coworkers have already constructed a very successful 3x3 array of detectors of this sort where each imager has a taper ratio of approxi-

25

mately 2:1. Phillips is constructing a 2x2 array of 4:1 tapers. Several commercial firms are preparing to produce similar detectors. It seems likely that detectors of this sort will be the principal instruments used at U.S. synchrotrons during the next few years.

ROUTINE DATA-COLLECTION -- DIFFICULT EXPERIMENTS

Iron-Carbonyl Chemistry in a Protein Molecule

In general, one makes measurements at these macromolecular crystallography beam-lines, one doesn't do experiments. That is to say, the "experimental" part of the work is done away from the synchrotron, and one hopes to be able to perform accurate measurements on these specimens without being faced with any partcular variability in the likely outcome. An exception to this rule has been the work done by Joel Berendzen of Los Alamos Nat'l Lab and colleagues, who have managed to trap a transient state in the binding of carbon monoxide to the oxygen-storage protein, myoglobin. Myoglobin (abbreviated Mb) is a fairly ordinary protein, except that it is coordinated with an iron-containing haeme group. This iron atom oxidizes easily, but it can be reduced and then reacted with CO to form a stable complex. The first part of the diffraction experiment involved construction of a cryostat capable of freezing a highly hydrated protein crystal quickly enough that only vitreous-water ice and no crystalline ice formed. Then the the crystal had to be held for several hours at a temperature near to 20K. While being held at this temperature, the Mb crystal was illuminated for an hour or so; this illumination stimulated a transition that led to a breaking of the Fe-CO bond. Finally, as quickly as possible, complete three-dimensional diffraction data were measured to determine the structure of this transient state, stable only at this low temperature. In this case, data were taken from two different crystals over a total elapsed time of approximately eight hours. The availability of the synchrotron and reliable equipment was an absolute requirement for the success of this delicate experiment.

The result was dramatic and satisfying. In Figure 4 one can see molecular structures in the region of the heaem group representing three states of the Mb molecule: in white is shown pure deoxy myoglobin with no O_2 or CO molecule bound; grey represents the state with CO firmly bound; black is the photolyzed state, frozen at 20K. In the photolyzed state the CO molecule can be seen, still in the binding pocket and nestled against hydrophobic residues at the rear. The conformational changes of both the haeme and the protein are at an intermediate state.

Very High-Resolution Data

A challenging set of measurements we made recently were from crystals of a synthetic peptide designed to fold into an α helix.[5] The investigators were faced with a crystal structure they couldn't solve. Standard macromolecular methods had failed -- isomorphous replacement with heavy-atoms had failed, as had the method by which a model structure, in this case a stretch of α helix, is fitted into the crystal's unit cell to give a starting solution. They were determined to attempt the direct methods of phasing that have revolutionized small-molecule crystallography. This required, however, that the diffraction data be measured to atomic resolution. They were unable to get adequate data at their home laboratory for various reasons. However, the flexibility of the X12-C hybrid diffractometer -- ability to tilt the detector upwards, short x-ray wavelength -- made it possible to get excellent data to 0.9 Å resolution and some data beyond. Finally, the structure was solved by the recent "Shake-and-Bake" method.[6,7] It finally was comprised of four independent helical molecules in the crystal assymetric unit: at total of over 400 C, N, and O atoms. This represents one of the larger structures solved *ab initio* from a single data set with no heavy atom scattering being included. The extraordinary

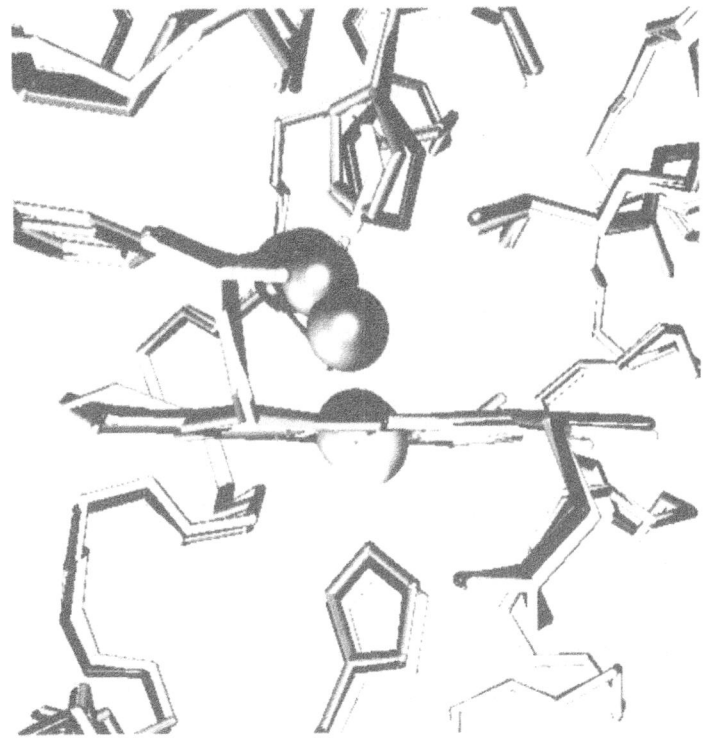

Figure 4. Superimposed models of three states of Mb: White -- deoxy; Grey -- CO; Black -- photolyzed CO.

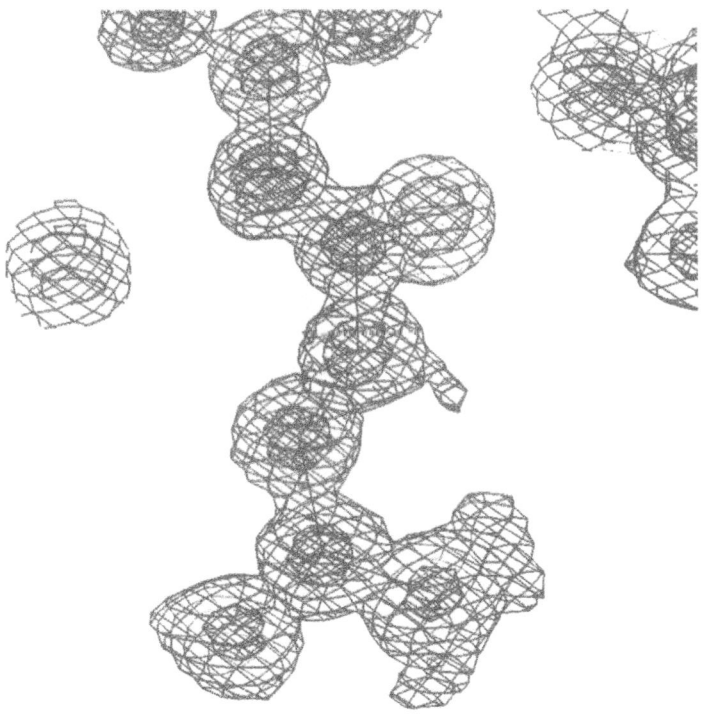

Figure 5. View of the 0.9-Å resolution map of Alpha-1 protein. The outer contours are at 1σ; the inner ones are at 2σ.

resolution of this structure is revealed in Figure 5, where one sees a portion of a leucine residue, the -CH$_2$-CH-(CH$_3$)$_2$ sidechain being quite evident in the electron density. On several of the Carbon atoms one even can see regions of density that may represent hydrogen atoms.

Virus Structure Determination

An important type of investigation where synchrotron x-ray sources have had a major impact is the study of virus crystals. Here the unit cells are very large, which leads to weak diffraction and closely spaced diffraction maxima. The brightness of the synchrotron beam is especially valuable in producing usable diffraction patterns. Such a pattern is shown in Figure 6 from a crystal of turnip yellow mosaic virus, investigated at beamline X12-C by Alex McPherson and colleagues[8]. One can see that individual reflections are well resolved, even though they are spaced by 1/500Å. Use of the Θ arm to tilt the detector has allowed measurement of the highest resolution data that the crystals could provide.

MULTIWAVELENGTH ANOMALOUS-DISPERSION (MAD) PHASING

The Physical Basis

The ability to choose the wavelength of radiation can be useful in determination of the phase of the diffracted ray, necessary for calculation of the structure. Electrons on heavy atoms, bound to molecules of the specimen, resonate with x-radiation of particular wavelengths. This resonance causes anomalous scattering, in which the diffracted rays have phases and amplitudes that are shifted slightly from rays of different wavelengths. These shifts facilitate nearly direct determination of the overall phase of the diffracted ray.

The nature of these effects is displayed nicely in curves of real and imaginary shifts in the scattering factors for heavy atoms as a function of wavelength. An example is shown in

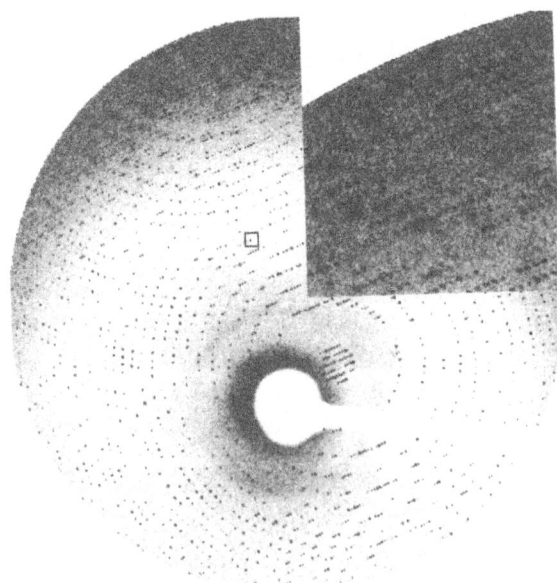

Figure 6. A diffraction pattern from a crystal of turnip yellow mosaic virus. The insert at upper-right is magnification of the upper-left portion of the pattern.

Figure 7 for selenium atoms in a protein of modest size.[9] Here the f" curve is derived directly from the x-ray absorption spectrum of the specimen, and the f' curve comes from the Kramers-Kronig inversion of f". One can see that the shifts in both curves, especially right at the transition, are dramatic. Phase information comes, then, from diffraction data that are taken at several wavelengths that exploit the differences in the spectrum. Commonly one wavelength chosen is at the point of maximum inflection of absorption (f") where f' will be a strong minimum. A second choice will be the maximum of absorption and f". This will enhance the difference between portions of the diffraction pattern related by inversion symmetry (Friedel mates). Finally, a third wavelength is often employed, chosen at a point where f" is still large but f' is less strongly negative, to enhance the f_3-f_1 difference.

Hendrickson and LeMaster[10] have defined a protocol for a structure-solving vehicle that depends on modern molecular biology and the use of synchrotron radiation. Their method, which has succeeded many times and is finding increasing popularity, is to clone the gene for the protein for which a structure is to be determined, express this gene in a species of organism that requires the sulfur-containing amino acid methionine for growth, and then to replace ordinary methionine in the growth medium with seleno-methionine (Se-Met). One finds that the biology is not too difficult and the protein often crystallizes under conditions identical to the normal crystals. When the protein is not too large, the crystals are good, and there is at least one Se-Met for each 100 amino acids in the protein, the structure usually can be solved directly by this MAD (multiwavelength anomalous dispersion) method.

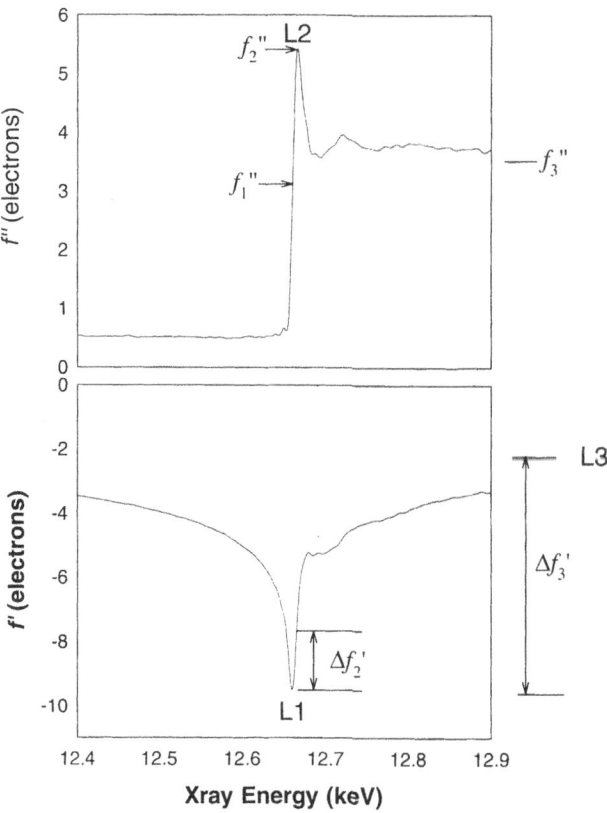

Figure 7. Plots of the imaginary (top) and real variation of scattering for Se atoms as a function of wavelength.

The Quality of Results

An example of such a structure is one solved recently by Brookhaven biologist Venki Ramakrishnan from data measured at beamline X12-C.[11] This is the structure of the C-terminal fragment of the ribosomal helper protein, Initiation Factor 3 (IF3-C). A representative portion of the electron density map is shown in Figure 8. The quality of the phases is so good that little imagination was required to build a model to fit this electron density. In the figure, an ion pair between the carboxylate of a glutamine side-chain and the guanidino group of an arginine appears in almost atomic detail.

There is prospect for the measurement of phases to produce extraordinarily accurate electron-density maps. The atomic transition that produces the anomalous-scattering effects used in this Se-Met-phased map involves the K-shell electrons of selenium. Use of this sort of atom and this transition is important because the energy of the transition is especially convenient for crystallographic use: something in the range of 8-15 keV (0.7 to 1.5 Å wavelength) is best. Higher atomic-number atoms, say the lanthanides and noble metals, have L-shell transitions that also lie in this range. Crystallographers have already worked out vehicles for binding of such atoms as gold, platinum, mercury, and lead to protein molecules in the context of conventional phasing methods. These also are useful for MAD phasing. Sometimes the lanthanides bind to proteins in ways that produce especially sharp transitions, producing amplitude shifts in the 20-electron range. The phasing from these atoms can be dramatic.[12]

RAPID DATA COLLECTION -- THE WHITE-BEAM LAUE METHOD

The Basis of the Method

Most x-ray diffraction data are recorded with a beam of x-rays that is monochromatic, or at least quasi-monochromatic. However, some of the original x-ray tubes produced x-radiation that was polychromatic or "white." Therefore, many of the original diffraction

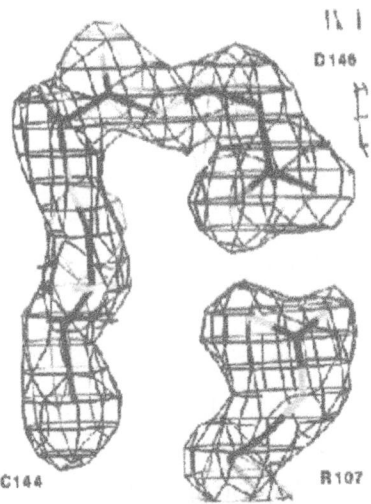

Figure 8. An electron-density representation from a MAD-phased map of IF3-C. A Glu-Arg ion pair is shown clearly.

patterns were taken with these tubes by Max von Laue and his co-workers.[13] These patterns were useful in understanding the symmetry of the crystals being studied, but little attempt was made to measure accurate diffraction intensities from them. This was owed, in part, to the supposition that many reflections that were energy harmonics of one another would appear superimposed on the x-ray film, and that their respective intensities could not be resolved.

However, driven in part by the abundance of polychromatic radiation available from modern synchrotron x-ray sources, workers have attempted to make some sense of these patterns. In two important works, Cruickshank and colleagues established the modern basis for the measurement of accurate diffraction itensities with a polychromatic beam of x-rays.[14,15] An important aspect of this was to have proven that really only a modest fraction of the data would be "harmonic overlaps."

An important feature of the white-beam Laue method is that a large number of reflections can be measured from a single, stationary crystal. The exposure time required to record these reflections depends on the spectral brilliance of the synchrotron source. To put it in perspective without making too much of the theory, if the x-ray beam from a given synchrotron will produce a complete data set from something like 100 difraction images with a total exposure time of 2-6 hours, then a nearly equivalent set of data might be measured from the same synchrotron from something like eight exposures in a total exposure of 100 milliseconds!

One might ask, why wouldn't one take all ones data that way? The answer is that although the data can be of reasonable quality, the specimen crystal must be quite perfect for the method to work at all, the signal-to-noise of the data is systematically lower than for monochromatic data, and a portion of the data, especially the lowest resolution data, are systematically missing from the set of Laue data.

Molecular Movies

However, there are experiments for which the Laue method can be extremely valuable. These are cases where one wants to trap an intermediate of some reaction in the crystals and a very short exposure time is required. A protocol might be established to trap such an intermediate: a reaction is triggered in a crystal of an enzyme by some outside stimulant, such as a flash of light or the diffusing into the porus crystal of a reactant, then x-ray exposures are taken, timed to correspond to a major event in the course of the reaction being triggered. In some circumstances one might take multiple exposures after an event had been triggered that would evolve relatively slowly. In others, the event might be triggered several times while the necessary diffraction patterns are recorded.

Protein cleavage by Trypsin. This strategy has been used several times in experiments performed at the beamline at the NSLS that has been fitted out for Laue crystallography, X26-C.[16] One of these involved study of the hydrolytic enzyme trypsin. In this case we were able to prepare a transient intermediate in hydrolysis, then to trigger the continuation of the reaction by a change in pH stimulated by simple diffusion of buffer into the crystals. The hydrolytic cycle of trypsin involves nucleophilic attack by a serine hydroxyl group on the carbonyl carbon atom of the scissile bond, either an amide or an ester. This attack is assisted by the basic side chain of a histidine residue. The reaction proceeds through a tetrahedral intermediate to form an acyl enzyme, with the leaving group, amine or alcohol, being cleaved away. The next step is attack by water, catalyzed in the same way. The intermediate we were able to prepare, the p-guanidino benzoate, forms at low pH, but its subsequent hydrolysis is so slow that there is time for crystals to be grown.[17]

The protocol for the experiment was firstly to mount a guanidino-benzoyl trypsin crystal in a flow cell, a thin walled capillary with the crystal held in place with cotton fibres and thin flexible tubing connected to allow a flow of buffer. The pH was held at 5.5, which is too low for hydrolysis of the ester bond. Then diffraction data were taken over the course of several minutes (six or seven exposures, each with an exposure time of several tens of milliseconds each). The pH was raised around the crystal by the changing of the solution passing through the flow cell, the crystals were allowed to incubate three minutes at this pH, and then another set of exposures was taken. Finally, an additional set of exposures was taken 90 minutes later.

The result of the experiment, shown in Figure 9, was dramatic and beautiful. The initial low pH transient structure showed the acyl group positioned firmly in the active site of the enzyme. There was a constellation of fixed water molecules in the region, but none was in a position either to attack the carbonyl carbon atom or to be polarized by the catalytic histidine residue. However, the structure determined after the three-minute incubation at higher pH revealed a rearrangement of water molecules, and one of these had moved in to lie in contact with the sensitive carbonyl and near to the histidine residue, in a perfect position for attack.[18] The structure determined 90 minutes later revealed substantially the same thing as the three-minute structure, except that the fractional occupancy of the guanidino benzoate moiety had dropped to approximately two thirds, a fraction confirmed by later chemical analysis. We judge that this state represents a point on the trajectory of a water molecule during hydrolytic attack.[19]

A Step in the Citric Acid Cycle. A more recent work by Barry Stoddard[20] involved the trapping of steady-state intermediates by the clever choice of site-directed mutants. The reaction being investigated was that catalyzed by isocitrate dehydrogenase (IDH). The reaction, goes in two steps:

$$E + NADP^+ + COOH-CH_2-CHCOOH-CHOH-COOH$$

$$E \cdot COOH-CH_2-CHCOOH-CO-COOH + NADPH \rightarrow$$

$$E + COOH-CH_2-CH_2-CO-COOH + CO_2$$

The first is the transformation of isocitrate to oxalosuccinate, an intermediate that had never actually been observed before this experiment. The second is conversion of this to α-ketoglutarate. Each of these is catalyzed by a different portion of the enzyme.

An important residue in catalysis of the first step of the reaction was found to be the tyrosine residue at position 160 in the protein. When this residue is changed by artificial mutation to a phenylalanine (removal of a single OH group) the catalytic activity decreases substantially but does not go to zero -- several lines of evidence suggest that the mechanism for catalysis of the mutant is the same as that of the natural enzyme, so structures one sees in the mutant may reflect the mechanism of the normal enzyme. Similarly, a lysine residue at position 230 was found to be important in catalysis of the second step. Conversion of this residue to methionine (about the same size side-chain, but with the active amino group missing) preferentially slows the second of the two reactions.

These workers determined the structure of the intermediate states of the enzyme through use of the Laue diffraction method. Crystals of both mutants of IDH were mounted in flow cells. Diffraction data were taken first on the apo-enzymes, that is with no substrate nor co-factors present in the solution. Then the reactants were flowed into the cell and full three-dimensional data were taken, in an approach similar to that used above for trypsin. In this case the species that were to be observed should represent a steady-state accumulation of intermediates of the reaction. The importance of ones using this rapid method for data collection was that the crystals themselves were not stable in the presence of the reactants -- only a short time was available to take the data before the crystals fell apart.

The result of the two experiments are revealed in Figure 10. On the left is shown the accumulation of the reactants NADP+ towards the bottom and isocitrate complexed to a magnesium ion at the top and towards the rear. This difference-electron density map was calculated from the difference in diffraction amplitudes obtained from the crystals of the

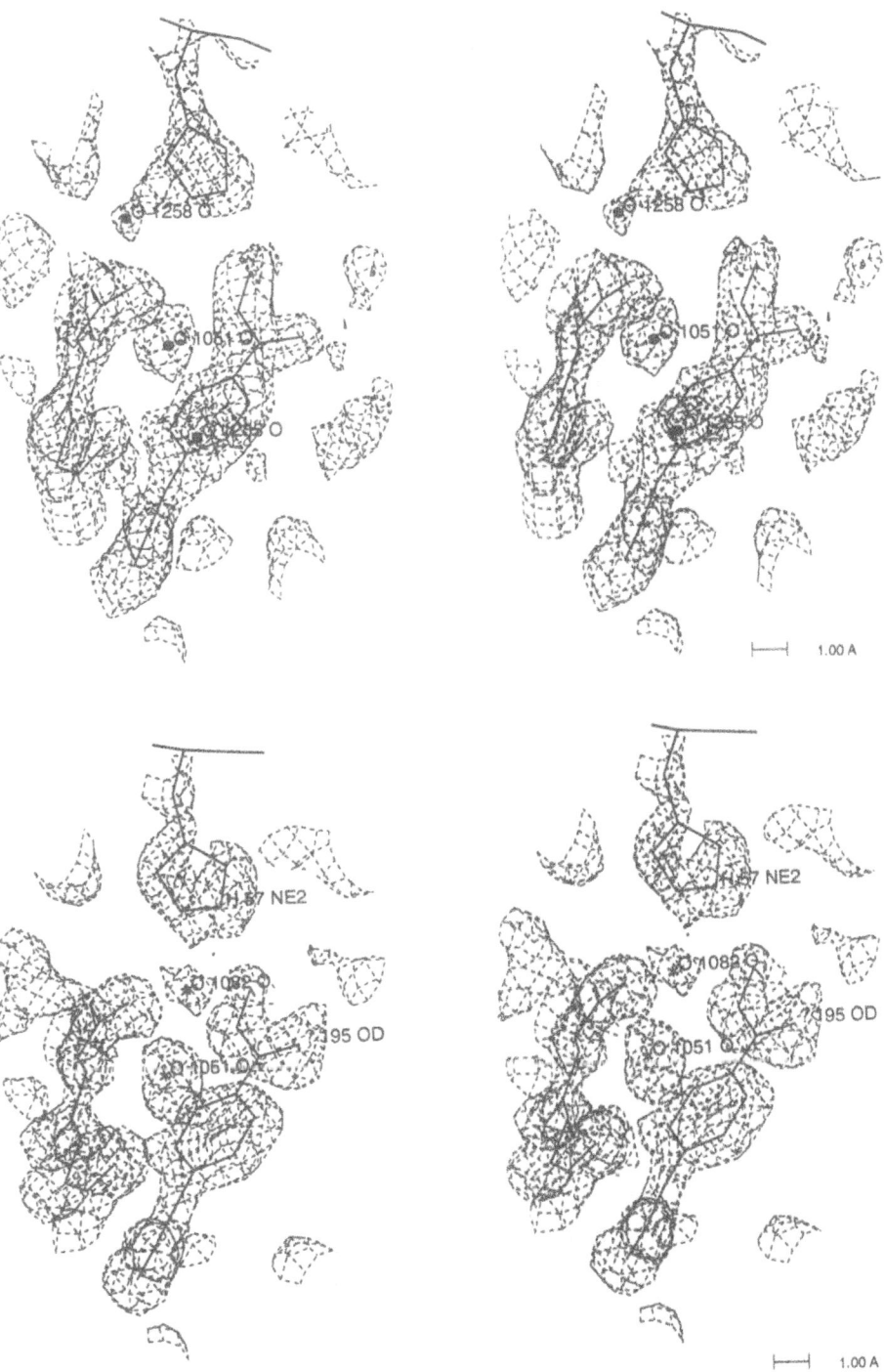

Figure 9. Guanidino-benzoyl trypsin. Low pH above; High pH below, three minutes later.

phenylalanine mutant after and before the substrate and cofactors were added. The arrangement of the nicotinamide ring with respect to the isocitrate substrate is completely consistent with all other information about how the catalysis might occur.

Accumulation of the supposed intermediate oxalosuccinate is revealed in the right hand side of the figure. The dramatic difference that is seen between the two halves of the figure arises from the loss of electron density in the portion of the NADPH molecule that includes the nicotinamide ring and its phosphate backbone. The adinine residue is seen clearly at the bottom, so it must be that the remainder of this coenzyme molecule is not bound firmly enough to the rest of the molecule for its electron density to be seen in this time-averaged crystal structure. This result suggests that the ordering of the positively charged, oxidized nicotinamide in the first case is driven by an ion-pair interaction between it and the acidic substrate. This interaction is missing in the second step. Nonetheless, this structure confirms nicely the appearance of the oxalosuccinate intermediate, which had only been proposed before this experiment.

CONCLUSION AND SUMMARY

Modern synchrotron x-ray sources provide a powerful weapon in the armamentarium of the modern structural biologist. The combined power of the intense and collimated beams from these sources and the well instrumented and run facilities allow measurement of data and performance of experiments that are not possible with ordinary laboratory facilities. They have provided an important proving ground for methods and instruments while producing both conventional and exotic results. The consequence of this is that diffraction methods in structural biology will continue to be driven by and to depend upon the synchrotron sources and their dedicated facilities well into the next century.

ACKNOWLEDGEMENT

Beamline X12-C and most of the work of this author is supported by the Department of Energy Office of Health and Environmental Research. Some support comes also from the National Science Foundation, and the Bristol-Meyers Squibb and Upjohn pharmaceutical companies.

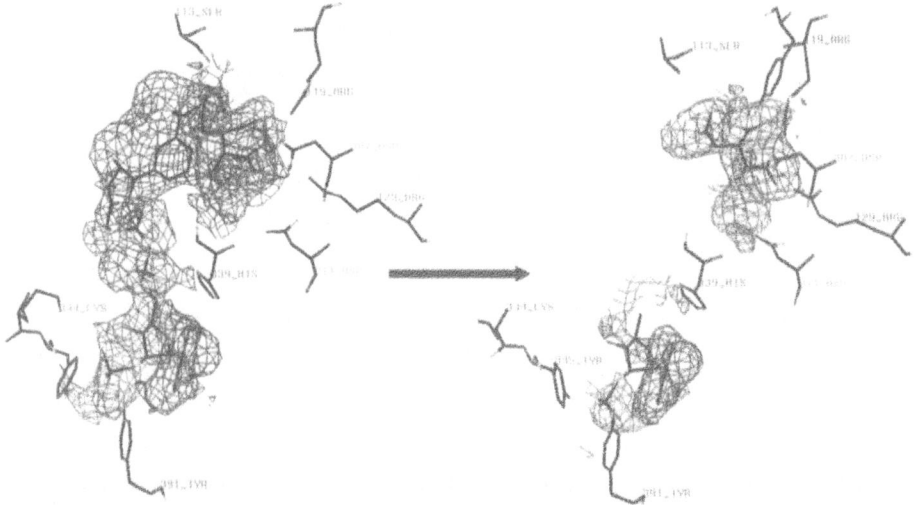

Figure 10. Left: NADP+ and isocitrate bound to IDH. Right: oxalosuccinate (top) and NADPH bound IDH.

REFERENCES

1. Strauss, M.G., Westbrook, E.M., Naday, S., Coleman, T.A., Westbrook, M.L., Travis, D.J., Sweet, R.M., Pflugrath, J.W., and Stanton, M. (1991). *Proc. Soc. Photo-Opt. Instr. Eng.* **1447**, 12-27.

2. Phillips, W., Stanton, M., O'Mara, D., Li, Y., Naday, I. and Westbrook, E. (1993). *Proc. Soc. Photo-Opt. Instr. Eng.***2009**, 133-138.

3. Stanton, M. (1993). *Nucl. Instr. and Meth.* **A325**, 550-557.

4. Stanton, M., Phillips, W.C., O'Mara, D., Naday, I. and Westbrook, E.M. (1993). Nucl. *Instr. and Meth.* **A325**, 558-567.

5. Prive, G. and Eisenberg, D. (1995) Personal communication.

6. DeTitta G.T., Weeks C.M., Thuman P., Miller R., Hauptman H.A. (1994). *Acta Crystallographica - A* **50**, 203-10.

7. Weeks C.M., DeTitta G.T., Hauptman H.A., Thuman P., Miller R. (1994). *Acta Crystallographica - A* **50**, 210-220.

8. Canady, M. A., Day, J., and McPherson, A. (1995). *Proteins: Structure, Functioin, and Genetics.* **21**, 78-81.

9. V. Ramakrishnan, personal communication.

10. Hendrickson W.A., Horton J.R., LeMaster D.M., (1990). *EMBO J.* **5,** 1665-72.

11. Biou, V., Shu, F., and Ramakrishnan, V. (1995). *EMBO J.* In press.

12. Brünger, A., Burling, F.T., Grost, P., Flaherty, K., and Weis, W.I., (1995) *ACA Annual Meeting* **Abstract 2m.1.A**, 51.

13. Friedrich, W., Knipping, P., Von Laue, M. (1912). *Proc. Bavarian Acad. Sci.* p. 303.

14. Cruickshank, D.W.J., Helliwell, J.R., and Moffat, K. (1987) *Acta Cryst.* A**43**, 352 - 373.

15. Cruickshank, D.W.J., Helliwell, J.R., and Moffat, K. (1987) *Acta Cryst.* A**43**, 656 - 674.

16. Getzoff, E.D., Jones, K.W., McRee, D., Moffat, K., Ng, K., Rivers, M.L., Schildkamp, W., Singer, P.T., Spanne, P., Sweet, R.M., Teng, T.-Y., and Westbrook, E.M. (1993). Nucl. Instr. and Meth. in Phys. Res. **B79**, 249 - 255.

17. Mangel, W. F., Singer, P. T., Cyr, D. M., Umland, T. C., Toledo, D. L., Stroud, R. M., Pflugrath, J. W., and Sweet, R.M. (1990). *Biochemistry* **29**, 8351-8357.

18. Singer, P.T., Smalås, A., Carty, R.P., Mangel, W.F., and Sweet, R.M. (1993). *Science* **259**, 669 - 673.

19. Perona, J.J., Craik, C.S., Fletterick, R.J., Singer, P.T., Smalås, A., Carty, R.P., Mangel, W.F., and Sweet, R.M. (1993). **261**, 620 - 622.

20. Bolduc, J.M., Dyer, D.H., Scott, W.G., Singer, P, Sweet, R.M., Koshland , D.E., Jr., and Stoddard, B.L. (1995). *Science* **268**, 1312 - 1318.

THE CRYSTAL STRUCTURES OF SOME NEW FORMS OF ALUMINUM FLUORIDE AS DETERMINED FROM THEIR SYNCHROTRON POWDER DIFFRACTION PATTERNS

Richard L. Harlow, Norman Herron, and David L. Thorn

Central Research and Development
E. I. DuPont de Nemours & Co., Inc.
Wilmington, DE 19880-0228

INTRODUCTION

AlF_3? -- You Can't Be Serious!

Why would anyone want to look at the structure(s) of such a simple material as AlF_3? The answer to this question is as clear as the hole in the ozone layer: AlF_3 phases have diverse catalytic properties where fluorochlorocarbons are concerned and could play an important role in the production of more benign Freons®. Some examples of their extensive diversity are given below:

Isomerization:	$CF_2ClCFCl_2 \rightarrow$	CF_3CCl_3
Disproportionation:	$2CF_2Cl_2 \rightarrow$	$CF_3Cl + CFCl_3$
Exchange:	$CF_3Cl + HF \rightarrow$	$CF_4 + HCl$
Addition:	$CCl_2{=}CCl_2 + HF \rightarrow$	$CHCl_2CFCl_2$

Ok, but surely the phase diagram and the structural details were worked out years ago? As it turns out, until the recent effort to make Freon® substitutes, most of the purported AlF_3 phases were known only from their x-ray powder diffraction signatures, very little had been done in the way of actually determining the structures of these materials or even to demonstrate that they actually were pure and unique phases. As the Freon® substitutes program geared up, all of the phases came under closer scrutiny. In addition, new methods of synthesis were yielding previously unknown phases of AlF_3 but only in microcrystalline, "powder" form. It was then clear that detailed characterizations of these materials would require *ab initio* structure determinations from their powder diffraction patterns, the best of which could only be obtained from synchrotron sources. This chapter

Synchrotron Radiation Techniques in Industrial, Chemical, and Materials Science
Edited by D'Amico *et al.*, Plenum Press, New York, 1996

37

discusses some of our experiences with this technique and outlines the structural results which we have obtained on a number of aluminum fluoride compounds.

The "Dark Ages" Were Not So Long Ago!

Before the mid-1980's, very little was actually known about the structures of the various AlF_3 phases. The structure of the most stable, condensed form, rhombohedral α-AlF_3, had been reported by numerous authors over a period of 60 years but had to wait until 1990 before the correct space group, R-3c, was assigned to it (1). The beta phase, the first of less dense forms, was patented in 1965 (2) but was incorrectly described in terms of an "hexagonal" unit cell based on its powder diffraction pattern. In the same year, another patent was issued for the rhombohedral gamma form (3) which the authors claimed had the VF_3 structure. Unfortunately, the powder pattern on which the claim was based did not match a calculated pattern based on the VF_3 structure with Al substituted for V. Moreover, attempts to repeat the synthesis of this phase have failed. In 1966, another "gamma" phase was reported (4); again, the structure was not determined but the powder pattern was indexed on the basis of a tetragonal unit cell. Unfortunately, the powder pattern is suspiciously similar to that of β-AlF_3 and thus is probably not a "new" phase but a poorly crystalline version of the beta form. Finally, a cubic phase with a 29.5 Å unit cell dimension, the delta form, was also patented in the mid-1960's (5) but, curiously enough, hasn't been seen since then.

Associated Synthetic Folklore

In amongst the various early patents and publications on AlF_3 are some rather crude syntheses for the various phases (2-5):

$$\alpha\text{-}AlF_3 \cdot 3H_2O \xrightarrow{\quad 400°C \quad} \beta\text{-}AlF_3$$

$$\beta\text{-}AlF_3 \cdot 3H_2O \xrightarrow{\quad 400°C \quad} \gamma\text{-}AlF_3$$

$$(NH_4)_3AlF_6 \xrightarrow{\quad 300°C \quad} \gamma'\text{-}AlF_3$$

$$\beta\text{-},\gamma\text{-},\gamma'\text{-}AlF_3 \xrightarrow{\quad >450°C \quad} \alpha\text{-}AlF_3$$

Only the fourth reaction, the high-temperature conversion of the less-dense forms of AlF_3 into α-AlF_3, is a universal truth. The thermal decomposition of α-$AlF_3 \cdot 3H_2O$ does generate the β-AlF_3 but it is poorly crystalline, a good property for catalysis but a poor property for structural analysis. The second and third reactions are more problematic. Industrially, AlF_3 is made by treating Al_2O_3 with HF at temperatures exceeding 450°C where the alpha form is virtually the only product.

Importance of Lower-Density forms of AlF_3

As the move to commercially manufacture Freon® substitutes gained momentum in the last half of the 1980's, one particular reaction was being investigated in detail at

DuPont, namely the conversion of perclene, $CCl_2=CCl_2$, into F-123, CF_3CHCl_2, and F-124, CF_3CHClF. Pure α-AlF_3, synthesized via the industrial route noted above, was not an active catalyst for this reaction. When, however, a few percent of a first-row transition-metal chloride, such as $CoCl_2$, was absorbed on the starting Al_2O_3, a small amount of β-AlF_3 was produced along with the α-AlF_3. This mixture was active. Interest in the less-dense forms of AlF_3 was thus heightened and the search for new, more open forms began. The early work generally involved thermal dehydration and/or decompositions of both known and novel hydrates, salts and acid salts (6). In most cases, only known phases were produced but occasionally new forms were at least suggested as impurities from weak peaks in the powder diffraction patterns.

The Structure of β-AlF_3

In late 1988, the structure of β-AlF_3 was reported (7). It appeared without the fanfare that it deserved: it was a classic piece of crystallographic work. The crystals of β-AlF_3 were grown in a RbCl flux containing α-AlF_3 with, curiously enough, a small amount of $CoCl_2$ and CoF_2. The crystals were discovered to be orthorhombic, but triply twinned to yield, in general, diffraction patterns with hexagonal symmetry in which all of the reflections exactly superimposed, making it impossible to deconvolve the intensities into their single crystal equivalents. The authors, however, managed to find a crystal which was less than perfectly twinned, i. e. the volumes of the three domains were not equal. In this situation, the superimposed intensities could be deconvolved and the structure determined and refined. For the first time, there was hard evidence for the structure of the most important of the less-dense forms of AlF_3.

COMPARISON OF THE α-AlF_3 AND THE β-AlF_3 STRUCTURES

α-AlF_3 crystallizes in rhombohedral space group R-3c (1). A view of the structure along the unique axis is shown in Figure 1a. From this vantage point, the structure appears to be quite "solid". Closer inspection, however, reveals that the structure actually consists of an infinite array of distorted cubes, where the aluminum atoms sit on the corners of the cube and the fluorine atoms sit along the edges (Figure 1b). Alternatively, the structure can also be described as three interpenetrating, 4-ring channels (based on the number of Al atoms in each ring). The Al atoms are octahedrally coordinated; the polyhedra are corner shared.

β-AlF_3 is similar to that of α-AlF_3 only in that the aluminum atoms maintain their 6-coordination and their octahedral environment, and in that the octahedra continue to be corner-shared. However, the β-AlF_3 structure consists of layers containing 3- and 6-rings (Figure 2a) which stack directly on top of one another to produce an open-framework structure with 6-ring channels running parallel to the c-axis of the orthorhombic cell (Figure 2b). These channels are large enough to contain Co^{+2} ions or water molecules, the types of species that seem to template the formation of the beta phase. Furthermore, the modestly open framework offers a possible explanation for the activity of the beta phase relative to that of the alpha phase.

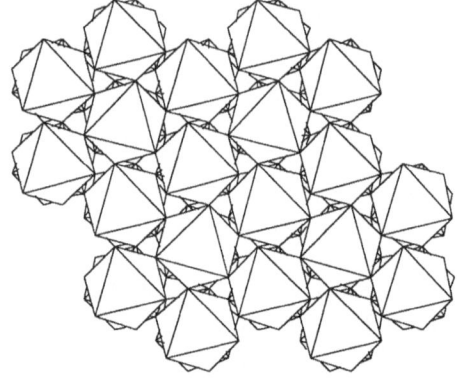

Figure 1a.   Polyhedral drawing of α-AlF$_3$ viewed parallel to the 3-fold axis of the rhombohedral cell.

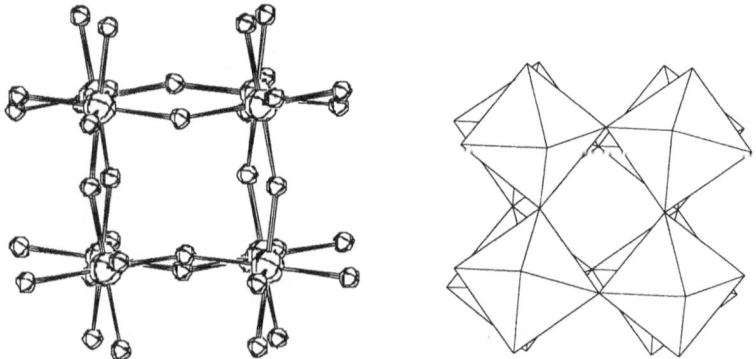

Figure 1b.   The basic building block of α-AlF$_3$, a distorted cubane-like structure. Left: atom
                drawing with the Al shown as larger spheres. Right: polyhedral version showing
                one of the 4-ring channels which interconnect at the center of the "cube".

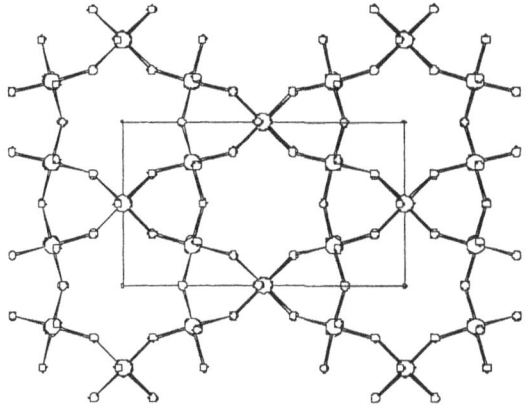

Figure 2a. One layer of the β-AlF$_3$ structure containing both 3- and 6-rings. The a-axis of the Cmcm orthorhombic cell is vertical; the b-axis is horizontal.

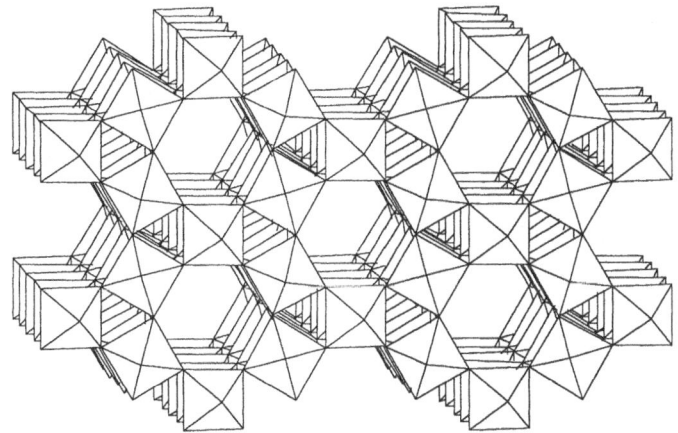

Figure 2b. The layers stack on top of each other to produce infinite 3- and 6-ring channels.

Figure 2c. The stacking of the layers in the β-AlF$_3$ structure produces distorted 4-rings between the layers.

41

ORGANIC ALUMINUM FLUORIDES

The synthesis of AlF_3 phases via dehydration and/or decomposition of inorganic compounds has severe disadvantages. First, the reactions are often unreliable, generally yielding more than one phase. Second, the products are usually not very crystalline and have broad diffraction peaks -- not very conducive to structural analysis. Third, the number of potential new phases appears (from experience) to be very limited. Finally, there is no sense of control when heating an inorganic compound until it decomposes, no way, for example, to guide the system to produce new compounds with larger channels that might be more active than the beta phase.

In an attempt to circumvent all of these disadvantages, it was decided to create a series of "organic" aluminum fluorides that would be soluble in common organic solvents. If some small aluminum fluoride species could be solubilized, then perhaps these species could be induced to join together in systematic ways (at low temperatures and in a controlled fashion) to produce new AlF_3 phases. From a reaction of one-part trimethylaluminum with four parts HF•pyridine adduct, a precipitate with a formula of $[pyridine-H]^+•AlF_4^-$ is produced in quantitative yield. While this microcrystalline solid is insoluble in most solvents, it was found that it could be rendered soluble by metathesizing the pyridinium cation by slurrying this solid into a solution containing an excess of 1,8-bis(dimethylamino)naphthalene, more commonly known as "proton sponge", PS. It was discovered that crystals of $[PSH•AlF_4]$ could be grown from a methylene chloride/toluene solution of suitable size so that an x-ray crystal structure analysis could be performed (Figure 3). Although the AlF_4^- in this structure was two-fold disordered, it was nonetheless the first structural characterization of a fluoroaluminate salt which contains discrete, tetrahedral anions (8). The structures of other salts, also made by the above metathesis reaction, have since followed (8,9).

The thermal decompositions of these salts proved to be far more interesting than their simple formulas might have suggested (10). Most salts, of course, simply decomposed in one step at relatively low temperatures. Curiously, however, they produced a variety of AlF_3 phases, including two new phases, depending on the exact conditions of

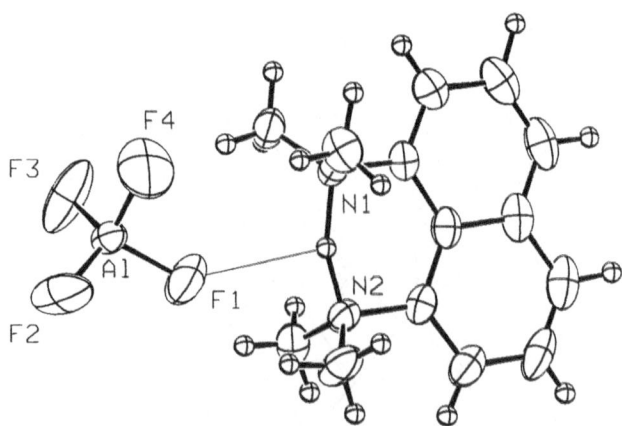

Figure 3.   Molecular structure of $[PSH•AlF_4]$. Only one of the two conformations of the AlF_4^- anion is shown: the anion is disordered about the F1--H hydrogen bond.

the decomposition:

$$PSH \bullet AlF_4 \quad \underset{300°C}{\rightarrow} \quad PS + HF + (\beta\text{-},\eta\text{-},\theta\text{-})AlF_3$$

The pyridinium salt, however, was found to decompose in two stages but essentially yielded only one product:

$$PyH \bullet AlF_4 \quad \underset{150°C}{\rightarrow} \quad Py + HAlF_4 \quad \underset{290°C}{\rightarrow} \quad HF + \eta\text{-}AlF_3$$

Decompositions via heating in high-boiling liquids were also attempted. Heating a concentrated solution of the pyridinium salt, for example, briefly in formamide yielded an unsuspected product, a new NH_4AlF_4 phase. Clearly, the salt had reacted with the solvent:

$$PyH \bullet AlF_4 + HCONH_2 \quad \rightarrow \quad Py + CO + \beta\text{-}NH_4AlF_4$$

Thermal decomposition of $\beta\text{-}NH_4AlF_4$ led to yet another previously unknown phase of AlF_3 which we have named the kappa form. A summary of the chemistry of these materials is given in Figure 4.

FLUOROALUMINATE CHEMISTRY

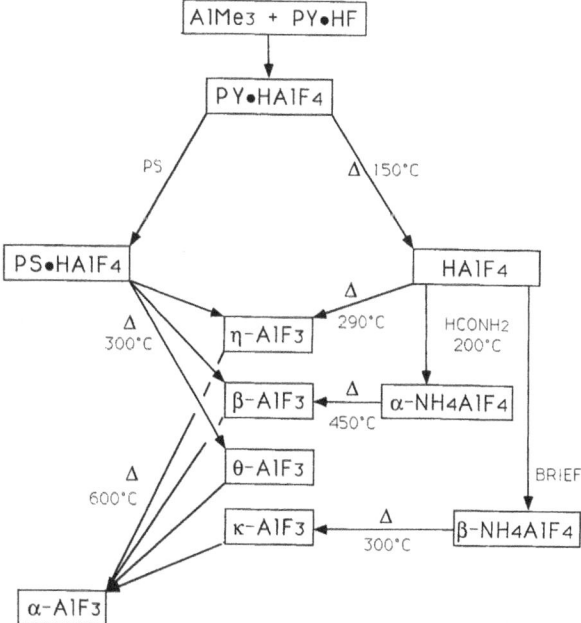

Figure 4.

AB INITIO STRUCTURE DETERMINATIONS

Refinement of the η-AlF$_3$ structure

Of the 10 compounds listed in Figure 4, only 4 had known structures: α-AlF$_3$ (1), β-AlF$_3$ (7), PSH•AlF4 (8) and α-NH$_4$AlF$_4$ (11). All attempts to grow crystals of PyH•AlF$_4$ and HAlF$_4$ large enough for x-ray, single-crystal structural analyses failed. In addition, there was no hope of obtaining single crystals of the new AlF$_3$ and NH$_4$AlF$_4$ phases, made as they were by decomposition techniques. Attempts were made to match the powder diffraction patterns against those in the ICDD-JCPDS database in hopes of finding known compounds with similar structures. Only one match was found: the cubic η-AlF$_3$ pattern was very similar to that of a series of known pyrochlore compounds, AlF$_x$(OH)$_{3-x}$ with 0.4<x<2.07 (12). A Rietveld refinement using synchrotron x-ray powder diffraction data quickly confirmed the structure shown in Figure 5 (13). As found for the alpha and beta forms, the structure consists of corner-shared octahedral AlF$_6$ units. Similar to the beta form, there are layers of octahedra (Figure 5b:right) which contain both 3- and 6- rings (compare with Figure 2a): in the beta structure, the layers stack on top of one another in only one dimension along the c axis. In the eta form, the layers stack along each of the body diagonals creating a 3-dimensional network of undulating 6-ring channels. Thus, η-AlF$_3$ has the lowest density of all the known aluminum fluoride phases. One feature which had not been previously seen is the formation of tetrahedral clusters at the points where these layers intersect (Figure 5b:left).

Details of the Solution and Refinement of θ-AlF$_3$

If the remaining phases were to be structurally characterized, it seemed obvious that the only method available was a full *ab initio* determination from their x-ray powder patterns. The "test case" for the technique was the theta phase of AlF$_3$. Its powder pattern was complicated by the fact that the compound has a tetragonal unit cell in which the c axis is roughly equal to the a axis divided by $\sqrt{2}$: many of the peaks, even at low two-theta values, were thus highly overlapped. Further complicating the pattern was the presence of peaks from other AlF$_3$ phases, most notably the eta phase. With all of these problems, it would have been foolhardy not to take advantage of the high resolution (for peak deconvolution), high signal/noise ratio (for observation of weak peaks) patterns available from the X7a and X3b1 beamlines at the National Synchrotron Light Source (NSLS) at Brookhaven National Laboratory (14). A low-angle segment of the diffraction pattern is shown in Figure 6: the lower tick marks represent contributions from the theta phase; the middle marks, the eta phase; the top marks, weak broad peaks of the beta phase. This figure emphasizes that the sample contains impurity phases (second major peak is from the eta phase) and that many of the low-angle peaks are overlapped (third major peak has two contributing hkl's).

An ab initio structure solution from powder data proceeds in the same five steps as would a structure determined from single-crystal data. Each step is at least more difficult in the case of powder data and sometimes is impossible. The structure of the theta phase can be used to illustrate several points. The first step is the determination of the unit cell. Many programs are available for the purpose but they all expect that the peak positions

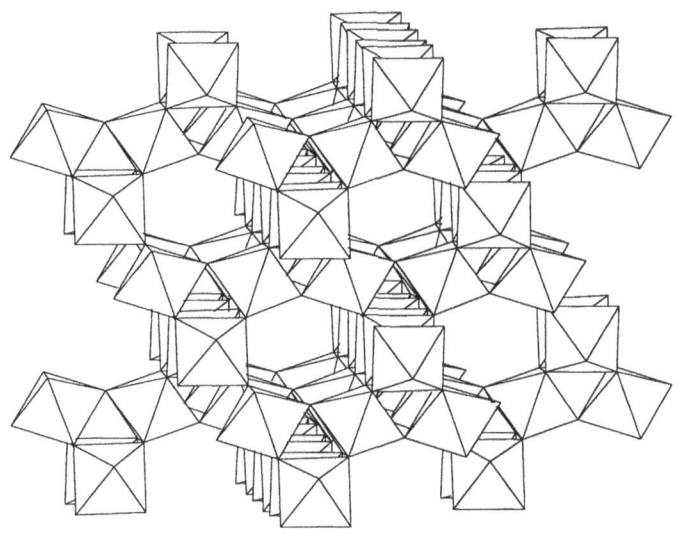

Figure 5a.   Polyhedral drawing of η-AlF$_3$ showing the undulating 6-rings. The view is approximately parallel to the [110] direction.

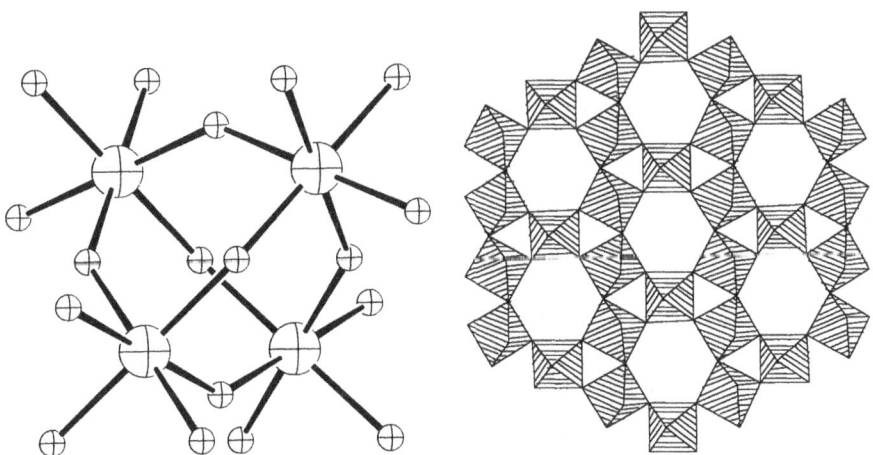

Figure 5b.   η-AlF$_3$ structural components. Left: tetrahedral cluster of four AlF$_6$ units. Right: one of four intersecting layers which stack perpendicular to the body diagonals of the cubic unit cell.

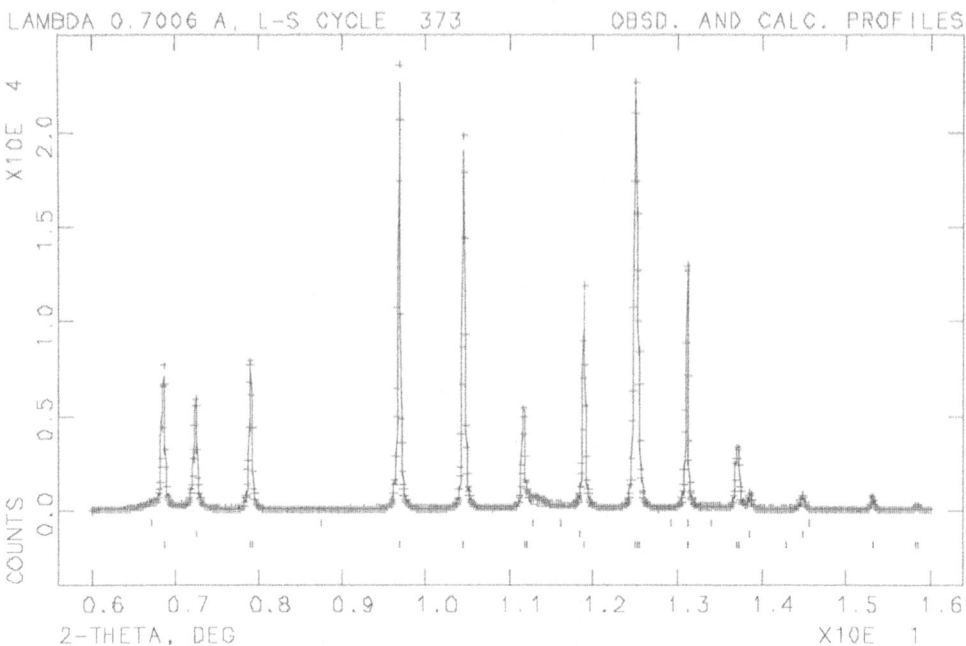

X10E 4

COUNTS

2-THETA, DEG   X10E 1

Figure 6.   A portion of the synchrotron x-ray pattern of the theta phase taken at beamline X7a
               at a wavelength of 0.70059 A.

represent a single phase. When all the peaks in the pattern, a mixture of eta and theta
peaks) were used as input, the programs were unable to find the correct unit cell. If the eta
peaks could not be identified and remove from the input, the structural solution would have
come to an end because the cell could not be determined. Suggestions concerning the
random removal of peaks in the hopes of finding a cell do not appear to be very practical in
the general case and have not been tried on this example. Removing the eta peaks from the
input allowed the program to determine the unit cell without any ambiguity. In the present
case, it found a tetragonal cell with a = 10.1844(1) and c = 7.1728(1) Å. With the unit cell
in hand, the d-spacings for all of the potential reflections could be calculated. The overlap
problem mentioned above was now obvious. The second step, the deconvolution of the
peaks, was now carried out with the aid of a computer fitting routine but was very carefully
monitored to make sure that the assignments of intensities to the various overlapped peaks
was reasonably correct. The third step was the determination of the space group. A review
of the systematic absences, always made difficult by the general overlap problem, pointed
to the possibility of an n-glide perpendicular to the c-axis: possible space grougs were thus
P4/n and P4/nmm. The ambiguity in the Laue group, 4/m vs 4/mmm, is a very serious
problem in *ab initio* powder diffraction studies. In either case, the d-spacings of the hkl
and khl reflections are identical, which means that most of the diffraction peaks consist of
two reflections which are exactly overlapped. If the structure conforms to space group
P4/nmm (Laue group 4/mmm), then the hkl and khl have equal intensities and thus the
intensity of the hkl peak can be readily assigned as half of the total intensity. If the
structure is P4/n (Laue group 4/m), then these two reflections are likely to have very
different intensities and there is no way of assigning an appropriate intensity to either of

46

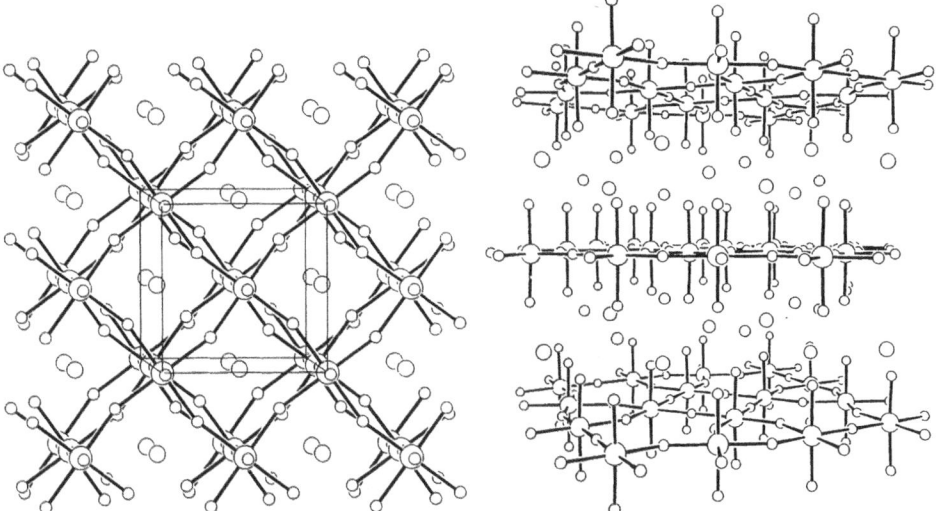

Figure 9.   Structure of α-NH$_4$AlF$_4$ with the unattached spheres representing the nitrogen atoms of the ammonium ions. Left: view approximately parallel to the c-axis. Right: perspective view approximately parallel to the a-axis.

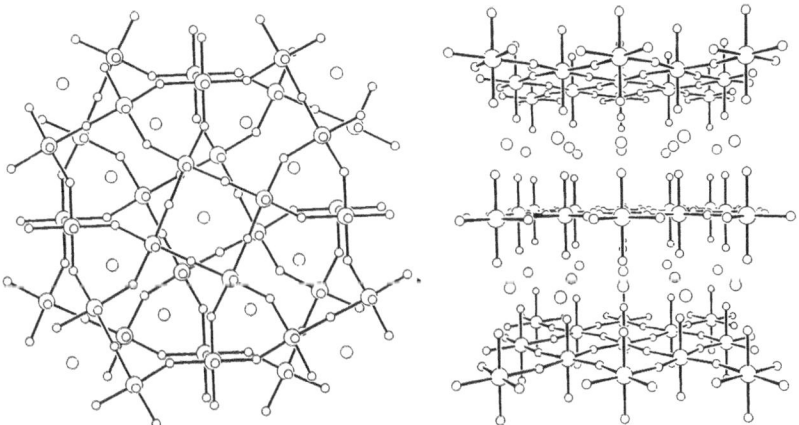

Figure 10.   Structure of β-NH$_4$AlF$_4$ with views similar to those in Figure 8. Left: view approximately parallel to the c-axis. The 4-rings (center) stack on top of each other but are staggered from one layer to another. A 3- and 5-ring pair in one layer sit above/below a 5- and 3-ring pair in adjacent layers. Right: perspective view approximately parallel to the a-axis.

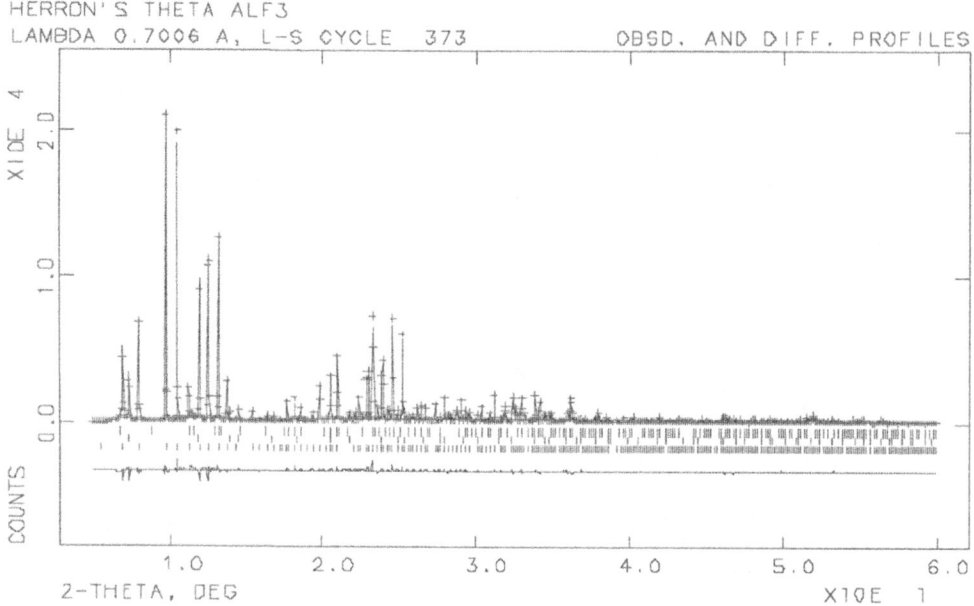

Figure 7a.   Observed (crosses) and calculated (line) intensities for the x-ray refinement of the theta phase of AlF_3 with inclusion of the impurity phases, eta and beta AlF_3. The difference intensity is plotted below the tick marks.

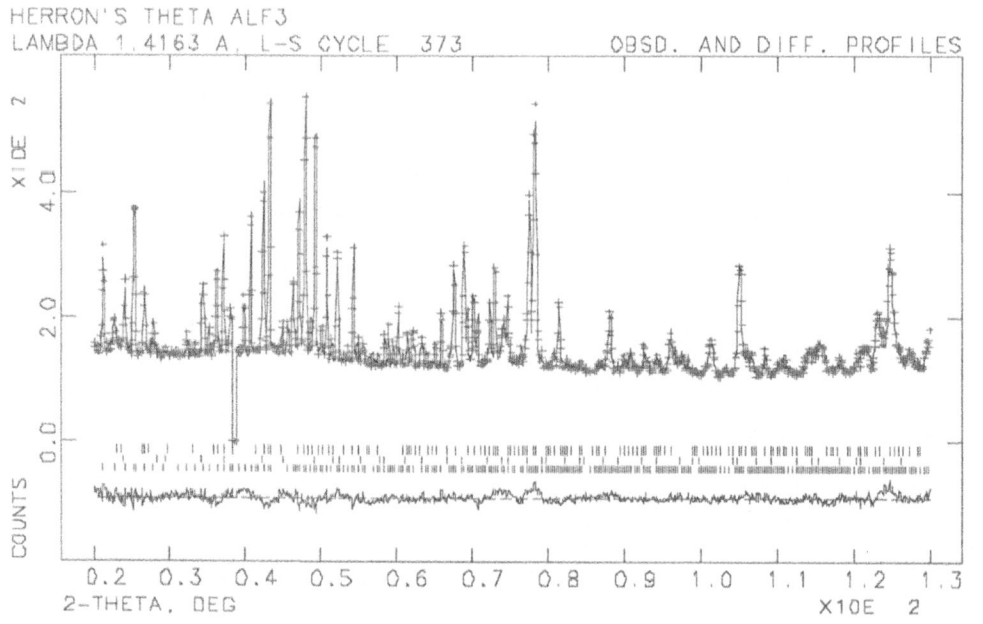

Figure 7b.   Same as Figure 7a but showing the fit to the neutron diffraction data.

these reflections and ultimately, no way to determine the structure directly from intensity information alone. This type of Laue group ambiguity occurs not only for tetragonal crystals but also hexagonal and cubic crystals and this can be a real impediment to the structure determination process. Assuming that the structure conformed to P4/nmm, a table of hkl's and their intensities were generated and fed into a direct methods program normally used for single-crystal studies. The E-maps which resulted had to be examined very carefully because both the figures-of-merit and the electron densities were much less reliable than in the single-crystal case. A structural solution was found and the final step, a refinement of the structure by the Rietveld technique was inititated (15). Eventually, a neutron diffraction pattern was also collected at Oak Ridge National Laboratory and the structure was then refined with both data sets in a joint fashion. Figure 7 shows the fits to the x-ray and neutron diffraction patterns (16).

The structure of the theta phase (Figure 8) consists of corner-shared AlF_6 octahedra assembled into rings of 3, 4 and 5 Al atoms bridged by F. The 5-rings (a new structural motif for AlF_3 phases) form an undulating 3-D interconnected channel system around the same Al_4F_6 tetrahedral clusters that were seen in the eta phase. The structure also contains "linear" $[-F-Al-F-Al-F-]_n$ chains running parallel to the c axis.

Details of the Solution and the Refinement of β-NH_4AlF_4 and κ-AlF_3

As mentioned earlier, the structure of α-NH_4AlF_4 is known (11). It consists of infinite layers of 4-rings stacked along the unique axis, separated by layers of ammonium ions packed into a tetragonal unit cell (Figure 9). The characterization of the new β-NH_4AlF_4 phase proceeded in much the same manner as that followed for the theta phase of AlF_3 except that this determination was initially much more straightforward. An x-ray powder pattern was obtained from the X3b1 beamline at the National Synchrotron Light Source, the peaks were fit, and the pattern was indexed with a tetragonal cell: a = 11.6390(1) and c = 12.6602(3) Å. The space group was assigned as I4/mcm; the presence of the c-glide eliminated the Laue group ambiguity. The structure was again solved by direct methods and refined by the Rietveld technique. The structure of β-NH_4AlF_4 is shown in Figure 10. It is very similar to that of the α-NH_4AlF_4 in that it contains layers of cornered-share Al octahedra interspersed with layers of ammonium ions. The beta phase

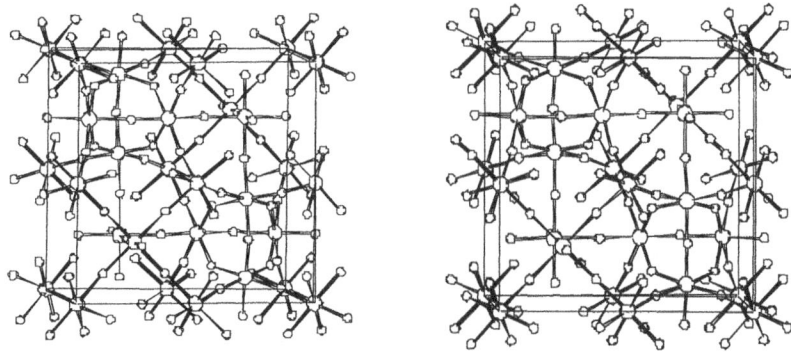

Figure 8. Stereodrawing of the structure of θ-AlF_3 as viewed perpendicular to the ab face. The "filled" Al atoms outline two of the tetrahedral clusters.

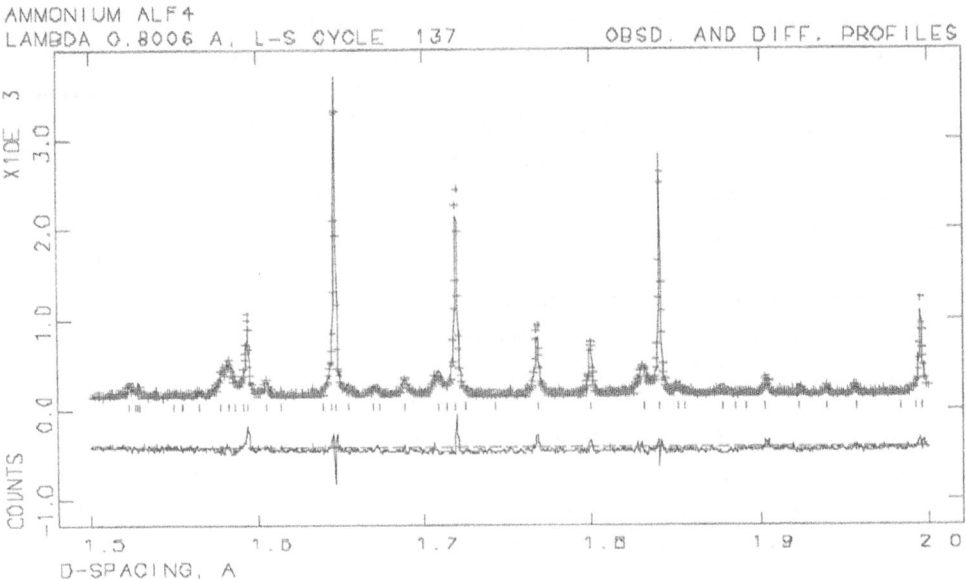

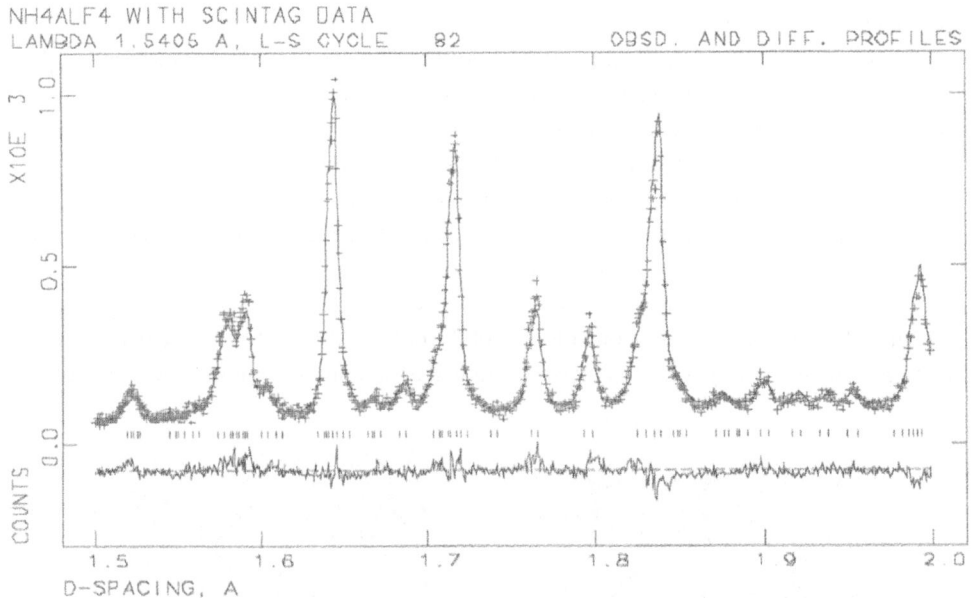

Figure 11.   Comparison of the high-angle regions for the two x-ray refinements of β-NH$_4$AlF$_4$:
top, synchrotron data; bottom, Scintag data. The patterns are plotted as a function
of d-spacing rather than 2θ because the two patterns were taken using very different
wavelengths: synchrotron, 0.8006 A; Scintag, CuKα radiation (the tick marks in
the Scintag pattern represent contributions from both Kα1 and Kα2).

distinguishes itself by having 3-, 4- and 5-rings within the layer as opposed to the pure 4-ring construction of the alpha phase.

During the refinement process, the isotropic thermal parameters of the two independent nitrogen atoms both went negative. As part of our effort to understand the reason for this, an x-ray powder pattern was collected on an in-house diffractometer. Although two refinements gave similar results as far as the nitrogen thermal parameters were concerned, it was interesting to see the differences between the synchrotron and in-house x-ray diffraction patterns as shown in Figure 11. The problem with the negative thermal parameters apparently arose because the refinement included only the nitrogen atoms of the NH_4^+ ions. The hydrogen atoms are necessarily disordered in this structure and so it seemed that their contribution to the structure factors would be very small. The refinement, however, obviously suggested that their contributions are important. At one point, the occupancy factor of the nitrogen atoms were refined and they immediately jumped to values which were equivalent to 11 electrons; at the same time, their thermal parameters became very large, an attempt to spread the 11 electrons over a large area -- not an unreasonable way to describe a disordered ammonium ion.

When heated, β-AlF_4 converts into κ-AlF_3. Although the peaks in this pattern were somewhat broad, indicating the presence of small crystallites, the structure was solved in an *ab initio* fashion without any apparent difficulties: tetragonal, P4/mbm with a = 11.4060(4) and c = 3.5443(2) Å. The structure is very similar to that of its parent: the individual layers are, in fact, exactly the same. The extraction of the ammonium ions and half of the out-of-plane fluorine atoms causes the layers to collapse onto one another. The reaction is not truly topotactic since adjacent layers in the parent compound had alternating orientations while all of the layers in the kappa phase are in identical orientations. The kappa phase (Figure 12) thus contains 3-, 4- and 5-ring channels which run parallel to the unique axis.

Because the fit between the observed and calculated intensities was not as good in this case as in the previous examples, a neutron diffraction pattern was obtained in order to

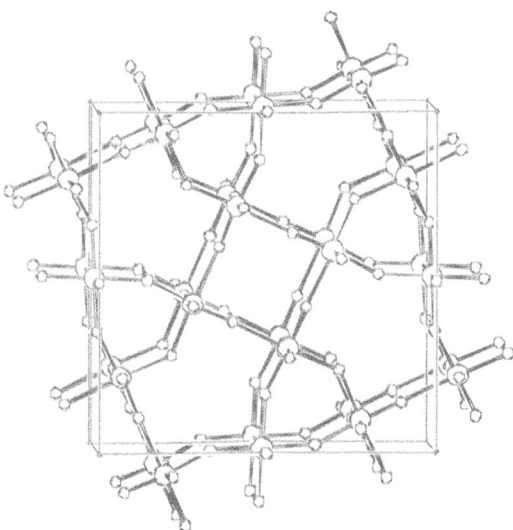

Figure 12. The structure of κ-AlF_3 viewed perperdicular to the ab plane.

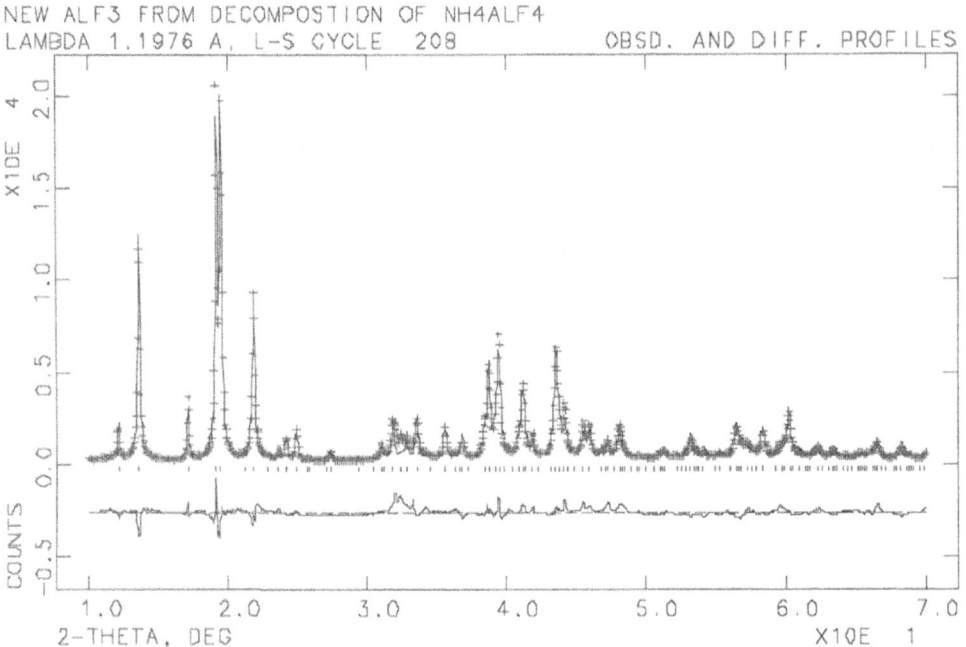

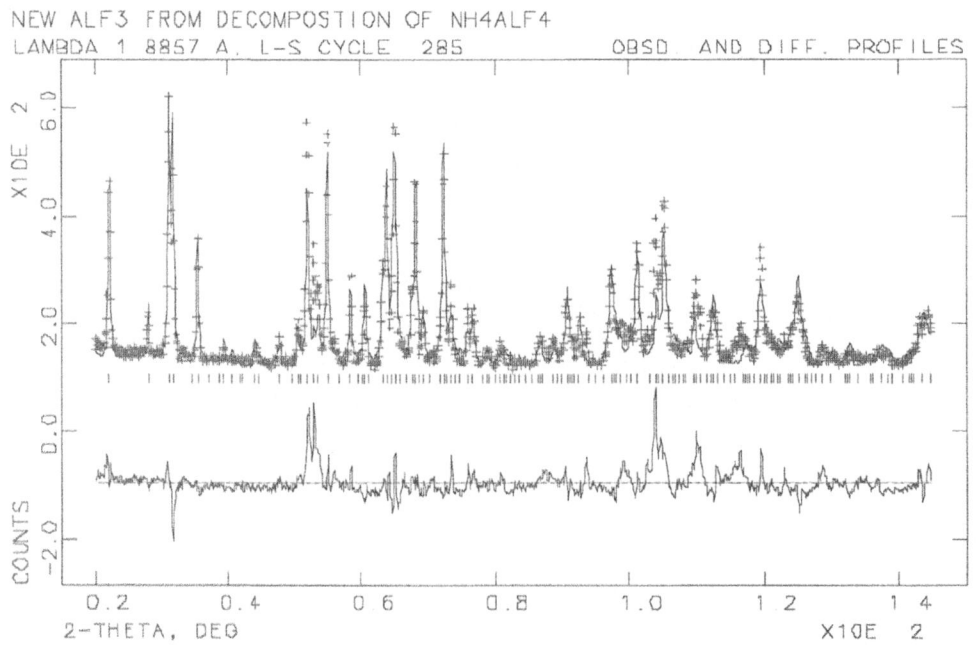

Figure 13. Results of the x-ray (top), 1.1976 Å, and neutron (bottom), 1.8857 Å, refinement of κ-AlF$_3$.

double-check the structure. The quality of the fit for the neutron pattern was, however, even worse . The two patterns are compared in Figure 13. Work to date has not elucidated the source of the problem. We believe that the structure is basically correct, but are concerned about two possiblities. First, the compound may contain an impurity of unknown structure. Second, there may be some hydrogen atoms still in the channels of the structure as suggested by the high background of the neutron pattern. The presence of these hydrogen atoms would affect the neutron intensities much more than the x-ray intensities.

CONCLUSIONS

It should be clear by now that the *ab initio* determination of structures from synchrotron powder diffraction patterns can be a powerful analytical technique. Figure 4, for example, now contains a majority of phases with known structures, 8 out of 10, and has given us a remarkable overview of the structural versatility of these aluminum fluorides (17). Our biggest disappointment is that we have been unable, so far, to determine the structure of the acid aluminum fluoride phase, $HAlF_4$.

We have also learned that the technique has many limitations. It requires a relatively pure sample (unless the impurity peaks can be identified). For some crystal classes, structures which conform to the lower of two Laue symmetries simply cannot be solved in a direct fashion. Accidental overlap of peaks will always be a problem in the assignment of intensities and will, in fact, become more of a problem as structures with larger unit cells are tackled. Even if a structure can be solved and refined, there still may be nagging questions as to whether it is a "uniquely" correct solution or only one of several possibilities. Joint refinements with both x-ray and neutron data can give added confidence (θ-AlF_3) or, as the kappa phase made clear, added concern (κ-AlF_3). No powder structure will ever be as "good" as a single-crystal structure, but surely it is better to have a crude structure than none at all.

ACKNOWLEDGMENTS

The authors are grateful to the Department of Energy for their general support of the neutron and synchrotron facilities at Brookhaven (HFBR and NSLS) and Oak Ridge (HFIR), and for their direct support of the beamlines mentioned in this chapter. We would also like to thank our colleagues at these facilities: J. Fernandez-Baca (HFIR) and T. Vogt (HFBR) who collected the neutron diffraction data on the theta and kappa phases, respectively; D. Cox and J. Hriljac for their operation of NSLS beamline X7a and for their technical expertise in *ab initio* structure determinations. We also owe thanks to P. Stephens (SUNY, Stony Brook) for bringing up the new powder station at X3b1 in a timely fashion. From a long list of DuPont colleagues, we would like to particularly mention the technical assistance of W. Marshall and our in-house x-ray facilities under G. Jones and C. Foris

REFERENCES

(1)   Ph. Daniel, A. Bulou, M. Rousseau, J. Nouet, J. L. Fourquet, M. Leblanc,
        and R. Burriel, A study of the structural phase transitions in AlF_3: x-ray powder
        diffraction, DSC and raman scattering investigations of the lattice dynamics and
        phonon spectrum, *J. Phys: Condensed Mat.* 2:5663 (1990).

(2)   F. J. Christoph and G. Teufer, U.S. Patent No. 3,178,483 (1965).

(3)   F. J. Christoph and G. Teufer, U.S. Patent No. 3,178,484 (1965).

(4)   D. B. Shinn, D. S. Crocket, and H. M. Haendler, *Inorg. Chem.* 5, 1927 (1966).

(5)   F. J. Christoph and G. Teufer, G. B. Patent No. 1,026,105 (1966).

(6)   D. R. Corbin and R. L. Harlow, unpublished.

(7)   A. Le Bail, C. Jacoboni, M. Leblanc, R. De Pape, H. Duroy, and J. L.
        Fourquet, Crystal structure of the metastable form of aluminum trifluoride β-AlF_3
        and the gallium and indium homologs, *J. Solid State
        Chem.* 77:96 (1988).

(8)   N. Herron, D. L. Thorn, R. L. Harlow and F. Davidson, Inorg. Chem.,
        Organic cation salts of the tetrafluoroaluminate anion. Yes, it does exist and yes, it
        is tetrahedral, *J. Am. Chem. Soc.* 115:3028 (1993).

(9)   N. Herron, R. L. Harlow, and D. L. Thorn, Novel coordination geometries in
        fluoroaluminate salts, *Inorg. Chem.*, 32:2985 (1993).

(10)   N. Herron, D. L. Thorn, and R. L. Harlow, Molecular precursors to
        functional materials, Proc. M. R. S. Spring 1994, Vol. V, "Molecularly
        Designed Ultrafine/Nanostructrued Materials", K. E. Gonsalves, G-M. Chow, T.
        D. Ziao and R. C. Cammarata, eds., in press.

(11)   A. Bulou, A. Leble, A. W. Hewat, NH_4AlF_4: Determination of the ordered
        and disordered structures by neutron powder profile refinement, *Mat. Res. Bull.*
        17:391 (1982).

(12)   J. M. Cowley and T. R. Scott, Basic fluorides of aluminum, *J. Am. Chem.
        Soc.* 70:105 (1948).

(13)   J. B. Parise, SUNY at Stony Brook, personal communication.

(14)   D. E. Cox, B. H. Toby, and M. M. Eddy, Acquisition of powder diffraction data
        with synchrotron radiation, *Aust. J. Phys.* 41:117   (1988).

(15)   A. C. Larsen and R. B. Von Dreele, GSAS, generalized structure analysis
        system, Los Alamos National Laboratory, LAUR 86-748 (1986).

(16)   After completion of this work, a similar pattern and x-ray powder refinement of the
        theta phase was reported by U. Bentrup, Thermal decomposition of
        $[(CH_3)_4N]M(III)F_4 \cdot H_2O$ compounds of iron and aluminum and about a new form
        of AlF_3, *Eur. J. Solid State Inorg. Chem.* t.29:51 (1992) and by A. LeBail, J. L.
        Fourquet, U. Bentrup, θ-AlF_3: Crystal structure determination from x-ray powder
        diffraction data. A new MX_3 corner-sharing octahedra 3D network, *J. Solid State
        Chem.* 100:151 (1992).

(17)   Structural details of the compounds discussed in this chapter can be found in the
        following reference: N. Herron, D. L. Thorn, G. A. Jones, J. B. Parise, J.
        Fernandez-Baca, T. Vogt, and R. L. Harlow, The structural characterization of
        three new phases of AlF_3 using powder diffraction techniques. Eta, theta, and
        kappa-AlF_3, *Chem. of Materials*, submitted.

SYNCHROTRON RADIATION - BASED RESEARCH AT THE DOW CHEMICAL COMPANY

R. A. Bubeck[1], S. R. Bare[1], B. M. DeKoven[1], M. D. Heaney[3], and
P. R. Rudolf[2]

The Dow Chemical Company
[1] Central Research and Development
[2] Analytical Sciences Laboratory
Midland, Michigan 48674

[3] Analytical Sciences Laboratory
Lake Jackson, Texas 77566

I. INTRODUCTION AND HISTORICAL PERSPECTIVE

The Dow Chemical Company has been conducting synchrotron-based experimentation through the user proposal system at various US facilities since 1983. The typical lag time between the submission of a user proposal and the allocation of beam time is about nine months. Given that high flux, tunability, polarization, time structure, and the ability to highly focus the radiation and of synchrotron sources vis-à-vis laboratory sources enable a wide range of experiments, it is clear that timely access to synchrotron beam time is key to our ability to develop synchrotron technology as a critical research tool. This experience is in tune, as well, with the findings of the Structural Biological Synchrotron Users Organization.[1] The time lag between proposal submission and obtaining beam time can be a major hurdle for being able to deliver the answers to research problems in a timely fashion.

Dow's Synchrotron Users Group, formed in December 1991, was the first attempt at bringing together key synchrotron researchers from different technology areas throughout the company. The task of bringing together the group was easy, however, accommodating all of the group needs was much more difficult based on the diversity of the synchrotron technologies needed to address the problems that are commonly encountered. Although a major portion of our early synchrotron studies were conducted on polymeric materials, from the outset it was understood that any formal synchrotron program would have to encompass everything from soft X-ray spectroscopy to non-destructive examination.

The evaluation of Dow's synchrotron needs and the formulation of both an interim and long term plan was conducted using four technical focus groups. These groups represented: (i) polymer applications, (ii) catalysis, (iii) surface science and soft X-ray spectroscopy, and (iv) crystallography and other applications. Findings from these groups indicated that the

Synchrotron Radiation Techniques in Industrial, Chemical, and Materials Science
Edited by D'Amico *et al.*, Plenum Press, New York, 1996

55

following techniques were essential to a synchrotron-based problem solving ability: small and wide angle X-ray scattering, fiber diffraction, powder and single crystal diffraction, reflectivity, EXAFS, NEXAFS, and SEXAFS. There is also much need for several different microprobe-related applications and soft X-ray microscopy.

Reviewing present (second generation) US synchrotron facilities lead to the belief that the equivalent of four or five different beamlines could probably cover the range of needed capabilities. However, opportunities for becoming Participating Research Team (PRT) members at highly subscribed beamlines are limited. It was also concluded that a certain amount of redesign/retrofitting to adapt beamlines to our specific requirements would be necessary even if specific PRT membership was available to us. The decision to join a collaboration at the third generation Advanced Photon Source (APS) at Argonne National Laboratory (ANL) was logical. In mid-1992 a long term strategy was finalized by conducting an evaluation of all the APS Collaborative Access Team (CAT) scientific proposals. Four of these overlapped somewhat with the envisioned program, however, the DuPont-Northwestern (DUNU-CAT) proposal was a good match. The resulting agreement provides 40% partnership for DuPont and Northwestern, and 20% for Dow. This share is in line with Dow's projected access needs for the long term synchrotron-based program.

The benefits of becoming a part of APS include access to a state-of-the-art third-generation synchrotron facility. It is also beneficial to be part of the design phase for a facility that will be a home base for at least ten years of operation. The ability to cost-share capital items, and conduct shared technical development also will allow us to be able to leverage our resources. There are also less tangible benefits for our scientific personnel. These include the personal interactions with world-class researchers from throughout the world. There will be opportunities to publish and gain scientific credibility in many developing fields. The collaborative access team alignment between Dow, DuPont, Northwestern, and Argonne National Laboratories is very important. The collaboration with DuPont is the first very broad technical alliance between Dow and another major corporation. The DuPont-Northwestern-Dow Collaborative Access Team (DND-CAT) comprises a total of about sixty principal investigators. Each of the partners is aligned into one of the four focus groups mentioned earlier. Each of these groups has the job of charting its course, determining joint needs such as ancillary equipment, and exploring unique opportunities such as shared capital purchases for use of the CAT. DND-CAT is governed by a management board comprised of a rotating non-voting chair, and two members each from DuPont and Northwestern University, and one member from Dow. This reflects the pro rata share in the CAT.

DND was the first CAT to sign a Memorandum of Understanding with ANL in mid-December 1993. One of main goals for 1993 was the evaluation of Lab-Office-Module (LOM) design to make them consistent with the operating procedure of our own institutions. A beamline group, located in Chicago is responsible for many facets of the collaboration including sector design and development, our management and safety plans, and the LOM evaluation. The beamline group took beneficial occupancy of the experimental hall floor at the end of April 1994. At the time of writing (June 1994) the structural steel for DND-CAT sector 5 LOM was being erected.

The interim synchrotron program to take Dow's efforts to the commissioning of APS has taken several different avenues. Consistent with early evaluations, no single beamline could fit the broad range of requirements. The interim (1993-1995) recommendations from Dow's synchrotron group indicated participation in a PRT for diffraction and scattering needs. Access to, and possible direct involvement at, ultra soft and soft X-ray facilities was also thought to be critical to our effort since ultra soft spectroscopies are key to the analysis of carbon, nitrogen, and oxygen which play a major role in many of Dow's products. To this end, an alignment with Oak Ridge National Laboratory has been negotiated and concluded, and Dow is presently a PRT member of the hard X-ray scattering beamline X-14A at the National Synchrotron Light Source (NSLS), Brookhaven National Laboratory (BNL). This has allowed the performance of a range of studies involving thin film reflectivity, residual stress, powder diffraction, and wide angle X-ray scattering. A Cooperative Research and Development Agreement (CRADA) between NIST and Dow allows PRT access to the NIST hard X-ray spectroscopy beamline X-23A2. Dow Chemical is also a founding member of the newly-formed U-7A PRT at BNL, which will satisfy our projected program for soft X-ray spectroscopies and surface science.

Dow researchers have also been involved recently in a variety of successful synchrotron-based studies. These include in-situ real-time X-ray scattering measurements of the tensile impact of engineering thermoplastics, in-situ studies of lyotropic liquid crystalline polymers, and in-situ fiber drawing experiments[2] conducted at the Cornell High Energy Synchrotron Source (CHESS). Studies of low energy (non-stick) anti-graffiti surfaces have also proved successful, as have studies in self-assembling monolayers[3] conducted at the VUV ring at NSLS. In-situ RTSAXS of the curing of polyurethane foaming have been carried out at NSLS,[4] as have some preliminary studies in the developing technology of X-ray synchrotron microscopy.[5] Catalysis studies at NSLS have also been a large part of our interim effort[6] (See also ref. 59). A CRADA with Sandia National Laboratory has also interfaced with our synchrotron program in the area of chlorinated polyethylene. Further details of several of these efforts are described in this review.

II. REAL-TIME IN-SITU X-RAY SCATTERING STUDIES OF THE DEFORMATION OF ENGINEERING THERMOPLASTICS

The first significant involvement by Dow in synchrotron-based X-ray scattering research began in the fall of 1983 as part of a collaborative effort between E. J. Kramer (Cornell University), H. R. Brown (IBM Almaden), and R. A. Bubeck (Dow Chemical). The research, which lasted over a seven year period, concentrated on real-time small-angle X-ray scattering studies of engineering thermoplastics during tensile and flex fatigue deformation. A special feature of the study is that these were the first ever in-situ X-ray scattering measurements during the deformation of polymers at near-impact deformation rates. Historically, the study of plastic deformation of polymers and materials in general has often included: (1) examination of the microstructure, (2) deformation of the specimen to the point of fracture, and (3) a post-mortem morphological analysis. Although the sequence of deformation events (i.e, crazing, shear banding, cavitation, etc.) can be determined sometimes by microscopy, the evaluation of the relative contribution of each deformation mode to the total plastic deformation as a function of time can not. Applying the technique of real-time small angle X-ray scattering (RTSAXS) to the study of relatively high strain rate deformation opens the way to the direct observation of changes in X-ray scattering associated with the modes of deformation. The combined RTSAXS mass thickness technique for studying deformation permits: (1) the observation of the sequence of deformation events and (2) the discrimination between crazing and other larger scale cavitation processes under controlled deformation rates.

The technique, which is reported in more detail elsewhere,[7,8,9] is summarized in the following brief description. High intensity X-radiation of wavelength $\lambda = 0.157$ nm was obtained utilizing the synchrotron source at CHESS. Samples were deformed in a tensile mode using a small hydraulic tensile fixture fitted with a load cell and displacement indicator. The frame was mounted on an undercarriage which was translated perpendicular to the X-ray beam by two stepping motors so that the area of maximum stress on the specimen was in the X-ray beam at the moment of deformation and fracture. The data acquisition system used to obtain the SAXS patterns is based on a Princeton Applied Research optical multichannel analyzer (OMA) equipped with a linear 1024-element X-ray sensitive Reticon array. The OMA records a one-dimensional X-ray scattering pattern with software-determined time resolution and angular spread. Time resolution is variable from milliseconds to hours while scanning the region between 2 and 20 milliradians from the primary beam. This range corresponds to microstructures of 90-900Å in size at the wavelength of X-rays used. The detector has a relatively high dynamic range (16,800), thus permitting acquisition of scattering patterns from tensile impact samples without saturation.

Simultaneously with the measurement of the SAXS pattern, the intensity of the transmitted X-ray beam (as attenuated by the sample) was measured. This measurement was done by using a beam stop thinned over a small area so that a fraction of the transmitted beam intensity was recorded on a few elements of the Reticon detector. Real-time measurement of the transmitted X-ray beam intensity permitted the measurement of changes in the mass thickness, and thus the total plastic deformation of the sample weighted over the area of the X-ray beam. An example of the line summed intensities for the in-situ RTSAXS deformation experiment obtained during the tensile impact of a high impact polystyrene (HIPS) engineering thermoplastic is shown in Figure 1.

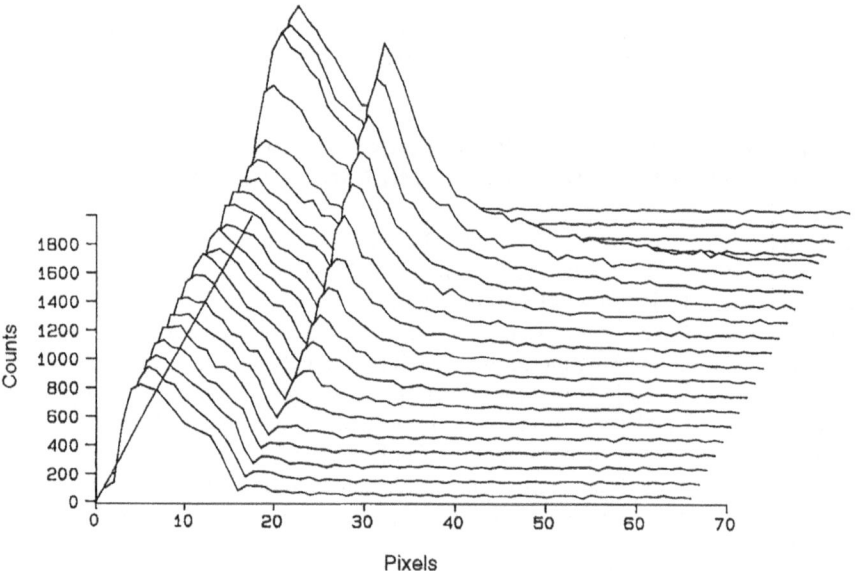

Figure 1. An example of line-summed intensities for RTSAXS during tensile impact of Styron™ (Trademark, The Dow Chemical Co.) 484 (HIPS-1) at a deformation rate of 42.cm⁻¹. Time resolution is 18.2 ms/scan.

The protocol for analysis of the data, such as that shown in Figure 1, involves the following steps: (1) the determination of total plastic strain ε_T from the decrease in X-ray absorption resulting from the sample decreasing in thickness and/or developing internal cavities as a consequence of deformation; (2) calculating the plastic strain due to crazing ε_{CR} from the analysis of the absolute scattering invariant Q(Abs.) resulting from the scattering from the craze fibrils; and (3) the subtraction of ε_{CR} from ε_T to obtain the plastic strain due to non-crazing mechanisms (i.e., shear yielding, particle cavitation, etc.). The determination of Q(Abs.) requires the calibration of the scattering experiment with a primary standard such as a Lupolen polyethylene in conjunction with an N_2 gas primary beam ionization monitor. Once Q is of sufficient magnitude, information on the evolution of the craze microstructure can be estimated from each scattering curve. The visual deformation of the sample is also video recorded using a fiber optic microscope. In cases for which there is significant ductile necking of the sample, the strains need to be corrected as a function of time to account for the decreasing amount of the material in the X-ray beam that results from the severe draw down of the sample. This correction is discussed in refs. 1 and 2.

An example analysis of RTSAXS results for a high impact polystyrene (HIPS-1) is shown in Figure 2. Included in the study were two materials (HIPS-1 and HIPS-2) each contained 7.5 weight percent butadiene rubber in the form of gel particles (i.e., "composite" rubber particles with polystyrene inclusions) as generated by mass polymerization.[10] The average particle diameters were 2.8 μm and 1.2 μm, respectively, for HIPS 1 and 2. The weight average molecular weights of the polystyrene matrix for both was about 240,000 with a polydispersity of 2.8. Other than a minimal amount of antioxidant, no other additives were present in the resins. The RTSAXS analysis for HIPS-2 is similar *in kind* to that of HIPS-1.

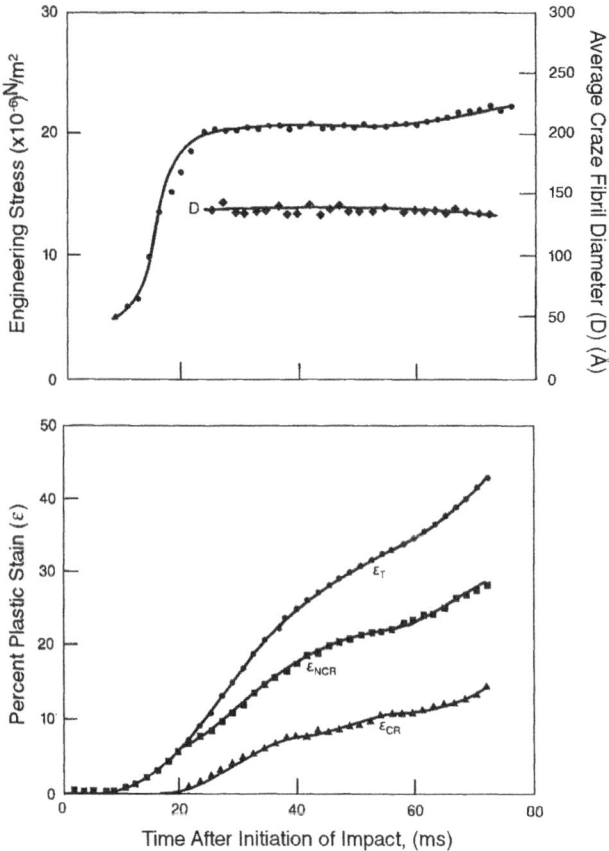

Figure 2. RTSAXS analysis of HIPS-1 at 7.1s. The upper group of plots shows engineering stress (●) and average craze fibril diameter(D)(◆). The lower group of plots show total plastic strain (e_T)(•), non-crazing strain (e_{NCR})(■), and strain due to crazing (e_{CR})(▲).

The only material parameter difference between HIPS-1 and HIPS-2 is the rubber gel particle size. For both HIPS-1 and HIPS-2, the deformation was found to be dominated by non-craze deformation. A decrease in the particle size by half increases the yield stress from 3000 psi. (20.7 MN/m^2) to 3300 psi. (22.8 MN/m^2) and the lag time for crazing from about 19 ms to about 25 ms. The time to fracture increased slightly from 73 ms to 78 ms. The decrease in particle size also results in a decrease in percent plastic strain at failure from roughly 45% to about 30%, in the tensile impact test in the strain rate range of 7/s to 35/s. The scattering center size (the average craze fibril diameter) was found to remain at about 14 nm independent of the change in particle size or the decade change in strain rate.

Many analyses of dilatation during deformation of rubber-modified thermoplastics attribute most of the contribution to this part of the deformation to crazing.[11,12,13,14,15] In this light, the two most striking (and unexpected) results of the real-time X-ray analyses are: (1) the substantial increase in noncrazing plastic strain that precedes the onset of crazing, and (2) that the noncrazing strain is, in each instance, greater than that due to crazing. The most likely source of this noncrazing strain is the cavitation of rubber particles and the bending (plastic or elastic) of the glassy polymer "ligaments" between such particles. Shear deformation on a microscopic scale in the ligaments between cavitated rubber particles can not be ruled out, however, by the absence of change in sample lateral dimensions.

Reported results (refs. 1 and 2) for acrylonitrile butadiene styrene (ABS) copolymers are consistent with the observations made by Breuer et al.[16] for ABS materials. It has been previously observed that shear yield, presumably associated with particle cavitation, precedes crazing in ABS materials with relatively high impact strength.[17] As with the HIPS samples, crazing occurs after the noncrazing strain is already large. Also, as with HIPS, we believe that the non crazing strain is most likely due to particle cavitation, resulting shear yield and/or ligament bending of the SAN ligaments between the cavitated particles in ABS, depending on the ABS composition. Crazing never contributes to more than half of the total plastic strain associated with the deformation.

RTSAXS analyses of the three point flex fatigue of polystyrene have also been reported.[18] When the minimum deformation during cyclic loading was 50% or more of the maximum, the load on the craze fibrils was found to remain tensile. When the minimum deformation was reduced below this, the load on the craze fibrils became compressive and they buckled. In this second regime, a decrease in minimum sample deformation caused a considerable decrease in fatigue lifetime associated with the severe craze fibril buckling.

Real-time small-angle X-ray scattering has been shown to be a powerful technique that clearly delineates the sequence and types of deformation events, and their absolute contributions to the total plastic deformation in rubber-modified thermoplastic polymers. Samples of realistic thicknesses can be studied at high rates of deformation without the inherent limitations of microscopy and its requirement of thin samples (i.e., plane strain constraint is maintained on sample morphology). Contrary to the conclusions of several previous dilatation-based studies, it has been demonstrated that noncrazing mechanisms, which are due mostly to rubber particle cavitation and associated ligament bending of the surrounding glassy matrix occurs before crazing in HIPS and ABS. Crazing accounts for roughly only half of the total plastic strain in HIPS, and at most half in ABS materials. The proportion of plastic strain attributable to crazing can be much less than half the total in ductile ABS systems. Because of the significant particle cavitation that is observed to precede crazing in both the HIPS and ABS systems using the RTSAXS experiments, there has been a recent reevaluation of the relative roles of cavitation and crazing in the deformation of rubber-toughened engineering thermoplastics by researchers in this field such as Bucknall,[19,20] Young,[21] and their coworkers.

III. IN-SITU X-RAY SCATTERING STUDIES OF LYOTROPIC LIQUID CRYSTALLINE MONOFILAMENT DRAWING

There has been considerable interest in fibers of very high modulus and tensile strength composed of semi-rigid rod molecules such as poly(trans-benzothiazole) (PBZT) and poly(cis-benzoxazole) (PBO).[22,23,24,25] The general scheme for the drawing of PBO fibers starts with a dry-jet wet-spinning process from a 13 - 15% solution of the polymer with a molecular weight greater than 10,000 in polyphosphoric acid (PPA). These solutions are lyotropic liquid crystalline. The fiber structure is then fixed by coagulation in H_2O, and can be further enhanced by a heat-set process under tension. In order to gain information needed to optimize the filament drawing process of these lyotropic systems, the first ever in-situ X-ray scattering experiments of the drawing process were performed as a function of basic processing conditions. The purpose of the study is to determine orientation and microstructure development in the draw zone as a function of shear rate (SR) in the capillary die and spin draw ratio (SDR).

Conventional and synchrotron source-based in-situ X-ray scattering studies of fiber drawing for semi-crystalline thermoplastics[26] have been pursued. In-situ X-ray scattering studies of the fiber formation of liquid crystalline lyotropes, which are mostly solvent by content, had not been previously attempted. The technological problems faced in the case of lyotropic liquid crystalline systems are a low polymer content, a solvent system which is X-ray absorbing, and a low coherent scattering. Subsequently, extended exposure times are required. In order to surmount these problems, a device to perform in-situ synchrotron-based X-ray scattering studies of PBO monofilament spinning was constructed. The details of its construction are to be found elsewhere.[27,28] The majority of the fiber orientation and subsequent structural changes, occur within a small draw zone typically several inches in length for our experimental unit. Many of the results of the work have been recently reported.[29]

The range of monofilament drawing conditions studied included shear rate, spin draw ratio, and barrel temperature. In-situ wide-angle X-ray scattering (WAXS) patterns were obtained in the draw zone at positions 0.25 cm (0.1 in) (zero point position), 2.54 cm (1.0 in.), and 3.81 cm (1.5 in.) below the capillary die face. An exposure time of approximately 30 s was used for the thickest extrudates and 5 min. for the thinnest extrudates. The 14% PBO/PPA solution had an inherent viscosity = 30 dl/g (using methylsulfonic acid as a solvent) for the polymer, and a 83.7% P_2O_5 level for the PPA. The solutions were filtered, degassed, and stored in Teflon™ (Trademark, DuPont) tubes prior to shipment to CHESS for their subsequent use. Monochromatic 8 KeV X-rays (wavelength $\lambda = 0.157$ nm) from the wiggler magnet were collimated with a 100 µm pinhole collimator resulting in an incident intensity of 7 x 10^{10} photons mm^{-2}sec^{-1}. The WAXS patterns were acquired by exposing Fuji phosphor image plates, which were laser-scanned after exposure with a Fuji BAS 2000 system. He-filled chambers were fitted on the optical rail before the collimator and between the extrudate and the imaging plate in an effort to minimize air scatter and absorption.

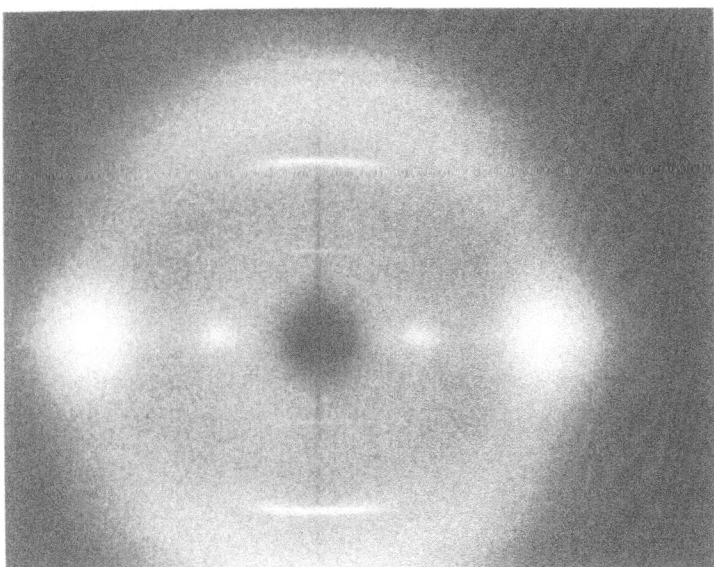

Figure 3. Typical wide-angle X-ray scattering (WAXS) pattern of a 14% PBO/PPA extrudate. The shadow of the filament can be seen aligned in the vertical direction.

A representative WAXS pattern obtained from 14% PBO/PPA extrudate during drawing is shown in Figure 3. The principal features are: (1) a prominent ring resulting from scattering from the PPA; (2) a broad equatorial reflection consisting of a doublet associated with the PBO and scattering from oriented PPA which obscures the doublet; (3) a second weaker equatorial [100] reflection; and successive layer lines stacked in the meridional direction associated with the molecular repeat spacing of the PBO molecules. These general features are consistent with what has been observed and reported for PBO/PPA solvates by Cohen et al.[30,31] For this study, the [100] equatorial reflection is the most important from a data analysis standpoint because an orientation parameter can be calculated from the azimuthal distribution of intensity in the peak.[32] The lateral and axial coherence lengths sizes, respectively, can be estimated from the width of the [100] equatorial reflection and the width of the [002] layer line.[33] Signal to noise ratios of about 3/1 were observed for the [100] reflection.

An azimuthal intensity distribution I(f) which included the equatorial [100] reflection was obtained. This profile was reformatted to enable data transfer into a curve fitting program. The orientation parameter (f) was then calculated by first fitting this intensity profile with a Split Pearson VII function. From the fitted profile the mean square cosine was determined as:

$$\langle \cos^2\phi \rangle = \frac{\int_0^{\frac{\pi}{2}} I(\phi)\sin\phi\cos^2\phi \, d\phi}{\int_0^{\frac{\pi}{2}} I(\phi)\sin\phi \, d\phi} , \tag{1}$$

where I is intensity as a function of subtended arc angle ϕ. The values for f were then determined (See ref. 25.), viz,

$$f = \frac{3\langle \cos^2\phi \rangle - 1}{2} , \tag{2}$$

where, f = 0 is the isotropic condition, and f = 1 is the perfectly ordered condition for orientation. The outlined method of analysis yields values for f with a 0.02 variation due to background selection. The coherence length, δ_{hkl},(which is often thought of as an average "effective crystallite size") for each WAXS pattern was calculated using the Scherrer method.[33]

A plot of orientation parameter (f) versus spin draw ratio (SDR) is shown for the 14% PBO/PPA solution in Figure 4 where the processing barrel temperature is 150°C, the capillary die SR is 600/s, and the draw zone position is 3.81 cm. For comparison, f for the 1.27 cm and the 2.53 cm draw zone positions are shown for a SDR of 10. An opaque to transparent transition in the extrudate occurs at a SDR of 3, and f is independent of the capillary shear rate above 2/s. Consequently, the majority of the obtainable orientation in the draw zone is attained at the same SDR as the opaque-transparent transition. Above an SDR of 3, f increases still further, although more gradually than from SDR of 1 to 3. At SDR of 20, f plateaus at 0.95. Between a SDR of 3 and 20 orientation was found to be an increasing linear function of measured line tension. At a fixed SDR, orientation increases as one proceeds down the draw zone.

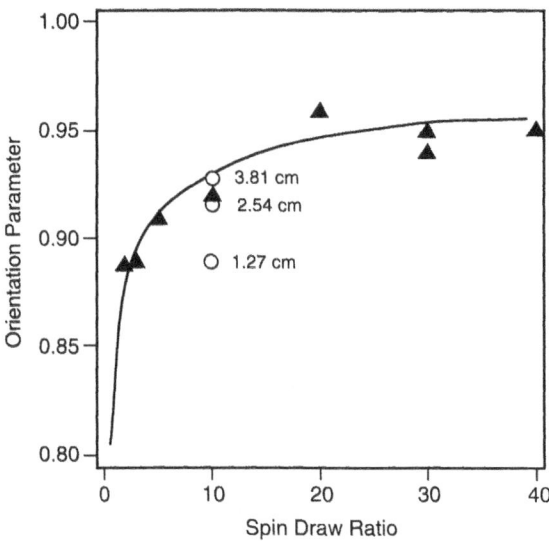

Figure 4. Orientation parameter f versus spin draw ratio (SDR) for 14% PBO/PPA. Orientation for three positions in the draw zone (1.27 cm, 2.54 cm, and 3.81 cm) is also shown at a SDR = 10.

The lateral and axial coherence lengths for the conditions plotted in Figure 4 are shown in Figure 5. The lateral coherence length decreases from 5.6 nm to 3.6 nm over the same SDR range. The nematic 14% PBO/PPA (83.7% P_2O_5 PPA) d-spacing in the [100] equatorial direction was found to remain constant at 10.6 Å for any SDR from 2 to 40. The equivalent values for the axial coherence length derived from the [002] reflection along the c-axis (molecular backbone) direction are also plotted in Figure 5. The axial coherence length increases from 17.5 nm to 20.0 nm over the same SDR range. The effective scattering center size elongates and narrows with increasing extension (SDR). The nematic 14% PBO/PPA (83.7% P_2O_5 PPA) d-spacings for the [001], [002], and [003] meridian (fiber axis direction) layer lines were found to be 12.55 Å, 6.20 Å, and 4.15 Å, respectively. These remain constant for any SDR from 2 to 40.

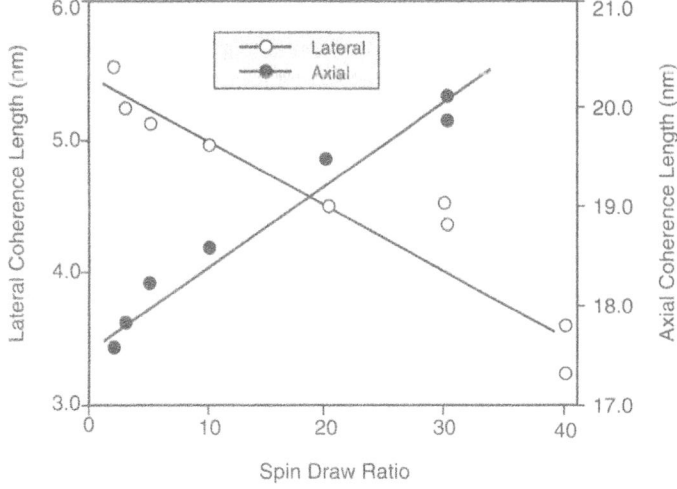

Figure 5. Axial and lateral coherence lengths as a function of SDR.

The coherence length determinations indicate the existence of "micro-domains" in the extrudate on the size order of about 200Å by 50Å before any further processing steps such as coagulation. This size is roughly that observed for the lengths of "crystallites" in as-spun PBO and PBZT fiber using transmission electron microscopy by Martin and Thomas.[34,35]

The results of a synchrotron-based in-situ X-ray scattering study of lyotropic PBO/PPA monofilament fiber drawing indicate that the molecular orientation depends strongly on spin draw ratio and line tension, but not shear rate in the capillary. With increasing SDR, the axial coherence length increased from 17.5 nm to 20 nm, and the lateral coherence length decreased from 5.5 nm to 3.5 nm. These micro-domain sizes are consistent with the paracrystalline domain sizes typically observed in coagulated fiber. The average values for the [002] (filament axis) and [100] (lateral) d-spacings are constant as a function of SDR above 2. The occurrence of the microdomains in the draw zone as a precursor to the microfibrilar structure is consistent with the great difficulty in improving filament compressive strength by post-draw processing techniques.

IV. FORMATION OF POLYURETHANE SLABSTOCK FOAM MICROSTRUCTURE AS DETERMINED BY REAL-TIME SAXS

The manufacture of polyurethane foam is an activity of considerable commercial importance (about 2.6 billion lbs. per annum).[36] The development of microstructure during the course of the cure of the foam is very rapid and usually takes place in the time span of a few minutes. The primary objective of this study was to gain detailed information on the in-situ formation of microstructure in polyurethane slabstock foams in real time during the foam reaction. SAXS has been used to examine the hard domain structure in polyurethane materials for many years.[37,38,39] Nevertheless, how the domain structure develops cannot be determined by static SAXS scans. Dynamic or real-time SAXS performed at BNL has allowed observation of the differences in microstructure formation with foam formulation changes.[40] A future objective is to correlate microstructural information to final macroscopic foam properties.

Prior work on polyurethane elastomers focused on comparing changes in the experimental SAXS patterns such as intensity versus time at a selected angle or volume fraction determination from the invariant.[40] Newer work shows that additional information not readily apparent from the experimental patterns alone can be obtained by applying the one-dimensional correlation function and distance distribution function analysis to real-time SAXS data.[41,42,43,44,45]

For this set of experiments the effect of H_2O content was determined by varying H_2O content from 2.2 to 4.8 parts per hundred parts (pphp) polyol. All other formulation parameters were kept essentially the same. Blank SAXS patterns were collected with the empty mold in place prior to filling the mold with the polyurethane foam mixture. The components of the foam mixture were then combined and transferred to the mold. SAXS patterns were then taken at 3 second intervals. An X-ray wavelength λ of 0.128 nm was used for all the experiments. A sample to detector distance of 78 cm was used for the 2.2 and 3.8 pphp water slabstock foams and 103 cm for the 4.8 pphp water slabstock foam. Additional formulation and experimental details have been reported.[46]

A major finding from this study is that the slabstock foam hard domain growth observed from SAXS patterns matches the bidentate urea formation as determined from infrared analysis.[46] The experimental SAXS scans versus time for the different formulations are shown in Figures 6a through 6c. Time is measured from the start of sample mixing. For each scan, the intensity is plotted versus the reciprocal space variable h where $h = (4\pi/\lambda)\sin\theta$ (where θ is one-half the scattering angle). As can be seen, the scans of the 2.2 pphp water formulation show strong intensity at the extreme low-angle portion of the pattern and with

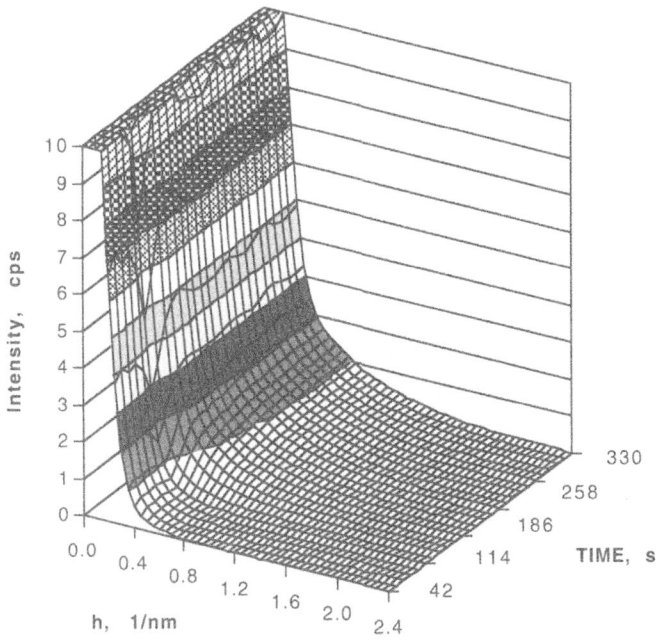

Figure 6a. Experimental SAXS scans versus time for the 2.2 pphp H_2O formulated polyurethane slabstock foam.

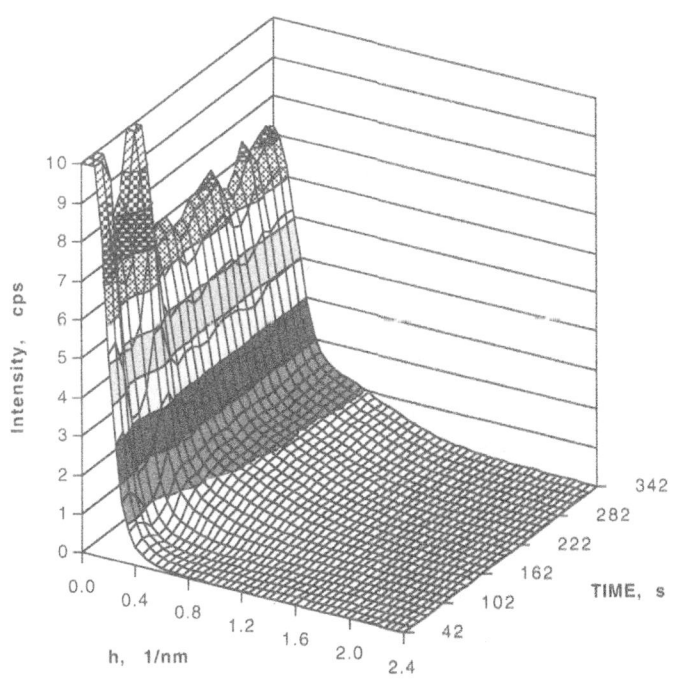

6b. Experimental SAXS scans versus time for the 3.8 pphp H_2O formulated polyurethane slabstock foam.

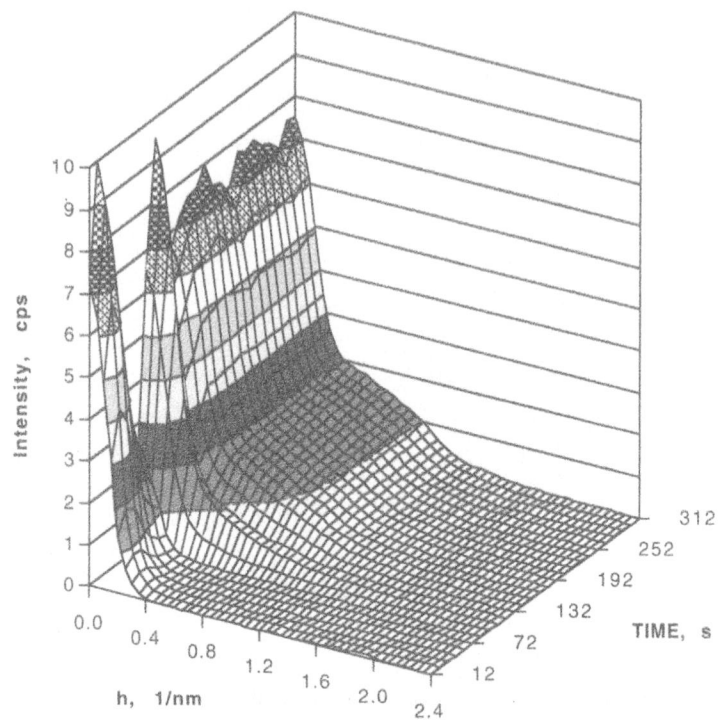

6c. Experimental SAXS scans versus time for the 4.8 pphp H$_2$O formulated polyurethane slabstock foam.

time a shoulder develops that is not very distinct compared to that for the higher water formulations. Nevertheless, the shoulder attributed to hard domain formation (h range of 0.4 to 1.2 nm^{-1}) can be observed at approximately 120 s. As water content is increased, the extreme low-angle portion of the pattern shows lower intensity and a more pronounced hard segment shoulder develops. In agreement with the infrared analysis, the time from sample mixing to the beginning of shoulder formation decreases with higher water content as well.

The 1D correlation functions were evaluated from the experimental SAXS scans using previously reported protocols.[41-43] Shown in Figures 7a and 7b are the 1D correlation function plots versus time for the 2.2 pphp and the 4.8 pphp H$_2$O formulations, respectively. The 1D correlation function plots for the 3.8 pphp H$_2$O formulation are not shown, but resembled an average of the 2.2 pphp and 4.8 pphp H$_2$O formulations. Correlation function plots before hard segment formation have excessive Fourier noise, and therefore microstructural information prior to hard segment formation is questionable. Once hard segment regions begin to form the domain microstructure becomes readily apparent. The 2.2 pphp H$_2$O formulation has no prominent maximum after 120 s, but shows an indication of some order with a spacing of approximately 6 nm. Definite domain order is observed for the 4.8 pphp H$_2$O formulation with a 7 nm spacing, consequently, higher water content affects the extent of hard domain order and spacing.

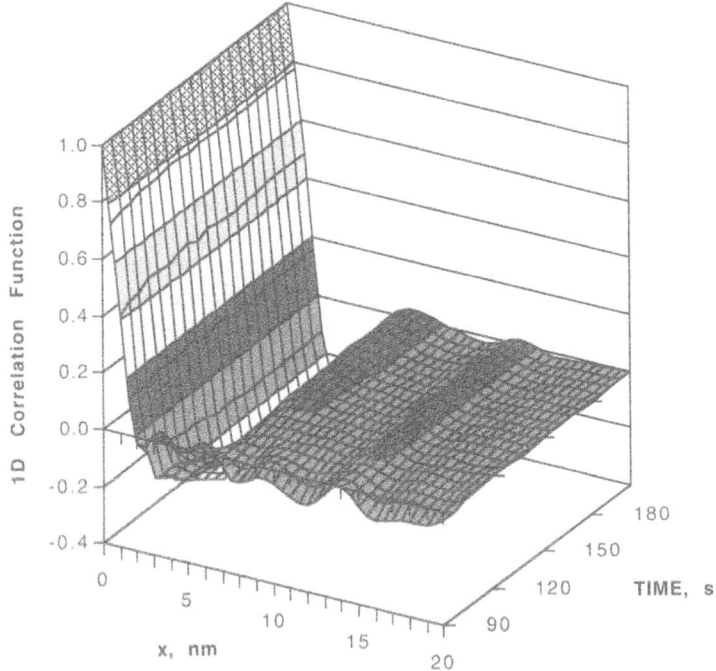

Figure 7a. 1D correlation function plots versus time for the 2.2 pphp H_2O formulated polyurethane slabstock foam.

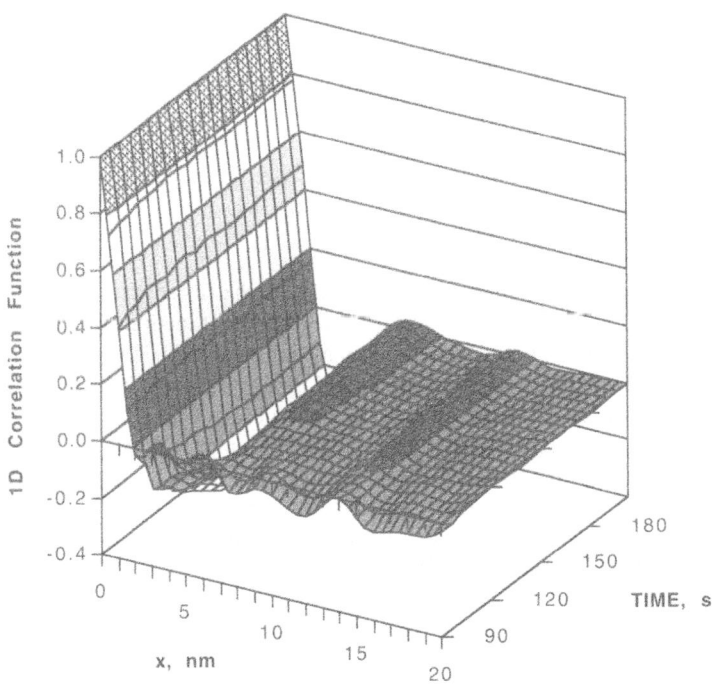

7b. 1D correlation function plots versus time for the 4.8 pphp H_2O formulated polyurethane slabstock foam.

Hard segment volume fraction was determined from the one-dimensional (1D) correlation function using the first minimum. In calculating the volume fraction it is assumed that the soft segment region is the larger volume fraction component of the polymer. The values obtained for the different formulations are plotted versus time in Figure 8. For the 4.8 pphp H_2O formulation, values before hard segment formation are also shown. In all cases, a sharp drop in the hard segment volume fraction estimate occurred at the onset of bidentate urea formation (hence hard segment formation). The estimates prior to the hard segment formation are an indication of a different microstructure or the result of Fourier noise in the 1D correlation function. At the onset of domain formation, the volume fraction increase follows the bidentate urea formation extremely well. The volume fraction estimates from the SAXS data at 300 s are greater than estimates based on the formulation, but they are in agreement with the trend of higher hard segment volume fraction with higher water content.

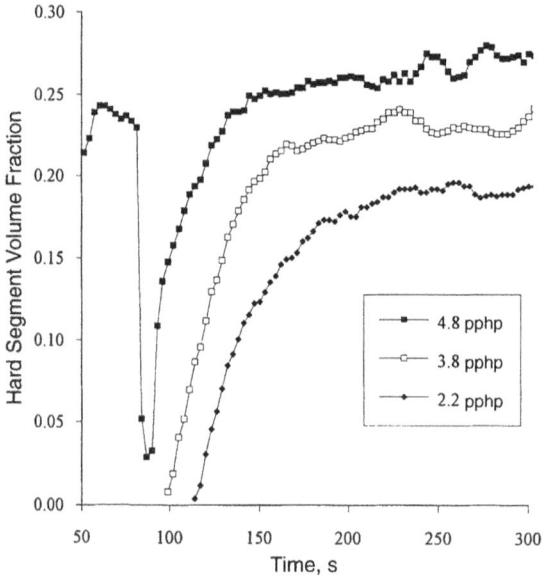

Figure 8. Effect of H_2O content on the hard segment volume fraction formation. Volume fractions estimated from minimum on 1D correlation function plots.

Additional information was obtained by calculating the first distance distribution functions from the SAXS scans. The first distance distribution function f(r) is defined as the three-dimensional correlation function multiplied by real-space r.[44] The overall hard domain shape is elusive because the close domain spacing results in scattering due to interparticle interference. Interparticle interference should be negligible at first if the hard domains form randomly throughout the polymer rather than in close groupings. First distance distribution function plots versus time are shown for time greater than 126 s for the 2.2 pphp H_2O formulation in Figure 9a. Plots are shown from 96 s on for the 4.8 pphp H_2O formulation in Figure 9b. As with the 1D correlation function plots, the distance distribution function plots for the 3.8 pphp H_2O formulation resembled an average of the 2.2 pphp and 4.8 pphp H_2O formulations. The f(r) plots show an initial shape comparable to an oblate (disk-like) ellipsoidal domain, as discussed by Glatter, with a large dimension greater than 20 nm and a small dimension of approximately 3 nm.[44,45] The exact shape and size cannot be determined from the f(r) plots alone since a flat plate-like structure can also give a similar f(r) pattern. Nevertheless, the observed pattern at the start of hard domain formation is definitely not a sphere, prolate ellipsoid, or rod-like shape. If the disk-like model is valid, the first f(r) max-

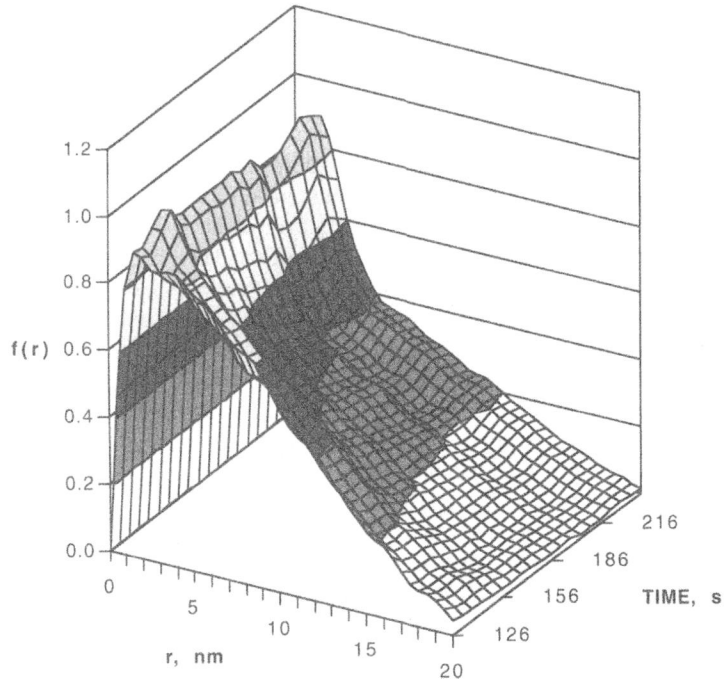

Figure 9a. First distance distribution function plots versus time for the 2.2 pphp H$_2$O formulated polyurethane slabstock foam.

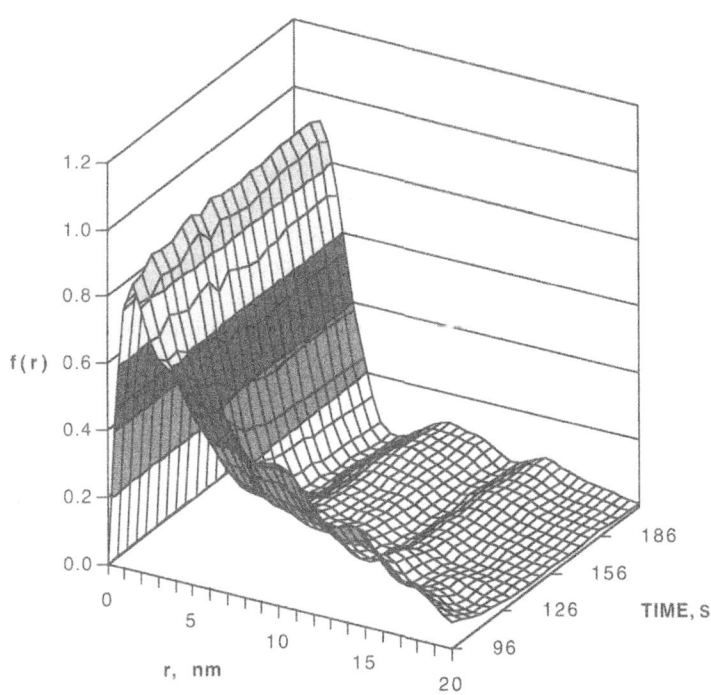

9b. First distance distribution function plots versus time for the 4.8 pphp H$_2$O formulated polyurethane slabstock foam.

imum position and inflection point after this first maximum correlate with the average domain thickness. Domain thickness growth was observed for the 4.8 pphp H_2O formulation by the first f(r) maximum position increase during the time range from 96 to 126 s. With time, additional maxima occur for the 4.8 pphp H_2O formulation which is in agreement with the maximum formation that occurs on the 1D correlation function plots. This again is evidence for interparticle interference when the hard domain concentration is high enough.

Results from the real-time SAXS experiments gave more detailed information on the evolution of the microstructure in reacting polyurethane slabstock foam microstructure than can be obtained by static experiments. Application of the 1D correlation function analysis and distance distribution function analysis to the SAXS patterns has been proven to give a clear indication of when the hard segment regions began to form and the approximate size and shape of the domains. The results of this study indicate that: (1) hard domain growth matches formation of highly organized urea (bidentate urea); (2) H_2O content controls hard domain volume fraction, spacing, and domain size; (3) isolated high density domains form first in a manner consistent with nucleation and growth theory; and (4) interparticle interference occurs as more domains form near existing domains.

V. CHARACTERIZATION OF POLYMER SURFACES AND INTERFACES USING TUNABLE SYNCHROTRON RADIATION-BASED ULTRA SOFT X-RAY ABSORPTION SPECTROSCOPY

This section addresses the application of ultra soft X-ray absorption spectroscopy (USXAS) for characterizing the surface and bulk properties of several model polymer systems. USXAS is an important extension of a traditional hard X-ray technique for structural determination of low Z elements (C,N,O,F) in materials.[47] The development of USXAS for characterization of polymeric materials will improve our ability to characterize both the bulk (top 40 nm) and surface (top 5 nm) of polymeric materials. USXAS offers unique advantages for polymer characterization.[48,49,50] These include direct determination of the surface concentration of many constituent functional groups and determination of the functional group orientation by comparing absorption cross sections for varying photon incidence angle.

Fundamental chemical and orientation information at surfaces can be obtained by using USXAS. For example, it is possible to detect the presence of specific carbon bonds in polymers because USXAS permits one to probe the carbon K near edge. USXAS involves the excitation of a core (usually 1s) electron into unoccupied valence level molecular orbitals. This renders the technique very sensitive to the elemental makeup and the chemical state. Orientation of functional groups can be determined from the angular dependence of the X-ray absorption cross sections using polarized X-rays.[47] The absorption event involves the interaction of the electric field of the X-ray photon with the transition dipole moment of the molecular bond. This results in high sensitivity to the orientation of the functional group involved. Further, since the initial state orbital for the electronic excitation is a 1s orbital (spherically symmetric) all the orientation dependence is conferred in a relatively simple way by the final state antibonding orbital symmetry to which the electronic excitation occurs.[47]

For the data presented herein, USXAS spectra were obtained at the U1A beam line at the NSLS. The USXAS were collected with a rapid sample interchange system equipped with facilities for simultaneous fluorescence yield and partial electron yield detection described fully in reference 50. The estimated total photon flux was about 1×10^{10}/s at 300 eV photon energy with a resolution of 0.6 eV. Typical storage ring currents were in the 400 - 600 mA range during these experiments. Partial electron yield measurements were performed with a channeltron located above the sample plane, negatively biased to deflect secondary electrons. Simultaneous fluorescence yield was measured using a high resolution detector optimized for carbon X-ray radiation, as described by Fischer, Colbert, and Gland.[51] The specimens were mounted on a sample manipulator so that the angle of the incident radiation to the film surface could be varied.

The enhanced chemical state information obtained for polymer surfaces using USXAS can be illustrated using results for polystyrene. Figure 10 shows a comparison between an X-ray photoelectron spectroscopy (XPS) spectrum and a fluorescence yield (FY) USXAS spectrum of a spin-cast film of polystyrene (~10 nm thick). The XPS spectrum is obtained by measuring the kinetic energy of photoelectrons using a non-resonant excitation X-ray (Al K_α at 1486 eV)[52], while the USXAS spectrum results when using tunable synchrotron radiation in the 270-310 eV range. In the XPS spectrum all of the carbon bonding information is contained in the intense spectral feature at 285 eV and the specific bonding contributions cannot be resolved. In contrast, there is enhanced chemical information in the USXAS spectrum. The intense C-C ($1s-\pi^*_1$)) resonance at 285 eV in the USXAS spectrum illustrates the sensitivity to carbon bonding. The π^* resonances involve the excitation of a 1s electron to unfilled π^* antibonding orbitals on the aromatic ring. Higher in energy, above the ionization potential for the 1s electron, are resonances excitations to the unoccupied molecular orbitals characteristic of σ bonds. Included in the σ^*_1) and σ^*_2) resonances are also contributions from the polystyrene backbone σ-bonds.[53]

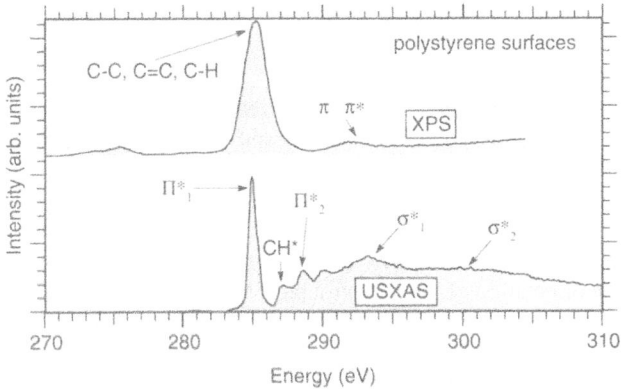

Figure 10. A comparison between an XPS spectrum (top trace) and a fluorescence yield (FY) USXAS spectrum (bottom trace) of a polystyrene film is shown. The XPS spectrum is obtained by measuring the kinetic energy of photoelectrons using non-resonant excitation X-rays while the USXAS spectrum results when using tunable synchrotron radiation in the 270-310 eV range.

Molecular bond orientation can be obtained by USXAS using linearly polarized synchrotron radiation. The polarized synchrotron X-rays actually function as a "searchlight" along the direction of the molecular bond. A schematic of this concept is shown in Figure 11. A hypothetical C_2H_2 molecule with C-C and C-H anitbonding orbitals on a surface with the polarized X-ray beam incident normal to the surface (parallel electric vector, $E_{||}$) is depicted in Figure 12a. In this configuration, the absorption to the C-C orbital will have a much higher transition probability. If the surface is rotated so the incident X-ray beam is nearly at glancing incidence (~25° with respect to surface plane), the spectral absorption cross section will now be intense for the C-H anti-bonding orbital and very weak for the C-C antibonding orbital (See Figure 12b for the perpendicular electric vector, $E \perp$.)

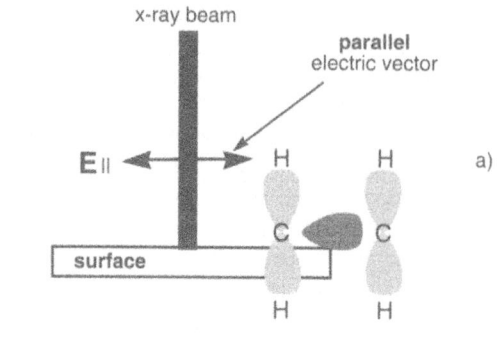

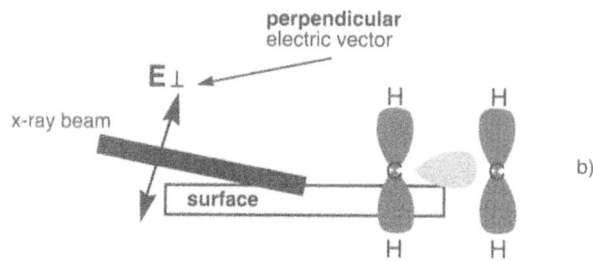

Figure 11. The concept of probing molecular bond orientation via USXAS using linearly polarized synchrotron radiation is shown. See text for further discussion.

The combination of enhanced chemical and orientation information that can be realized using USXAS for carbon in organic materials is shown in Figure 12 for a self assembled monolayer of hexadecylsilane on Si[54] and a stretched polyethylene film.[55] In Figure 12, partial electron yield (EY) spectra for both $E_\parallel$ and $E\perp$ for hexadecylsilane (HDS) on Si (a) and a stretched polyethylene (PE) film (b) are shown. Clearly, there is high spectral sensitivity in that the C-H and C-C bonds can easily be distinguished. Furthermore, there are dramatic differences in the orientation of the CH_2 groups at the surface and these are reflected when comparing spectra obtained in the $E_\parallel$ and $E\perp$ configurations. For HDS on Si, the σ^* (C-H) transition is most intense for $E_\parallel$, while for polyethylene this transition is most intense in the $E\perp$ configuration. A similar observation can be made for the σ^* (C-C) in comparing HDS on Si and polyethylene. The orientation of the CH_2 chains are shown pictorially in Figures 12a and 12b. These observations indicate that in the case of HDS the carbon chains are aligned perpendicular to the surface, while in the PE the carbon chains are aligned parallel to the surface.

There is a considerable demand for materials having surfaces to which unwanted substances will not easily adhere or stick. A new approach has been described for obtaining a new class of water-based, ultralow-surface-energy, "non-stick" coatings by self-assembling and immobilizing (via chemical crosslinking) reactive perfluoroalkyl-polymeric-surfactants (RPPS), which contain both pendant perfluoroalkyl groups (R_F) and reactive ionic functionality.[56] The RPPS adjust their chain configurations to orient R_F groups at the gas interface. Immobilization by crosslinking prevents molecular interdiffusion and reorientation and dramatically improves non-stick properties. Surface characterization using wettability, USXAS, XPS, and atomic force microscopy suggests the immobilized RF groups are assembled and oriented in layers about 8 nm thick. The USXAS characterization of these polymer films are briefly summarized below.

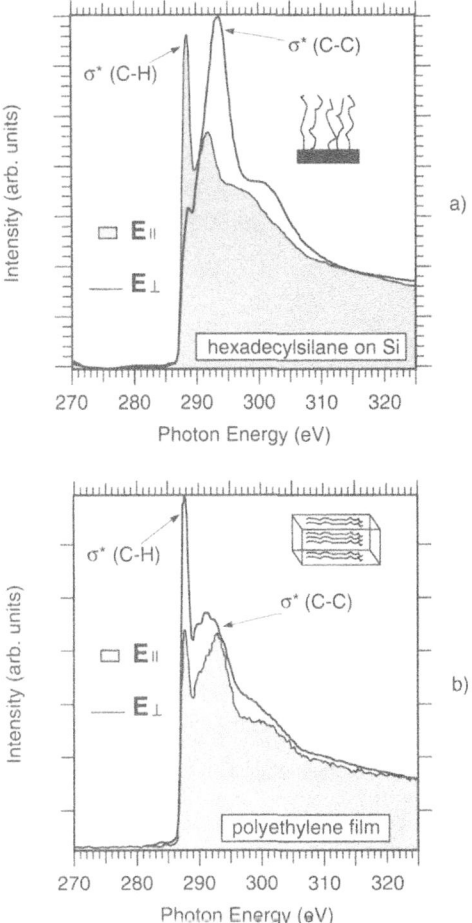

Figure 12. Partial EY USXAS spectra for both $E_{\parallel}$ and $E_{\perp}$ for hexadecylsilane (HDS) on Si (a) and a stretched polyethylene (PE) film (b) are shown. There is high spectral sensitivity in that the C-H and C-C bonds can easily be distinguished. There are also dramatic differences in the orientation of the CH_2 groups at the surface and these are reflected when comparing spectra obtained in the $E_{\parallel}$ and $E_{\perp}$ configurations. In the case of HDS the carbon chains are aligned perpendicular to the surface, while in the PE the carbon chains are aligned parallel to the surface.

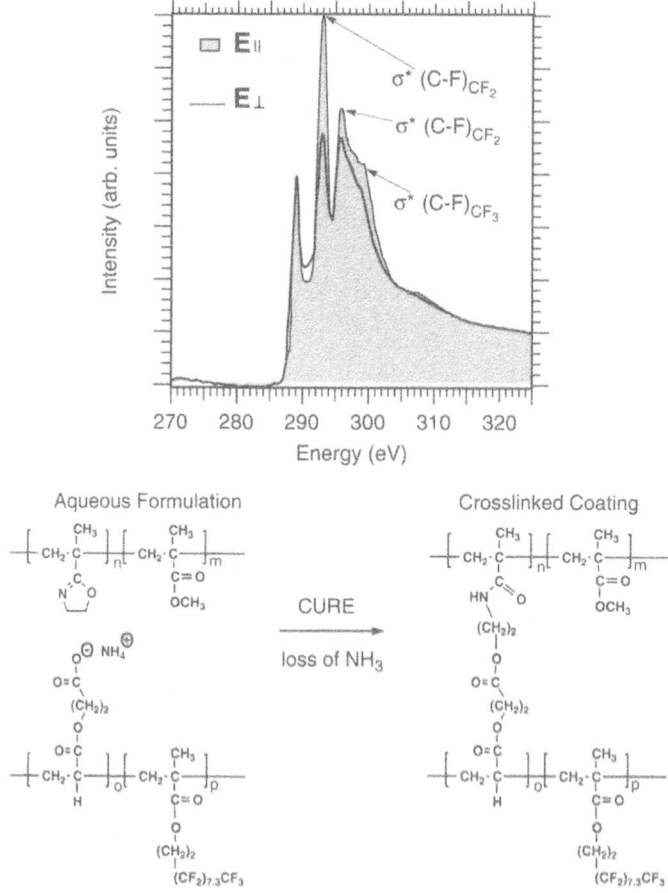

Figure 13. The USXAS carbon K-edge spectral region obtained for the surface of the water based, non-stick hydrophobic coatings is shown. The crosslinking reaction for several RPPSs is also shown. A dramatic change in the σ* (C-F) transition localized in the CF$_2$ groups (near 293.0 eV) is observed in comparing the obtained for **E**$_\parallel$ and **E** $_\perp$. This implies that more of the C-F bonds (in the CF$_2$ groups) in the first 3 nm of the surface are oriented parallel to the surface.

With USXAS it is possible to observe the orientation of the R$_F$ groups at the polymer surface of crosslinked fluorocarbons. Orientation was determined from the relative angular dependence of the X-ray absorption cross section by monitoring the partial electron yield.[47] The USXAS carbon K-edge spectral region obtained for the polymer surface is shown in Figure 13. The crosslinking reaction for several RPPS's is also shown in Figure 13. A dramatic change in the σ* (C-F) transition localized in the CF$_2$ groups (near 293.0 eV) is observed in comparing the results obtained for E$_\parallel$ and E $_\perp$. This implies that more of the C-F bonds (in the CF$_2$ groups) in the first 3 nm of the surface are oriented parallel to the surface. Hence, the R$_F$ groups are oriented preferentially perpendicular to the surface which leads to greatly reduced wettability. We determined from the polarization angular dependence of the C-F resonances (in the CF$_2$ groups) that the R$_F$ groups are oriented from the surface at an average angle of ~58°. Less self assembly of the R$_F$ groups (~15 % smaller ratio of the CF2 spectral features for E$_\parallel$ and E $_\perp$) was observed in films having higher crosslink density. The combination of the near edge X-ray absorption fine structure (NEXAFS) results with the ultra-low surface energy of these films provides strong evidence that the perfluoralkyl groups are self assembled and CF_3 terminated.

The bulk sensitivity of fluorescence yield (200 nm) has been utilized to study a buried metal/polymer interface, highlighting the use of USXAS to provide new detailed chemical information which will be useful in predicting and optimizing the adhesion in metal/polymer interfaces. DVS bis-BCB (divinyl siloxane bis-benzocyclobutene) is used for dielectric layers in new generation multilayer interconnect devices (multi-chip modules).[57] For this application, an understanding of the nature of the bonding and complexing which occurs at the metal/polymer interface is needed. Fluorescence yield is particularly useful in studying the buried interface between DVS bis-BCB coated with Al (10 to 100 nm)[48,58] and other metals because the fluorescence from the buried interface is able to penetrate the overlying metal film. In contrast the relatively short depth sensitivity (5 nm) of traditional electron yield USXAS makes it useless for the buried interface study, since it would only probe the outer 5 nm of the Al overlayer. We have collected FY carbon K near edge spectra for evaporated aluminum coated (50 nm thick) DVS bis-BCB (50 nm thick) as well as the identical bare polymer. To enhance the metal/polymer interface sensitivity we examined difference plots (coated polymer minus bare polymer) and found a dramatic enhancement in polarization dependence after the formation of the metal/polymer interface. This polarization enhancement at the interface region implies that the aromatic ring planes of DVS bis-BCB are highly oriented towards the surface normal. These preliminary results highlight the use of USXAS to provide new detailed chemical information which may be useful in predicting and optimizing the adhesion in metal/polymer interfaces.

Other systems examined using USXAS include the characterization of flame treated model acrylic automotive coatings and spin cast polystyrene films.[49] The chemical and surface sensitivity of the technique are emphasized by the ability of USXAS to detect an increase in the trigonally coordinated carbon at the surface after treatment with a reducing flame. Spin coated polystyrene films were characterized as a function of molecular weight, film thickness and casting solvent. The polystyrene pendant phenyl groups were found to be preferentially oriented towards the surface normal, independent of casting solvent, molecular weight, and film thickness.

VI. IN-SITU XANES AT THE Mo L3-EDGE OF MAGNESIUM OXIDE-SUPPORTED MOLYBDENUM OXIDE CATALYSTS

In order to develop structure-function relationships in heterogeneous catalysts for the purpose of optimizing chemical reaction processes, it is necessary to determine the spatial arrangement of the atoms comprising the active catalytic phase. These catalysts are typically composed of highly dispersed solid multiphase chemical agglomerates in low concentration in which the different phases are closely associated or interact with one another. In addition, the nature of the catalyst is usually a function of the environment of reactants and products in a dynamic equilibrium. Thus, in situ studies of the working catalyst are preferable to ex-situ analytical techniques. It is not surprising that no one technique can provide a detailed understanding of such materials. Rather, groups of complementary techniques are employed. This section addresses the application of X-ray absorption near edge spectroscopy (XANES) as an in-situ probe of the local structure of a highly dispersed catalytic phase.

A brief overview of in-situ determination of the local site symmetry of catalysts comprising nanodispersed molybdenum oxide phases supported on magnesium oxide using XANES at the Mo L_3-edge is illustrated by three examples: (a) an in-situ calcination study, (b) the effect of CaO impurity in the MgO support, and (c) the stability of the molybdate phases under butane dehydrogenation. A detailed study of the magnesium oxide-supported molybdenum oxide catalyst system using in situ laser Raman spectroscopy (LRS) and ex-situ XANES at the Mo $L_{2,3}$-edges has been reported previously.[59,60]

It has long been recognized that different pre-edge peak (1s to 3d transitions at K absorption edges) intensities are observed in the spectra of octahedral and tetrahedral transition metal complexes with incompletely filled d shells, and that this feature allows site symmetry determination.[61,62] However, at the Mo K-edge (20.0 keV) the effective resolution of ca. 10 eV hinders detailed structural characterization. An improved resolution of ca. 0.5 eV can easily be obtained at the Mo L_3-edge (2.5 keV), which greatly facilitates structural characterization. In addition, XANES at the L-edges probes molecular orbitals of d character which are primarily involved in bonding.

The Mo XANES data were recorded at the NSLS, on beam line X19A. The storage ring operated at 2.5 GeV with a current between 110 and 230 mA. The X-ray photons were monochromated with a NSLS boomerang-type flat crystal monochromator with Si(111) crystals. The slit width of the monochromator was fixed at 3 mm, estimated to give a resolution of 0.5 eV at the Mo L-edges. The harmonic content was reduced by detuning the monochromator crystals by approximately 90%. The in-situ X-ray absorption edges were measured as fluorescence yield excitation spectra using a Stern-Heald-Lytle detector with Ar as the detector gas and the XANES of the reference compounds were measured as electron yield spectra. To minimize absorption by the air the path length from the end of the beam pipe to the sample chamber was made as short as possible. Prolene™ (Trademark, Chemplex Industries, Inc.) windows (4 μm thick) were used on the I0 chamber and entrance window to the in-situ cell.

A commercially available X-ray absorption fine structure (EXAFS) cell was used for the in-situ XANES experiments, the cell for which has been described in detail elsewhere.[63] Briefly, the device is comprised of a water cooled, helium flushed aluminum block into which a cylindrical insert for soft X-ray work can be inserted. This cylindrical insert consists of the sample holder and cylindrical housing. The sample holder is made of stainless steel and supports a disk shaped sample which is heated by a Kanthal resistance heater. The sample-holder has a gas inlet and outlet in order to control the gas environment around the sample. The cylindrical housing is water cooled and has a 5 mm aluminized Mylar™ (Trademark, DuPont) window. The gas inlet is connected to a versatile portable feed gas system equipped with electronic mass flow controllers and switching valves. The catalysts were prepared by impregnating magnesium oxide powder with ammonium heptamolybdate solution.[62] Self-supporting catalyst disks (approx. 0.7 g) were loaded into the sample holder. They were then heated to the desired temperature in a flow of dry 20% oxygen in helium. Data were recorded with the catalysts at a specific temperature.

The XANES spectra are normalized to a unit edge jump according to conventional methods after subtraction of a polynomial fit over the pre-edge region. The energy of the edge is referenced by setting the first inflection point of the L_3-edge of Mo foil to 0.00 eV. In this manner the absorption edge of all of the catalyst samples falls in the range of 4.0 - 5.0 eV, as expected for Mo(VI) compounds.[64,65]

The Mo L_3-edge XANES, shown as electron yield signals, of a series of reference compounds, $CoMoO_6$, MoO_3, $(NH_4)_2Mo_2O_7$ and Na_2MoO_4, are shown in Figure 14. The prominent feature in the spectra is the intense white line. At the Mo L_3-edge this white line is a result of transitions from the dipole allowed 2p to 4d transition. In addition, splitting of the line is observed, reflecting the ligand field splitting of the final state d-orbitals.[64] The magnitude and relative intensity of the splitting can be understood using simple ligand field concepts. In a tetrahedral field the magnitude of the splitting of the d-orbitals is smaller than in an octahedral field (e, t_2 vs. t_{2g}, eg). The number of available orbitals is also consistent with the relative intensities observed (e:2 and t_2:3; t_{2g}:3 and eg:2).[64] In $CoMoO_6$ and MoO_3 the Mo is octahedrally coordinated to six oxygens, whereas in Na_2MoO_4 the Mo is tetrahedrally coordinated. $(NH_4)_2Mo_2O_7$ has Mo atoms both tetrahedrally and octahedrally coordinated to oxygen. Both the magnitude of the splitting, and relative intensity of the peaks of the compounds shown in Figure 14 are consistent with their symmetry. We have previously shown fluorescence yield data for these compounds, but here the magnitudes are free of thickness effects.[62] The inset of Figure 14 contains the second derivatives of the XANES data, which serve to highlight the differences between the spectra. The range of values for the splitting of tetrahedrally coordinated Mo oxides is 1.8-2.4 eV, whereas for octahedrally coordinated Mo oxides it is 3.1-4.5 eV.

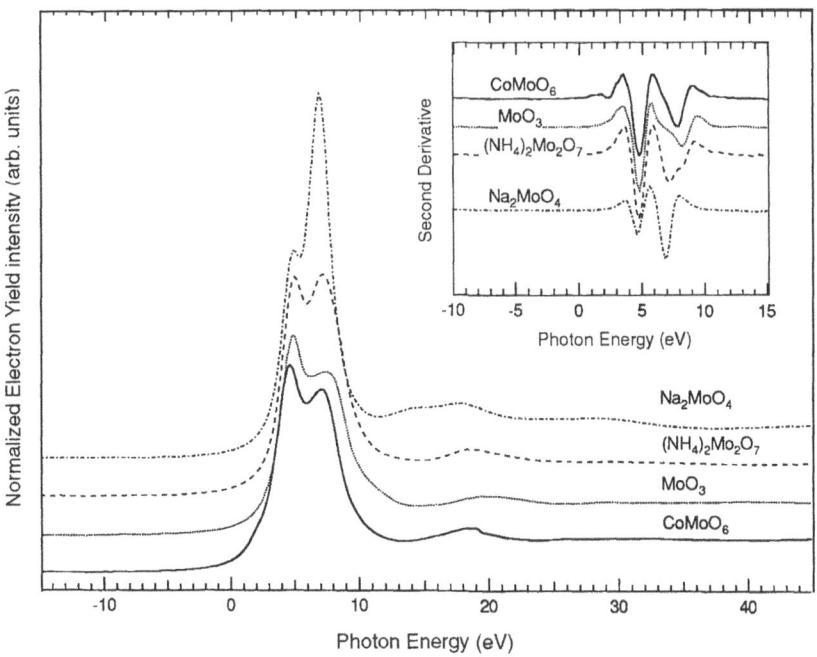

Figure 14. Electron yield Mo L_3-edge XANES spectra of a series of Mo(VI) reference compounds. The vertical scale is offset for clarity. The inset shows the second derivatives of the spectra.

The in-situ Mo L_3-edge XANES of two different loadings of magnesium oxide-supported molybdenum oxides (MoO_3 on MgO) are shown in Figures 15a and 15b as a function of calcination temperature in a flow of 20% O_2 in He. The upper panel shows the data from 5 wt % MoO_3/MgO, and the lower panel that from 20 wt % MoO_3/MgO. The spectra are offset in the energy axis with the spectra at the highest temperature on the unshifted scale. The initial spectrum in each case is for an initially prepared catalyst in its hydrated state. The Mo L_3-edge XANES of the catalysts in this state indicate that the Mo is tetrahedrally coordinated in both cases. Our previous data also showed this to be true for a 15 wt % MoO_3/MgO catalyst (Shown for comparison in Figure 15c.).[62]

As the 5 wt % MoO_3/MgO catalyst is heated in situ the XANES spectra remain similar up to ca. 400°C, with only a small decrease in intensity of the white line. The second derivatives of the spectra show a splitting of 2.0 eV, indicating tetrahedrally coordinated molybdenum. Above this temperature the relative intensity of the features changes, with the peak at lower energy now more intense, and the splitting between the peaks increases to 3.5 eV. Both of these observations indicate the presence of octahedral molybdate. The spectrum at 448°C shows the presence of both tetrahedral and octahedral species, while the spectra above this temperature are essentially constant up to the final calcination temperature of 650 °C. The same variation is also followed for the 20 wt % MoO_3/MgO catalyst. In this case, the species remain tetrahedral up to 441°C (a splitting of 2.5 eV). At 478°C, tetrahedral and

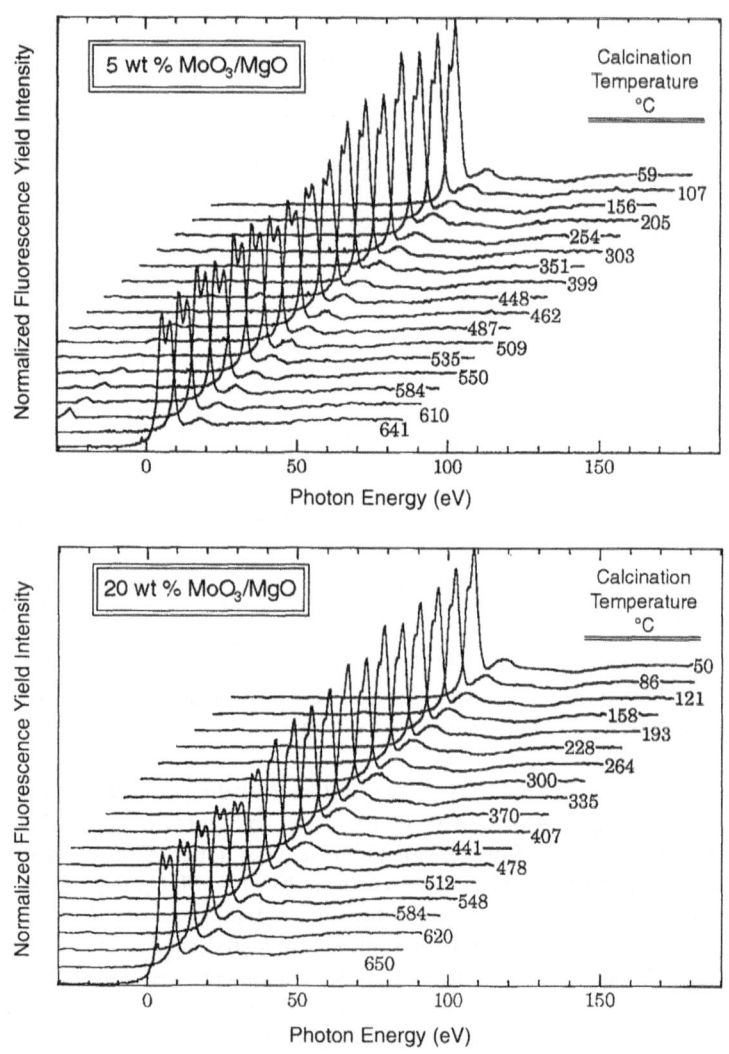

Figure 15a. Fluorescence yield Mo L$_3$-edge XANES spectra of 5 wt % MoO$_3$/MgO (top panel), and 20 wt % MoO$_3$/MgO (bottom panel). The spectra were acquired in situ at the temperatures indicated in a flow of 20% O$_2$ in He. The spectra are offset for clarity.

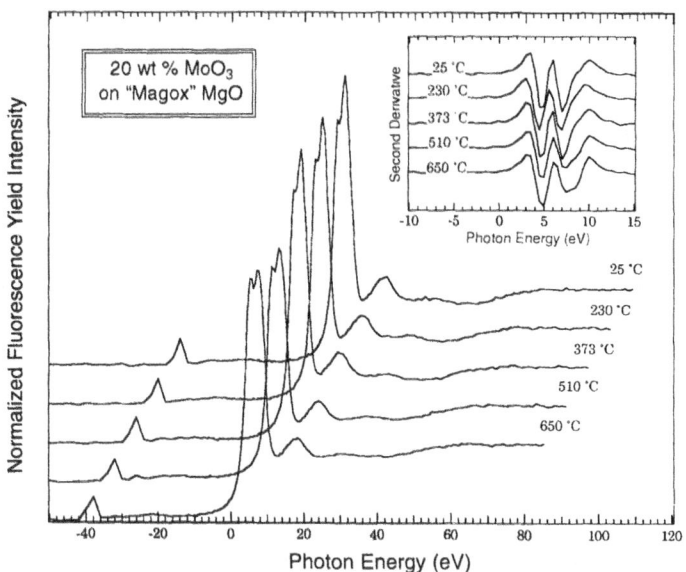

15b. Fluorescence yield Mo L_3-edge XANES spectra of 20 wt % MoO_3 on MAGOX MgO. The spectra were acquired in situ at the temperatures indicated in a flow of 20% O_2 in He. The spectra are offset for clarity. The inset shows the second derivatives of the spectra.

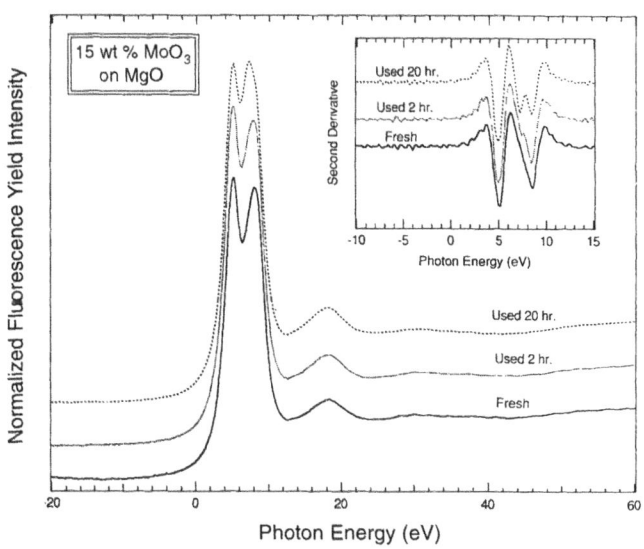

15c. Fluorescence yield Mo L_3-edge XANES spectra of 15 wt % MoO_3/MgO as a function of time on stream in butane oxydehydrogenation. The catalysts were recalcined in dry air prior to accumulation of the data. The spectra are offset for clarity. The inset shows the second derivatives of the spectra.

octahedral species are present, while at 512°C - 650°C all of the molybdate is octahedral. This change of Mo coordination from tetrahedral (for hydrated catalysts) to octahedral coordination (for dehydrated MoO_3/MgO catalysts) can be understood by a simple acid-base chemistry description of the molybdate and MgO support.[62,66] In particular, these data can now be combined with a detailed in situ laser Raman study for these catalysts.[61] The temperature range where the species change coordination correlates with the disappearance of the Mg-OH vibrations in LRS. A complete in-situ XANES study of weight loadings of MoO_3 from 1.5 wt % to 40 wt % on MgO has been performed, and discussed in detail elsewhere.[67]

The effect of CaO impurity is indicated by a similar in-situ Mo L_3-edge XANES calcination study of 20 wt % MoO_3 supported on a different grade of MgO (MAGOX Super Premium Grade supplied by the Premier Refractories and Chemicals Company), as shown in Figure 15b. The inset shows the second derivatives of the spectra. The spectra at 25°C - 373°C indicate the presence of only tetrahedral molybdate species. This is in agreement with the data illustrated in Figure 15a with catalysts prepared from ultra-pure MgO (Puratronic Grade supplied by Johnson-Matthey), however, at both 510°C and 650°C, the spectra show a mixture of both tetrahedral and octahedral molybdate. The LRS for the MAGOX MgO-supported catalyst contains an additional peak at 880 cm^{-1}.[61] This peak is indicative of the formation of calcium molybdate, which is tetrahedrally coordinated. One of the main impurities in MAGOX MgO is CaO (ca. 0.5 wt%). This impurity thus reacts preferentially with the molybdenum oxide to form calcium molybdate at temperatures above 400°C. Thus, the trace impurity of CaO has a major effect on the species formed in this catalyst. The small peak at -40 eV relative to the Mo L_3-edge in Figure 15b is the sulfur K-edge of impurity sulfur which is also present in the MAGOX MgO.

MoO_3/MgO catalysts of this type are used for the oxydehydrogenation of butane to butadiene in a transport bed reactor. In order to investigate the long term stability of these octahedrally coordinated molybdate species under reaction conditions, we have recorded the Mo L_3-edge XANES of a 15 wt% MoO_3/MgO catalyst (pure MgO) as a function of time on stream. Details of the reaction condition can be found reported elsewhere by Murchison and Vrieland.[68] The data in Figure 15c show results for a fresh calcined (dehydrated) sample, and those after 2 and 20 hr use. The inset contains the second derivatives of the spectra. A gradual transformation of the Mo species on the fresh catalyst from all octahedral to a mix of tetrahedral and octahedral is clearly indicated. In addition, X-ray powder diffraction data of the 20 hr sample show some increased background intensity in the region expected for diffraction peaks due to magnesium molybdate, $MgMoO_4$, in which the Mo is tetrahedrally coordinated. This change in structure of the molybdate species correlates with a decrease in activity of the catalyst.

To summarize, the use of Mo L_3-edge XANES has proven to be a most useful characterization probe of the structure of MgO-supported MoO_3 catalysts. The local site symmetry of the Mo can be determined at sub-monolayer concentrations of Mo using basic ligand field concepts to interpret the data. Examples are given of: (i) monitoring the symmetry of the Mo species as a function of calcination temperature, (ii) the effect of CaO impurity in the MgO, and (iii) the stability of the highly dispersed molybdate species under butane oxydehydrogenation.

ACKNOWLEDGMENTS

The real-time small-angle deformation and in-situ X-ray scattering fiber drawing studies were performed at the Cornell Energy Synchrotron Source (CHESS). The RTSAXS measures of polyurethane slab stock curing, and USXAS surface and interfacial studies were performed at the National Synchrotron Light Source (NSLS) at Brookhaven National

Laboratory. The XANES measurements were also performed at the NSLS, which is supported by DOE under contract number DC-AC02-76CH00016. The authors also wish to thank Fred Corson, Dow Vice President of Research and Development, for his support of Dow's synchrotron-based research efforts.

REFERENCES

[1] "Structural Biology and Synchrotron Radiation: Assessment of Resources and Needs," Report sponsored by BioSync The Structural Biological Synchrotron Users Organization, July 1991. BioSync, Dept. of Biological Sciences, Purdue University, West Lafayette, IN 47907.

[2] S. J. Nolan, C. F. Broomall, R. A. Bubeck, B. G. Landes, P. R. Rudolf, M. J. Mills, M. J. Radler, *Mat. Res. Soc. Symp.Proc.* **305** 111 (1993)

[3] D. Schmidt, C. Coburn, B. DeKoven, G. Potter, G. Meyers, D. Fischer, *Nature* **368**,39 (1994)

[4] M. D. Heaney, W. R. Willkomm, J. V. McClusky, R. E. O'Neill, M. Capel, National Synchrotron Light Source, 1992 Annual Report, 1992, p 282.

[5] H. Ade, A. P. Smith, S. Cameron, R. Cieslinski, B. Hsiao, G. Mitchell, E. Rightor, Manuscript submitted to *Polymer*.

[6] S. R. Bare, J. J. Maj, G. E. Mitchell, J. L. Gland, National Synchrotron Light Source, 1992 Annual Report, 1992, p 320.

[7] R. A. Bubeck, D. J. Buckley, Jr., E. J. Kramer, and H. R. Brown, *J. Mater. Sci.*, **26** 6249 (1991)

[8] D. J. Buckley, Jr., "Toughening mechanisms in the High Strain Rate Deformation of Rubber-Modified Polymer Glasses", Ph.D. Thesis, Cornell University, 1993

[9] R. A.Bubeck, J. A. Blazy, E. J. Kramer, D. J. Buckley , Jr., and H. R. Brown, *Polymer Comm.* **27**, 357 (1986)

[10] "Styrene Polymers", E. R. Moore, ed., in the Encyclopedia of Polymer Engineering and Science, 2nd edition, **16**, 1-246 (1989), John Wiley and Sons, Inc.

[11] C. B. Bucknall and D. Clayton, *J. Mater. Sci.* **7** ,202 (1972)

[12] C. B. Bucknall, D. Clayton and W. E. Keast, *J. Mater. Sci.* **7** ,1434 (1972)

[13] C. B. Bucknall, D. Clayton and W. E. Keast, *J. Mater. Sci.* **8** ,514 (1973)

[14] C. B. Bucknall, P. Davies, and I. K. Partridge, *J. Mater. Sci.* **22** ,1341(1987)

[15] C. B. Bucknall, "Toughened Plastics", Applied Science Publishers, London (1977), Ch. 7 and 8.

[16] H. Breuer, J. Stabenow, and F. Haaf, *International Conference on Toughened Plastics*, Paper 13 (1978), Plastics and Rubber Institute, London

[17] R. A.Bubeck, J. A. Blazy, E. J. Kramer, D. J. Buckley , Jr., and H. R. Brown, *Mat. Res. Soc. Symp. Proc.* **79** ,293(1987)

[18] H. R. Brown, E. J. Kramer, and R. A. Bubeck, *J. Mater. Sci.* **23** ,248 (1988)

[19] C. B. Bucknall, C. A. Correa, V. L. P. Soares, and X. C. Zhang, 9th International Conference on Deformation, Yield, and Fracture, Churchill College, Cambridge, 11-14 April, 1994, Paper 9.

[20] C. B. Bucknall, P. S. Heather, and A. Lazzeri, *J. Mater. Sci.* **28** 6799 (1993)

[21] P. A. Lovell, A. J. Ryan, M. N. Sherratt, and R. J. Young, 9th International Conference on Deformation, Yield, and Fracture, Churchill College, Cambridge, 11-14 April, 1994, Paper 3.

[22] Y. Cohen and E. L. Thomas, *Macromolecules* **, 21**, 433 (1988)

[23] Y. Cohen and E. L. Thomas, *Macromolecules*, **21**, 436 (1988)

[24] J. F. Wolfe, "Polybenzothiazoles and Oxazoles" in The Encyclopedia of Polymer Science and Technology, 2nd ed.,**11**, John Wiley and Sons, 1988, pgs. 601 - 635

[25] J. F. Wolfe, P. D. Sybert, and J. R. Sybert, U. S. Patent No. 4,533,692, Aug. 6. 1985

[26] H. G. Zachmann, "Studies of the Mechanisms of Crystallization by Means of WAXS and SAXS Employing Synchrotron Radiation", p403 in Crystallization of Polymers, edited by M. Dosiére, NATO ASI Series C, (Kluwer Academic Publishers, 1993)

[27] S. J. Nolan, C. F. Broomall, R. A. Bubeck, M. J. Radler, and B. G. Landes, Paper submitted to *Review of Scientific Instruments*.

[28] S. J. Nolan, C. F. Broomall, R. A. Bubeck, B. G. Landes, P. R. Rudolf, M. J. Mills, and M. J. Radler, *Mat. Res. Soc. Symp. Proc.* **134**, 195 (1989)

[29] M. J. Radler, B. G. Landes, S. J. Nolan, C. F. Broomall, T. C. Chritz, P. R. Rudolf, M. E. Mills, and R. A. Bubeck, In press with the *J. Polym. Sci.: Part B: Polym. Phys.*

[30] Y. Cohen, *Mat. Res. Soc. Symp. Proc.* **305**, 111 (1993)

[31] Y. Cohen, S. Buchner, H. G. Zachmann, and D. Davidov, *Polymer*, **33**, *18*, 3811 (1992)

[32] P. H. Hermans and P. Platzek, *Kolloid Z.*, **88**, 68 (1939)

[33] P. Scherrer, *Goettinger Nachrichten*, **2**, 98 (1918)

[34] D. C. Martin and E. L. Thomas, *Mat. Res. Soc. Symp. Proc.*, **134**, 465 (1989)

[35] D. C. Martin and E. L. Thomas, *Macromolecules*, **134**,9, 2450 (1991)

[36] *Modern Plastics*, January 1993, pg. 87

[37] J. T. Koberstein, B. Morra, and R. S. Stein, *J. of Appl. Cryst.* **13**, 34-45 (1980).

[38] L. M. Leung and J. T. Koberstein, *J.of Polym. Sci. Part B: Polym.Phys. Ed.*, **23**, 1883-1913 (1985).

[39] D. Tyagi, J. E. McGrath, and G. L. Wilkes, *Polym.Eng.Sci.*, **26**, 1371-1398 (1986).

[40] A. J. Ryan, W. R. Willkomm, T. B. Bergstrom, C. W. Macosko, J. T. Koberstein. C. C. Yu, and T. P. Russell, *Macromolecules*, **24**, 2883-2889 (1991).

[41] F. J. Baltá-Calleja and C. G. Vonk, *X-ray Scattering of Synthetic Polymers*, Elsevier, N.Y. (1989).

[42] C. Santa Cruz, N. Stribeck, H. G. Zachmann, and F. J. Baltá Calleja, *Macromolecules*, **24**, 5980-5990 (1991).

[43] C. G. Vonk, in *Small Angle X-ray Scattering*, edited by O. Glatter and O. Kratky, Chapter 13, Academic Press, New York (1982).

[44] O. Glatter, *J. of Appl.Cryst.*, **12**, 166-175 (1979).

[45] O. Glatter, in *Small Angle X-ray Scattering*, edited by O. Glatter and O. Kratky, Chapter 5, Academic Press, New York (1982).

[46] J. V. McCluskey, R. D. Priester. W. R. Wilkomm, M. D. Heaney, and M. A. Capel, Presentation at the Proceedings of the SPI, 35th Annual Technical/marketing Conference, (1993)

[47] J. Stöhr, *NEXAFS Spectroscopy* (Springer-Verlag, Berlin, 1992).

[48] D. A. Fischer, G. E. Mitchell, B. M. DeKoven, A. T. Yeh, J. L. Gland, and A. R. Moodenbaugh in "Applications of Synchrotron Techniques to Materials Science", *Mat. Res. Soc. Proc.* **307**, (eds. D. Perry, R. Stockbauer, N. Shinn, K. D'Amico, and L. Terminello, 1993).

[49] G. E. Mitchell, B. M. DeKoven, D. R. Speth. M. E. Jones, J. J. Curphy, D. L. Schmidt, A. T. Yeh. J. L. Gland and D. A. Fischer in "Polmer/Inorganic Interfaces", *Mat. Res. Soc. Proc.* **304** (eds. R. Opila, F. Boerio, and A. Czanderna, 1993).

[50] J. L. Jordan-Sweet, C. A. Kovac, M. J. Goldberg, J. F. Morar, *J. Chem. Phys.* **89**, 2482 (1988)

[51] D. A. Fischer, J. Colbert, and J.L. Gland, *Rev. Sci. Instr.* **60**, 1596(1989)

[52] Briggs, D. & Seah, M., Eds., *Practical Surface Analysis, Vol. 1 - Auger and X-Ray Photoelectron Spectroscopy* (John Wiley and Sons, New York, NY, 1990).

[53] D. Outka and J. Stöhr, *J. Chem. Phys.* **88** 3539 (1988)

[54] B.DeKoven, D. Fischer, and M. Chaudhury, to be submitted to *Langmuir*.

[55] G. Mitchell, A. Yeh, D. Fischer, and J. Gland, to be submitted to *J. Phys. Chem.*.

[56] D. L. Schmidt, C.E. Coburn, B.M. DeKoven, G.F. Meyers, and D.A. Fischer, *Nature* **368**,39 (1994)

[57] T. M. Stokich, Jr., W.M. Lee and R.A. Peters, *Mat. Res. Soc. Proc.* **227** (1991) 103

[58] B. DeKoven, D. Fischer, D. Parker, J. Curphy, G. Mitchell, C. Wedelstaedt, and J. Gland, to be published.

[59] S.-C. Chang, M. A. Leugers, S. R. Bare, *J. Phys. Chem.*, **96**, 10358 (1992)

[60] S. R. Bare, G. E. Mitchell, J. J. Maj, G. E. Vrieland, J. L. Gland, *J. Phys. Chem.*, **97**, 6048 (1993)

[61] F. W. Kutzler, C. R. Natoli, D. K. Meisemer, S. Doniach, K. O. Hodgson, *J. Chem. Phys.*, **73**, 3274 (1980)

[62] J. Wong, F. W. Lytle, R. P. Messmer, D. H. Maylotte, *Phys. Rev. B: Condens. Matter*, **30**, 5596 (1984)

[63] F. W. Lytle, R. B. Greegor, E. C. Marques, *Proc. 9th Int. Congr. Catal.*, Calgary, **5**, 54 (1988)

[64] B. Hedman, J. E. Penner-Hahn, K. O. Hodgson, in *EXAFS and Near Edge Structure III*, K. O. Hodgson, B. Hedman, J. E. Penner-Hahn, Eds.; Springer-Verlag: Berlin, 1984, p. 64

[65] G. N. George, W. E. Cleland, J. H. Enemark, B. E. Smith, C. A. Kipke, S. A. Roberts, S. P. Cramer, *J. Am. Chem. Soc.*, **112**, 2541 (1990)

[66] G. Deo, I. E. Wachs, *J. Phys. Chem.*, **95**, 5889 (1991)

[67] S. R. Bare, to be submitted to *Catalysis Letters*.

[68] C. B. Murchison, G. E. Vrieland, to be submitted to *J. Appl. Catal*.

POWDER CRYSTALLOGRAPHY OF USEFUL MATERIALS USING SYNCHROTRON RADIATION

James A. Kaduk[*], John Faber, and Shiyou Pei

Amoco Corporation
150 W. Warrenville Road, P.O. Box 3011
Naperville IL 60566

INTRODUCTION

Although the title seems innocuous, it is intended to be deliberately provocative. When chemists think of powder diffraction, it is generally in the traditional sense of a tool for phase identification, and perhaps the determination of lattice parameters, semiquantitative phase analysis, or average crystallite size. Recent advances in software (particularly the Rietveld method) and the availability of synchrotron radiation have facilitated the extraction of accurate and precise structural information from powder diffraction data, information comparable to that obtained from a typical single crystal experiment. It is now possible to do "powder crystallography", and not merely the traditional tasks of powder diffraction.

Synchrotron radiation has several characteristics which make it useful for powder crystallography. Most important are its intensity (which permits achieving high resolution and peak/background) and its tunability, allowing one to use the optimum wavelength for the problem at hand. We have applied synchrotron radiation to a wide variety of structural problems. In this paper we describe several recent structural studies, to illustrate the possibilities and some of the problems of powder crystallography

HYDRATED SODIUM ALUMINATE, $NaAlO_2 \cdot 5/4H_2O$

Sodium aluminate is an important industrial inorganic chemical. It is used in water treatment, and also as a convenient source of aluminum in synthetic applications. Sodium aluminate is often used as the aluminum source in the preparation of zeolites and other catalytic materials.

The most commonly encountered form of sodium aluminate is the solid obtained by crystallization from concentrated aqueous solutions[1]. At low temperatures (5-45°C),

Synchrotron Radiation Techniques in Industrial, Chemical, and Materials Science
Edited by D'Amico *et al.*, Plenum Press, New York, 1996

NaAlO$_2$· 3/2H$_2$O crystallizes, while at higher temperatures (60-140°C) NaAlO$_2$· 5/4H$_2$O is produced.

Although powder patterns have been reported for NaAlO$_2$· 3/2H$_2$O (Powder Diffraction File entry 2-1025), NaAlO$_2$·3H$_2$O (29-1165[2]), and NaAlO$_2$· 5/4H$_2$O (41-638[3]), the crystal structures have never been reported. The unit cell of the sesquihydrate has been reported[2] as tetragonal, with a = 10.53 and c = 11.40 Å. For NaAlO$_2$· 5/4H$_2$O, ^{27}Al and ^{1}H NMR and infrared spectroscopic studies[3] indicate the presence of highly polymerized anions with tetrahedrally-coordinated Al. The unit cell of NaAlO$_2$· 5/4H$_2$O has been reported[4] as tetragonal, with a = 10.530(5) and c = 5.300(5) Å.

Experimental

The material used in this study was E. Merck NaAlO$_2$·xH$_2$O reagent, lot # 8352. Thermogravimetric analysis indicated that this material contained 21.55 wt% water. Since powder diffraction demonstrated that the dehydrated material is NaAlO$_2$, the stoichiometry of the hydrated material is NaAlO$_2$·5/4H$_2$O. From the observed density and the unit cell dimensions, the cell contents are Na$_8$Al$_8$H$_{20}$O$_{26}$.

The sample used for the data collection was mixed with NIST 640b silicon internal standard at a concentration of 4.14 wt%. The sample was milled in a McCrone micronising mill using n-hexane as the milling liquid.

Data collection was carried out at beamline X3B1 at the National Synchrotron Light Source (NSLS) at Brookhaven National Laboratory. The pattern was measured from 7.000-64.695° 2θ in 0.005° steps, counting 4 sec/step. The sample was rocked 1° in ω at each data point. The wavelength used was 0.699341(5) Å (determined using the SRM 1976 alumina plate and the Si internal standard), and the diffractometer zero was -0.002(1)° 2θ.

A powder pattern measured on a Scintag PAD V diffractometer in our laboratory could be indexed[5] on a primitive tetragonal cell having a = 10.5358(3) and c = 5.3366(2) Å; the figure of merit was 149. These lattice parameters were used as the initial values for the refinement of the synchrotron pattern. The systematic absences were consistent with the acentric space groups P$\bar{4}$2$_1$m (#113) and P42$_1$2 (#90). P$\bar{4}$2$_1$m was selected, and confirmed by successful solution and refinement.

Initial data processing was carried out using GSAS[6]. A 3-term cosine Fourier series background function, a scale factor, the lattice parameters, and the cauchy profile terms X and Y were refined. The Le Bail extraction procedure incorporated into GSAS was used to extract 357 individual observed structure factors for 2θ < 50° (d > 0.83 Å).

These structure factors were used to create a SHELXTL Plus[7] data file. The structure was solved in P$\bar{4}$2$_1$m using direct methods. (Attempts to solve the structure in P42$_1$2 yielded much poorer combined figures of merit.) The strongest peak in the E-map corresponded to a reasonable Al position. Successive cycles of least-squares refinement and difference Fourier synthesis revealed the positions of the remaining heavy atoms. The final SHELX refinement of 27 variables using 322 observations yielded the residuals R = 0.3379 and wR = 0.4241.

Final refinement of the structure was carried out using GSAS. The nonhydrogen atoms were refined independently, subject to a soft constraint of 1.74(1) Å on the Al-O-Al bridge bonds. The Al and two independent Na were refined anisotropically. The oxygen atoms were refined isotropically, with independent displacement coefficients.

The hydrogen atoms could not be located in difference Fourier maps, and were included in the structure factor calculations in calculated positions, 0.85 Å from the oxygens to which they are bonded covalently. The water molecule oxygen O4 was 2.63 Å from two framework oxygens O2, and the hydrogen H4 was positioned on the O4-O2 vector. The water molecule oxygen O5 was located 2.74 Å from two framework oxygens O3, and the hydrogen H5 was located on the O5-O3 vector. The shortest O1-O1 distances were 3.20 Å. The hydroxyl hydrogen was placed in two half-occupied positions along the shortest O1-O1 vectors. Attempts to refine the hydrogen atom positions subject to a soft constraint were unsuccessful. The hydrogen isotropic displacement coefficients were fixed at 0.05 Å^2. Refinement of the water molecule occupancies yielded values insignificantly different from unity.

Included in the refinement were a scale factor and the lattice parameters. The peak profiles were described by a pseudo-Voigt function. The refined coefficients were the gaussian U, cauchy X and Y, asymmetry, sample displacement, and anisotropic strain and size broadening terms (unique axis = [001]). The background was described by an 8-term real-space pair correlation function, which used 3 characteristic interatomic distances. The isotropic displacement coefficient of the Si of the silicon internal standard was refined, as well as the X, Y, and asymmetry profile coefficients, and a scale factor.

Two impurity phases were identified and included in the refinement: α-quartz (SiO$_2$) and thermonatrite (Na$_2$CO$_3 \cdot$ H$_2$O). At least one additional unidentified crystalline phase is present. For each of the impurity phases a scale factor and a cauchy X profile coefficient were refined.

The final refinement of 64 variables using 11541 observations yielded the residuals wRp = 0.1085 and Rp = 0.0747. The final reduced χ^2 was 7.488. The Bragg R(F) was

Table 1
Refined Structural Parameters of NaAlO$_2 \cdot$ 5/4H$_2$O

Space Group P$\bar{4}2_1$m, a = 10.53396(5), c = 5.33635(3) Å at 27°C				
Atom	x	y	z	U_{iso}, Å^2
Al	0.81175(13)	0.09450(13)	0.10055(23)	0.0248
Na1	0.60138(14)	0.10138(14)	0.59605(35)	0.0284
Na2	0.80414(15)	0.30414(15)	0.67529(37)	0.0418
O1	0.81046(23)	0.09109(22)	0.4388(4)	0.0328(8)
O2	0.93909(20)	0.18870(24)	0.0043(4)	0.0205(6)
O3	0.66923(20)	0.16923(20)	0.0129(6)	0.0172(10)
O4	0.38579(20)	0.11421(20)	0.7128(6)	0.0288(10)
O5	1/2	0	0.2176(9)	0.0354(14)
H1A, frac = 1/2	0.83624	0.01713	0.47151	0.05
H1B, frac = 1/2	0.88490	0.11773	0.47151	0.05
H4	0.32216	0.09678	0.80539	0.05
H5	0.55240	0.05240	0.15466	0.05

Anisotropic thermal parameters T = exp(h$^2a^{*2}$U$_{11}$ + ... + 2hka$^*b^*$U$_{12}$ + ...)

Atom	U$_{11}$	U$_{22}$	U$_{33}$	U$_{12}$	U$_{13}$	U$_{23}$
Al	0.0355(11)	0.0261(9)	0.0129(7)	-0.0022(8)	0.0034(8)	-0.0026(7)
Na1	0.0303(10)	0.0303(10)	0.0246(15)	0.0005(14)	-0.0055(9)	-0.0055(9)
Na2	0.0470(10)	0.0470(10)	0.0314(16)	-0.0041(15)	0.0054(11)	0.0054(11)

NaAlO2.5/4H2O, X3B1, 28 Feb 93, PLATE
Lambda 0.6993 A, L-S cycle 281

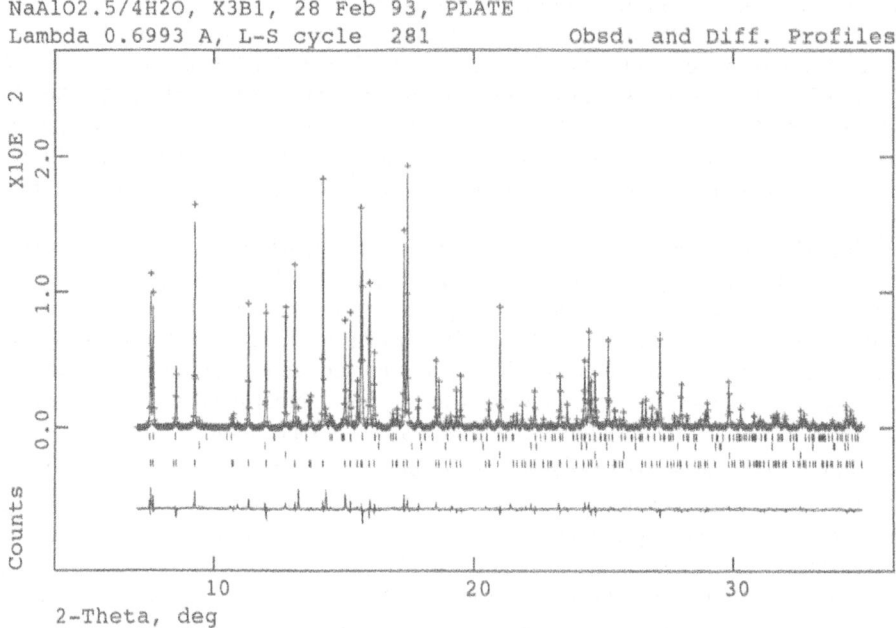

NaAlO2.5/4H2O, X3B1, 28 Feb 93, PLATE
Lambda 0.6993 A, L-S cycle 281

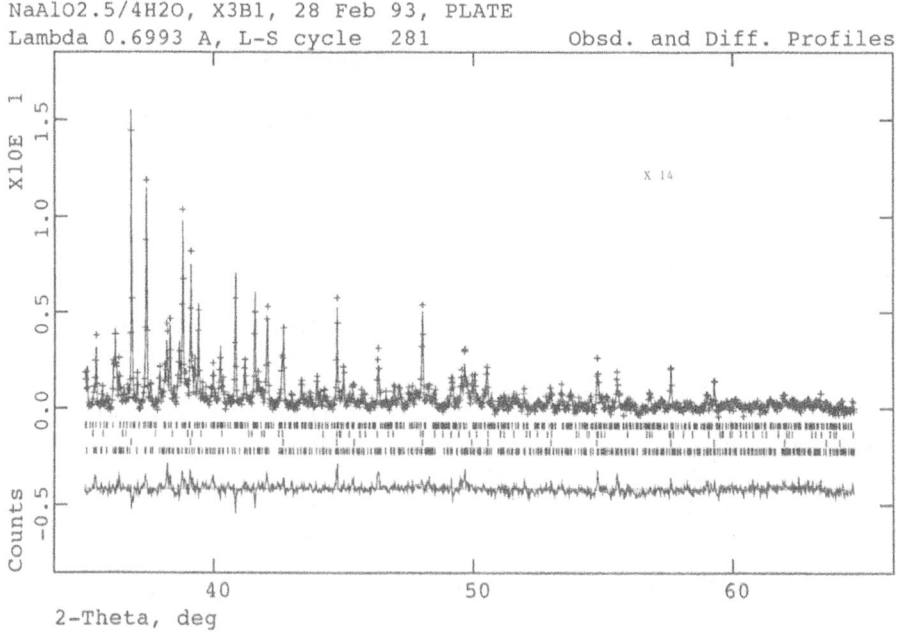

Figure 1. Observed, calculated, and difference diffraction patterns of $NaAlO_2 \cdot 5/4H_2O$. The crosses represent the experimental points, and the solid line the calculated pattern. The difference curve is plotted at the same scale as the other patterns. The bottom row of tick marks represents the positions of the $NaAlO_2 \cdot 5/4H_2O$ peaks. Successive rows upward represent the Si, SiO_2, and $Na_2CO_3 \cdot H_2O$ peak positions. The vertical scale of the high-angle portion has been expanded by a factor of 14.

0.0849. The agreement between F_o and F_c is poorest for the weak, high-angle reflections. The observed, calculated, and difference profiles are illustrated in Figure 1. The largest errors result from the peaks of the unidentified phases and the silicon internal standard. The slope of the final Wilson plot suggests that the ESDs are underestimated by a factor of 2. The refined structural parameters are reported in Table 1.

Results and Discussion

The basic unit of the crystal structure of $NaAlO_2 \cdot 5/4H_2O$ is a single layer parallel to (001) made up of corner-sharing AlO_4 tetrahedra. The tetrahedra are joined into sheets of 4-rings and 8-rings, in contrast to the double chains of 6-rings proposed by Gessner[3].

Three of the four oxygens (two O2 and one O3) lie in the plane of the layers, while the fourth (O1, a hydroxyl group) points into the interlayer regions. The hydroxyls alternate UDUD around the 4-rings, and are UUDDUUDD around the 8-rings.

The geometry of the AlO_4 tetrahedron is regular; the largest deviation from the ideal O-Al-O angle is 6.9°. Both O2 (general position) and O3 (site symmetry m) form the Al-O-Al bridges. The Al-O2, Al-O2A, and Al-O3 distances of 1.746(2), 1.730(2), and 1.758(2) Å are close to the average value of 1.74 Å typically found in framework structures. O2 forms the "UD" Al-O-Al bridge, and O3 forms the "UU" bridge. The twist angles[8], defined as the angles between the O1-Al-Al and O1A-Al-Al planes in the Al_2O_7 units centered on O2 and O3, are 179.9 and 0.0°, respectively. The Al_2O_7 units are almost exactly eclipsed or staggered.

The Al-O1 bond of 1.806(2) Å is considerably longer than the Al-O-Al bridge bonds. Bond valence and charge balance considerations suggest that this oxygen represents a hydroxyl group.

The crystal structure is illustrated by a polyhedral rendering in Figure 2. There are two crystallographically-independent Na ions. The coordination of Na1 is octahedral (Figure 3). The average deviation from the ideal angles is 4.6°; the largest deviations involve the water molecule O5. Coordinated to Na1 are two hydroxyl groups, one bridging oxygen, and three water molecules. Na2 is 7-coordinate (Figure 3). The Na2 coordination is best viewed as distorted octahedral, with an additional water molecule (O5) capping one of the octahedral

Figure 2. A polyhedral rendering[11] of the structure of $NaAlO_2 \cdot 5/4H_2O$. The Na are represented by the solid circles, and the water molecule oxygens by open circles. Na1 is near the long side of the 8-ring, and Na2 near the "end" of the 8-ring. O5 lies on $mm2$ through the center of the 8-ring, and O4 lies on m displaced from the ring center. The view is approximately down the c-axis, with a down and b to the right.

faces. The average Na1-O distance is 2.40 Å, while the average Na2-O distance is 2.62 Å.

The aluminate layers are held together by coordination to the Na ions, and by an extensive network of hydrogen bonds involving the water molecules and the hydroxyl groups. Na1 is coordinated to two hydroxyl groups from one layer, one bridging oxygen from the adjacent layer, and three water molecules (Figure 3). Na2 is coordinated to two hydroxyls from one layer, three bridging oxygens from the adjacent layer, and two water molecules (Figure 3). Bond valence sums[9] for the heavy atoms suggest that the Al is in a normal environment. Na1 appears to be crowded, and Na2 to be loosely coordinated. The bonding power of O1 is not satisfied, and it is clearly the location of the hydroxyl group. The valences of the framework oxygens O2 and O3 are not completely satisfied, which makes them plausible hydrogen bond acceptors.

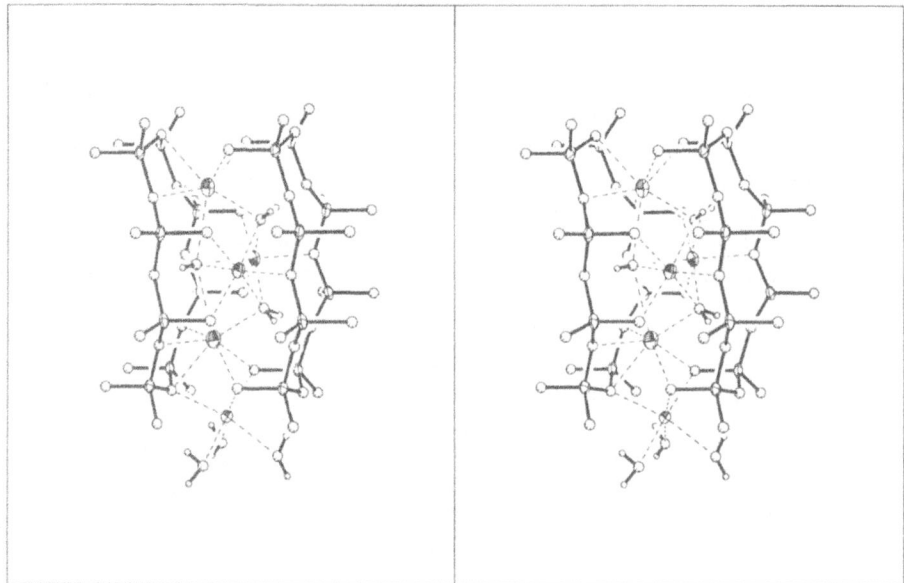

Figure 3. A stereo view of the coordination polyhedra of the Na ions in $NaAlO_2 \cdot 5/4H_2O$; 50% probability ellipsoids for the Na and Al atoms. Na-O bonds are indicated by dashed lines. The view is approximately down [110], with c to the right.

The refined lattice parameters of the primitive tetragonal unit cell (space group $P\bar{4}2_1m$) are: $a = 10.53396(4)$ and $c = 5.33635(3)$ Å at 27°C. A search of the Crystal Data database[10] for sub- and supercells from 1/4 to 4 times the volume of the observed cell yielded no plausible isostructures.

COPPER ALUMINUM BORATE, $Cu_2Al_6B_4O_{17}$

The unusual copper aluminum borate $Cu_2Al_6B_4O_{17}$ is useful as a dehydrogenation catalyst[12]. The average structure (I4/m, $a = 10.586(1)$, $c = 5.688(2)$ Å) has been known for some time[13], and has been redetermined recently using single crystal techniques[14]. Structure determination has been hampered by the difficulty of preparing homogeneous materials. Recent advances in sol-gel preparative chemistry[12] have led to the synthesis of uniformly-green

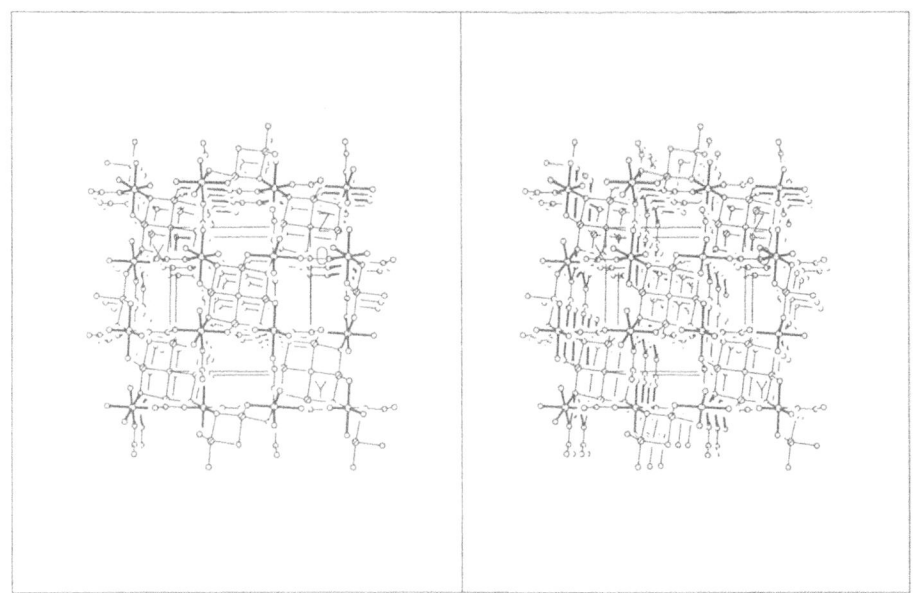

Figure 4. A stereo view of the crystal structure of $Cu_2Al_6B_4O_{17}$. The view is approximately down the tetragonal *c*-axis. The AlO_6 bonds are highlighted.

material, permitting a more-detailed structural study.

The crystal structure (Figure 4) is made up of edge-sharing chains of octahedral Al atoms parallel to the tetragonal *c*-axis. The AlO_6 chains are joined in the *a*- and *b*- directions by trigonal planar BO_3 groups. There is a 5-coordinate site, 50% occupied each by Cu and Al, which shares a face with the AlO_6 octahedron. These trigonal bipyramidal sites share equatorial corners at a square planar oxygen, O1.

Trigonal bipyramidal coordination is relatively unusual for both Cu^{2+} and Al^{3+}. We considered it unlikely that the Cu and Al ions would occupy exactly the same position within the coordination sphere, and that a split site model might be appropriate. Attempts to refine such a model using laboratory powder data did not yield improved residuals compared to a unified-site model. To study this site in more detail, we carried out a resonant scattering experiment, exploiting the tunability of synchrotron radiation.

To first approximation, X-ray scattering is an elastic process, and the atomic scattering factor is independent of the wavelength of the incident radiation. When the frequency approaches an atomic absorption edge, inelastic processes become significant, and lead to a dependence of the scattering factor on the wavelength of the incident radiation[15]. Both the real and imaginary parts of the anomalous scattering vary, and their variation can be exploited to alter the contrast between atoms. Simultaneous refinement of one structural model using several datasets collected at different wavelengths near and absorption edge can yield information about the populations of different atoms occupying the same site[16]. In this study, we sought to vary the scattering contrast between Cu and Al, to permit a more detailed look at the 5-coordinate site.

Experimental

The material studied was prepared from $Cu(C_2H_3O_2)_2$, $Al(C_3H_7O)_3$, and $B(C_3H_7O)_3$ using sol-gel techniques and a slow programmed drying and calcination. The product was ground in a McCrone micronising mill using ethanol as the milling liquid.

Three powder diffraction patterns were measured:

1. on a Scintag PAD V diffractometer in our laboratory, using Cu K_{α} radiation ($\lambda(K_{\alpha1/\alpha2} = 1.540629, 1.544451$ Å), from 10-140° 2θ in 0.02° steps, 10 sec/step.

2. on beamline X-7A at NSLS, using a wavelength of 1.293661 Å, from 9-70° 2θ in 0.02° steps, counting for 5-8 sec/step.

3. on beamline X3A2 at NSLS, using a wavelength of 1.37957 Å, from 30-75° 2θ in 0.01° steps. This wavelength is very close to the Cu K edge.

The variation in the scattering contrast produced significant differences in the relative intensities of the lines of the powder diffraction patterns.

All three powder patterns were included in a refinement of the crystal structure using GSAS[6]. The single crystal structure model[14] was used. The Cu1/Al1 site was refined anisotropically. All other atoms were refined with isotropic displacement coefficients. The U_{iso} of O2, O3, and O4 were fixed at the average of the single crystal U_{iso}. The occupancies of Cu1, Al1, Al2, and Cu2 were refined, subject to the constraint that the overall Cu:Al ratio was 1:3.

The powder patterns indicated the presence of a trace of CuB_2O_4 (PDF entry 25-268). This phase was included in the refinement using a fixed structural model[17]. Phase fractions

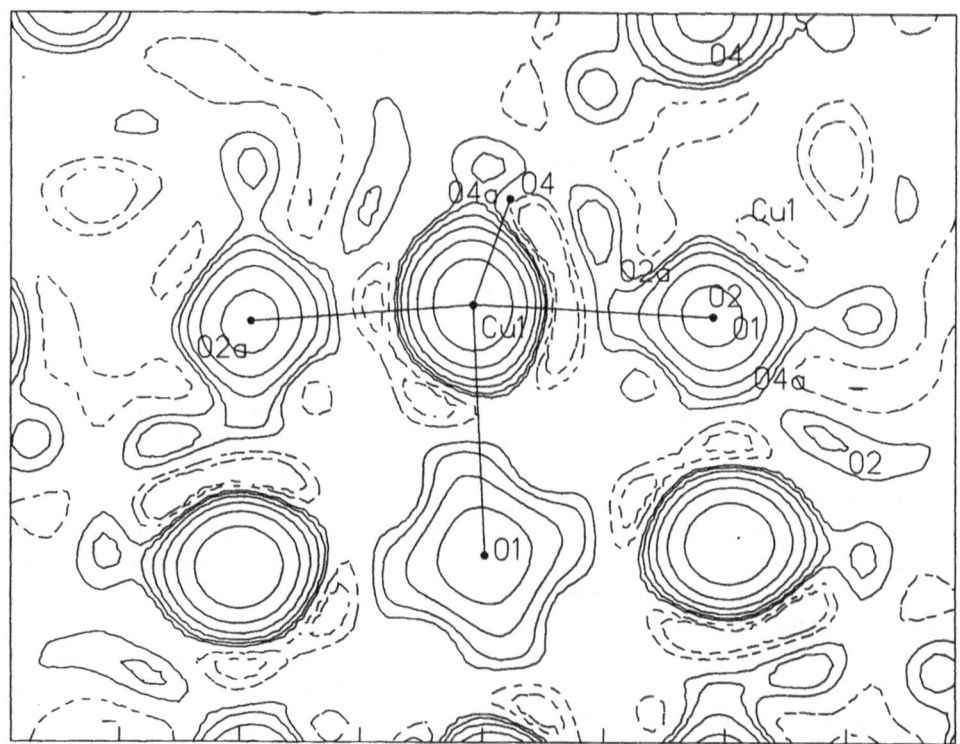

Figure 5. An Fobs map of $Cu_2Al_6B_4O_{17}$, calculated through the *xy0* plane, which contains Cu1, O1, and O2. The contours represent electron densities of 32, 16, 8, 4, 2, 1, -1, and -2 eÅ^{-3}. Negative contours are dashed.

for both phases were refined, subject to the constraint that the sum of the phase fractions was unity.

Included in the refinement were histogram scale factors for each pattern, and the lattice parameters of the two phases. A preferred orientation term was refined for $Cu_2Al_6B_4O_{17}$ (unique axis [001]). To minimize the effects of surface roughness, microabsorption, and incomplete interception of the beam, the low-angle portion of each histogram was excluded from the refinement.

The peak profiles were described by a pseudo-Voigt function. For the synchrotron data sets, a pure cauchy profile was used, while for the in-house data a gaussian component was required. Sample transparency and displacement terms were included for the laboratory pattern. All three profiles included an anisotropic size broadening term. The background of each histogram was modelled adequately by a 3-term cosine Fourier series.

An F_{obs} Fourier map (using either the multiple powder data sets or the single crystal data) calculated in the $xy0$ plane, which contains the Cu1/Al1 site and O1 and O2, exhibited no evidence for a split Cu1/Al1 site (Figure 5), but shows clear evidence for multiple O1 sites displaced from the origin. Refinement of a unified Cu1/Al1 site yielded the lowest residuals, but ordered and disordered O1 refinements yielded equivalent refinements. Non-crystallographic evidence can be used to distinguish between the models.

The final refinement of 53 variables using 13099 observations yielded the residuals wRp = 0.0409 and Rp = 0.0221; the final reduced χ^2 was 1.750. The largest peak in the final difference Fourier map was 0.80 eÅ^{-3}, 1.85 Å from Cu1/Al1, 2.03 Å, and 2.00 Å from O4, in the channels parallel to the c-axis. All of the largest difference peaks are in the $xy0$ plane, and are near framework atoms. The largest difference hole was -0.67 eÅ^{-3}, 0.90 Å from the B. The structural parameters from the combined refinement are reported in Table 2.

Table 2
Refined Structural Parameters of $Cu_2Al_6B_4O_{17}$

\multicolumn{6}{c}{Space Group I4/m, a = 10.57945(1), c = 5.67357(6) Å at 27°C}
Atom	x	y	z	Uiso	frac
Cu1	0.02424(12)	0.19110(12)	0	0.0267	0.487(2)
Al1	0.02424(12)	0.19110(12)	0	0.0267	0.513(2)
Al2	1/4	1/4	1/4	0.0050(5)	0.988(2)
Cu2	1/4	1/4	1/4	0.0050(5)	0.012(2)
B	0.2504(6)	0.5054(7)	0	0.0079	
O1	0.018	-0.018	0	0.0081(19)	1/4
O2	0.2086(3)	0.1501(3)	0	0.0098	
O3	0.2477(4)	0.3742(3)	0	0.0098	
O4	0.4319(2)	0.2407(3)	0.2068(3)	0.0098	

For Cu1/Al1, U_{11} = 0.0364(13), U_{22} = 0.0148(10), U_{33} = 0.0291(11), U_{12} = 0.0001(9), U_{13} = U_{23} = 0

Results and Discussion

The *average* crystal structure of $Cu_2Al_6B_4O_{17}$ has been confirmed, at higher precision. There was no evidence in the powder patterns for a different cell or space group. The octahedral sites reside on centers of symmetry, and are completely occupied by Al. The Al2-O3 and Al2-O4 distances of 1.934(2) and 1.943(2) Å respectively fall within the normal range for octahedral Al-O distances, but the Al2-O2 distance of 1.822(2) Å is relatively short. The average deviation from the ideal octahedral angles is 4.3°. The B-O3 distance of 1.388(6) Å two B-O4 distances of 1.352(4) Å fall within the expected range. The O-B-O angles differ insignificantly from the expected 120°.

The trigonal bipyramidal Cu1/Al1 site is half occupied each by Cu and Al. The axial distances to two O2, and are long and short (1.998(3) and 1.854(3) Å). Two of the equatorial distances (to O4) are short (1.872(2) Å) and one (to O1) is long (2.038(1) Å). The central Cu1/Al1 site is displaced 0.24 Å from the center of the coordination polyhedron.

The atomic valences, calculated from the sums of bond valences[9], of the Cu and Al are 2.63 and 2.44, far from the nominal values of 2 and 3. The calculated valence of O1 is only 1.54, reflecting the relatively long bonds. These differences are indications that the refined structure represents an average.

Analysis of 81 $Cu^{2+}O_5$ coordination spheres located in the Inorganic Crystal Structure Database[18] indicates that the typical CuO_5 coordination sphere contains four bonds in the range 1.90-2.05 Å, and one longer bond, averaging 2.2-2.3 Å. The average Cu1 coordination sphere is therefore very unusual, in that all five bonds are shorter than 2.04 Å. The Cu-O2 bond of 1.85 Å is among the shortest Cu-O bonds ever reported.

EXAFS experiments[19] provide evidence for Cu clustering. Each Cu has at least one Cu in the second coordination sphere. This observation, and the appearance of the F_{obs} map, suggest a new model for the local structure.

Consider the four 5-coordinate sites surrounding an individual O1. Stoichiometry mandates that there are two Cu and two Al in the average "4-ring" around O1, and that there is only one oxygen in the center of the "4-ring". If, according to the EXAFS results, the Cu ions occur in "cis" pairs, a displacement of the central oxygen away from the two Cu in the xy plane would result in two long Cu-O1 bonds and two short Al1-O1 bonds (Figure 6). A displacement of approximately 0.27 Å along [1$\bar{1}$0], to 0.018,-0.018,0, permits the bonding requirements of all atoms to be better-satisfied, is consistent with the EXAFS data, yields comparable residuals to the ordered model for O1, and describes the same average structure. The combination of crystallographic and spectroscopic information has resulted in a new model for the local structure, a model consistent with all observations and with the catalytic properties of this material. This model could be tested by carrying out diffraction anomalous fine structure (DAFS) experiments on reflections containing large contributions from the Cu1/Al1 site.

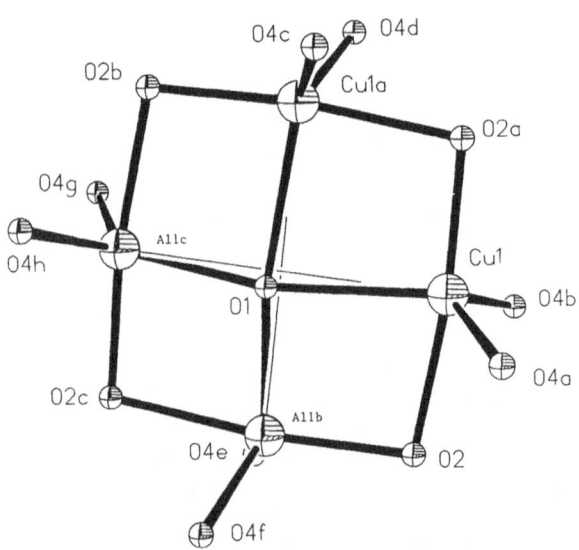

Figure 6. The proposed model for the local environment of the Cu1/Al1 sites in $Cu_2Al_6B_4O_{17}$. The true position of O1 is displaced approximately 0.27 Å from the average position. 50% probability ellipsoids.

FLUORIDE OPTICAL MATERIALS - BaY_2F_8 and $Ba(Y_{1/2}Yb_{1/2})_2F_8$

Materials having the BaY_2F_8 structure show promise as hosts for new lanthanide-doped laser materials. The only published report[20] ($C2/m$, $a = 6.935(1)$, $b = 10.457(2)$, $c = 4.243(1)$ Å, $\beta = 99.667(33)°$) of this structure type, that of $BaTm_2F_8$, contains unreasonable (negative) isotropic displacement coefficients. A desire to place the understanding of the optical properties of such materials on a firmer structural basis has led us to refine the structures of two materials, BaY_2F_8 and $BaYYb_{0.99}Tm_{0.01}F_8$. For the purposes of this work, the Tm content of the latter material will be ignored.

Study of such highly-crystalline materials demonstrates that synchrotron radiation is not an unmixed blessing. To carry out a powder diffraction experiment, a *powder* sample is required. By this we mean a sample containing a suitably large number of randomly-oriented crystallites. Achieving this condition is more difficult than commonly realized[21]. To sample a sufficiently large number of grains, the grain size needs to be < 10 μm, finer than is normally obtained. The size and divergence of the X-ray source also determine the number of grains sampled. It is here that synchrotron radiation makes life more difficult, for the beam divergence is much smaller than that of a typical laboratory source, resulting in "grainy" diffraction patterns, which can be impossible to fit.

An example of the potential magnitude of the problem is provided in Figure 7, which presents an ω-scan (rocking curve) of the (214) reflection of another optical material, Zn-doped $LiNbO_3$, measured at a wavelength of 0.65312 Å. The diffracted intensity varies by nearly 100% during the scan, as individual grains are sampled. Even rocking this sample by 2° in ω during each data point did not yield a true powder average, and a reasonable Rietveld refinement could not be obtained. Such effects make it imperative to check the granularity of each sample at the synchrotron before data collection begins.

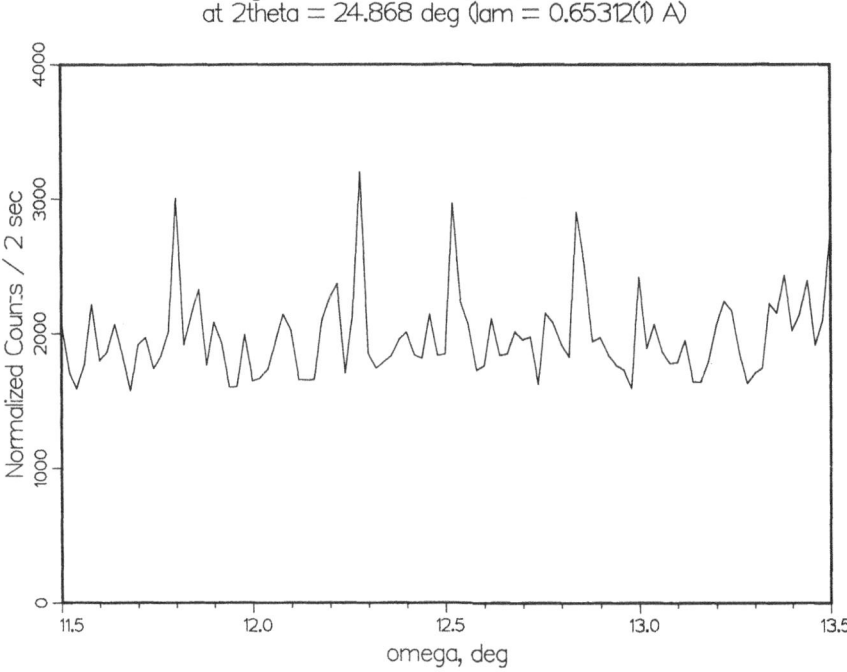

Figure 7. An ω-scan (rocking curve) of the (214) peak of Zn-doped $LiNbO_3$, measured at $2\theta = 24.868°$ and a wavelength of 0.65312(1) Å at beamline X3B1 at NSLS. The sharp spikes indicate the scattering of individual grains, the result of too-large particle size and the small beam divergence.

Just as in a single crystal experiment, all the structural information is ultimately derived from the structure factors, which are in turn derived from the observed integrated intensities. Accurate integrated intensities can only be obtained if the reflection profiles and the background are described accurately. The very sharp lines obtained from highly-crystalline materials at a synchrotron can present problems in fitting the profiles. The studies described here demonstrate two different approaches to overcoming these problems.

Experimental - BaY_2F_8

The sample was synthesized by A.P. Wilkinson. It was ground in a McCrone micronising mill twice for 30 minutes using ethanol as the milling liquid. After each grinding, the coarsest fraction was allowed to settle, and only the fine solid remaining in suspension was processed further. Such extreme measures were necessary to obtain a true powder specimen.

Powder patterns were measured both in-house and at NSLS. The laboratory pattern was measured on a Scintag PAD V diffractometer, equipped with an intrinsic Ge detector. The pattern was measured from 5-120° 2θ in 0.02° steps, counting for 12 sec/step. The sample was spun about the diffraction vector during data collection. The synchrotron pattern was measured on beamline X3B1 at NSLS, using a wavelength of 1.198808 Å. The pattern was measured from 11-66.975° 2θ in 0.005° steps, counting for 2 sec/step. The sample was rocked from ω-1 to ω+1° during each point. An ω-scan of the (200) peak indicated acceptable powder averaging.

All data processing was carried out using GSAS[6]. Both histograms were included in the refinement. The initial structure model was that of Izotova and Aleksandrov[20]. The Ba and Y were refined anisotropically; the three F were refined with a common isotropic displacement coefficient.

Some difficulty was encountered in describing the peak profiles of the synchrotron pattern. Although the instrumental pseudo-Voigt profile function contains principally gaussian components, the best profile function was pure cauchy. Even this function could not reproduce the profiles accurately; the peaks appeared to be "hyperlorentzian". To model the profiles, a

Table 3
Refined Structural Parameters of BaY_2F_8

Space Group C2/m, $a = 6.97862(3)$, $b = 10.51164(5)$, $c = 4.26118(2)$ Å, $\beta = 99.6751(4)°$

Atom	x	y	z	Uiso, $Å^2$
Ba	0	0	0	0.0117
Y	1/2	0.17659(21)	1/2	0.0113
F1	0.1839(3)	0.1404(2)	0.5585(5)	0.0113(5)
F2	0.3928(4)	0	0.2243(7)	0.0113(5)
F3	1/2	0.2414(3)	0	0.0113(5)

Anisotropic thermal parameters $T = \exp(h^2 a^{*2} U_{11} + ... + 2hka^* b^* U_{12} + ...)$

Atom	U_{11}	U_{22}	U_{33}	U_{12}	U_{13}	U_{23}
Ba	0.0120(7)	0.0170(6)	0.0066(7)	0	0.0031(5)	0
Y	0.0143(7)	0.0125(6)	0.0072(7)	0	0.0022(5)	0

second phase (having a different cauchy profile function) was included. All structural parameters were constrained to be the same for the two phases. A much better fit was obtained, although no structural parameter varied significantly as the refinement strategy changed.

Included in the refinement were scale factors for both histograms and the monoclinic lattice parameters. Both preferred orientation (unique axis [110]) and anisotropic size broadening terms (unique axis [010]) were refined. The backgrounds of both histograms were described by a 3-term cosine Fourier series.

The final refinement of 44 variables using 16656 observations yielded the residuals wRp = 0.1054 and Rp = 0.1076. The final reduced χ^2 was 2.674. The R(F) was 0.0410 for the laboratory data, and 0.0240 for the synchrotron pattern. No significant trends were noted in an analysis of variance. The slopes of the Wilson plots indicate that the ESDs are underestimated by a factor of 1.13 for the laboratory data, and 1.63 for the synchrotron data. The largest peak in a difference Fourier map was 3.6 electrons, at the Ba position. The largest hole was -0.9 electrons.

The refined structural parameters are reported in Table 3. The agreement of the observed and calculated patterns for the laboratory data is very good, while that of the synchrotron data is poorer, indicating residual profile errors (Figure 8).

Experimental - BaYYbF$_8$

Analysis of BaYYbF$_8$ began as a single crystal study. The available sample consisted of trimmings from an optical crystal. One of these fragments, approximately 0.17 x 0.17 x 0.15 mm, was the subject of a normal single crystal data collection (Nicolet/Siemens R3m, Mo K$_\alpha$ radiation) and refinement. Since there were no well-defined crystal faces, an empirical absorption correction, based on Ψ-scans of 28 strong reflections well-distributed in reciprocal space, was carried out. The maximum and minimum transmission were 0.029 and 0.001. The resulting systematic errors led to the presence of many large features ($\pm$ 10 eÅ^{-3}) in a difference Fourier map.

Since for a flat plate sample in symmetrical reflection geometry the absorption correction becomes negligible[22], we collected powder diffraction data, and hoped to overcome the effects of absorption by carrying out a combined refinement of the powder and single crystal data. A portion of the sample was ground in a McCrone micronising mill, using ethanol as the milling liquid.

Three X-ray powder patterns were measured: one in the laboratory (Scintag PAD V, Cu K$_\alpha$ radiation, 10-140° 2θ, 0.02° steps, 10 sec/step), and two synchrotron patterns - one on beamline X7A (λ = 1.29366(4) Å, 10-70° 2θ, 0.005° steps, 3 sec/step), and the other on beamline X3B1 (λ = 1.12706(6) Å, 11.000-79.470° 2θ, 0.005° steps, 4 sec/step). All data processing was carried out using GSAS[6]. The single crystal structural model was used to begin the refinement. All atoms were refined anisotropically. The Y and Yb were constrained to lie at the same position, and to have identical anisotropic displacement coefficients.

The profiles were described using only the cauchy terms of a pseudo-Voigt function, including both anisotropic size and strain broadening terms (unique axis [010]) and an asymmetry parameter. Although a good fit could be obtained for the laboratory pattern, some difficulty was encountered in fitting the synchrotron profiles and backgrounds. As with BaY$_2$F$_8$, the peaks appeared to be "hyperlorentzian". Most of the difficulties occurred in fitting the overlap of the peak tails in the regions containing strong peaks. A much-improved fit could be obtained by using a real space pair correlation function with two characteristic

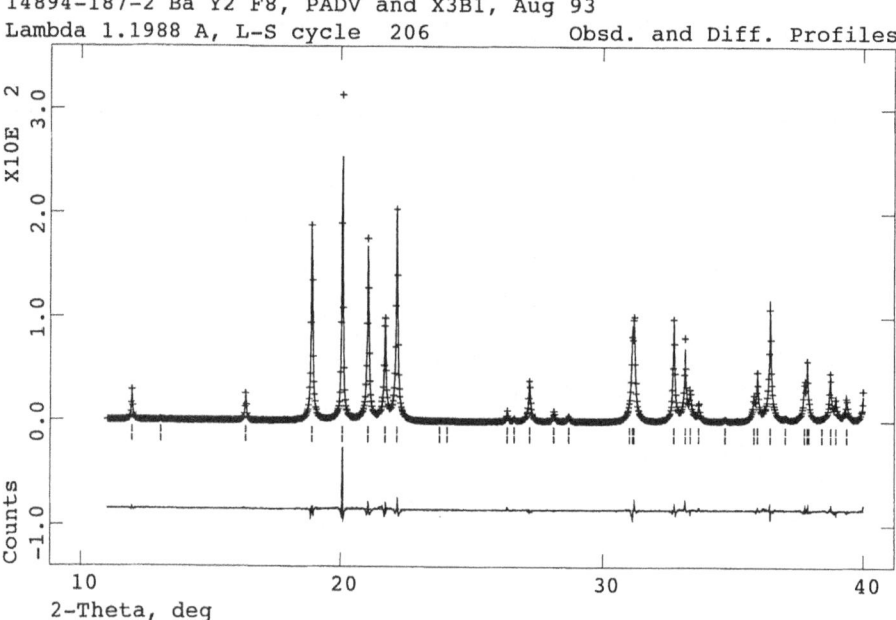

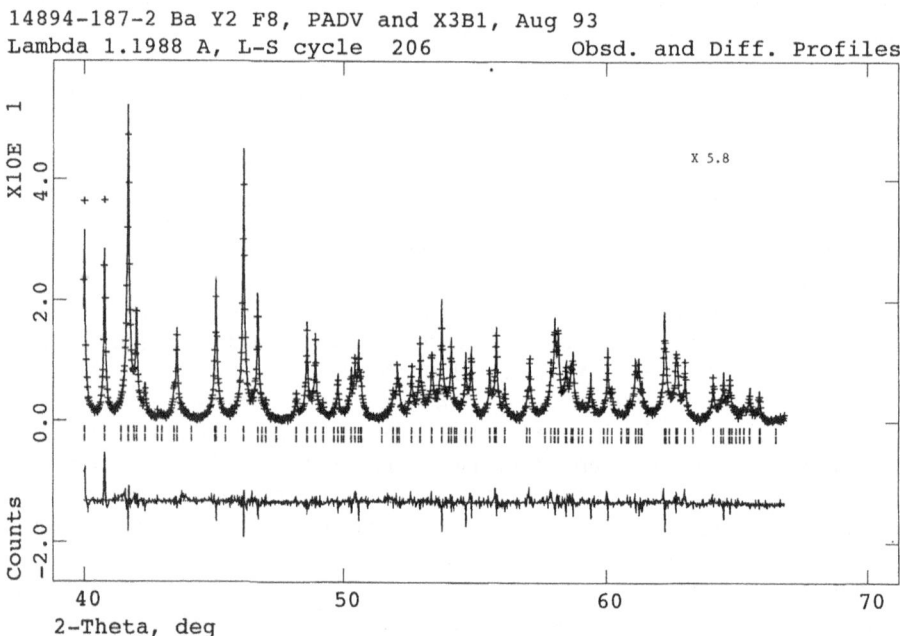

Figure 8. Observed, calculated, and difference diffraction patterns of $BaY_2F_8.$ The synchrotron pattern is presented on the left, and the laboratory pattern on the right. The crosses represent the experimental points, and the solid lines the calculated patterns. The difference curve is plotted at same scale as the observed and calculated patterns. The rows of tick marks represent the positions of the peaks. The vertical scale of the high-angle portion of the synchrotron pattern has been multiplied by 5.8, and the high-angle portion of the laboratory pattern by 13.5.

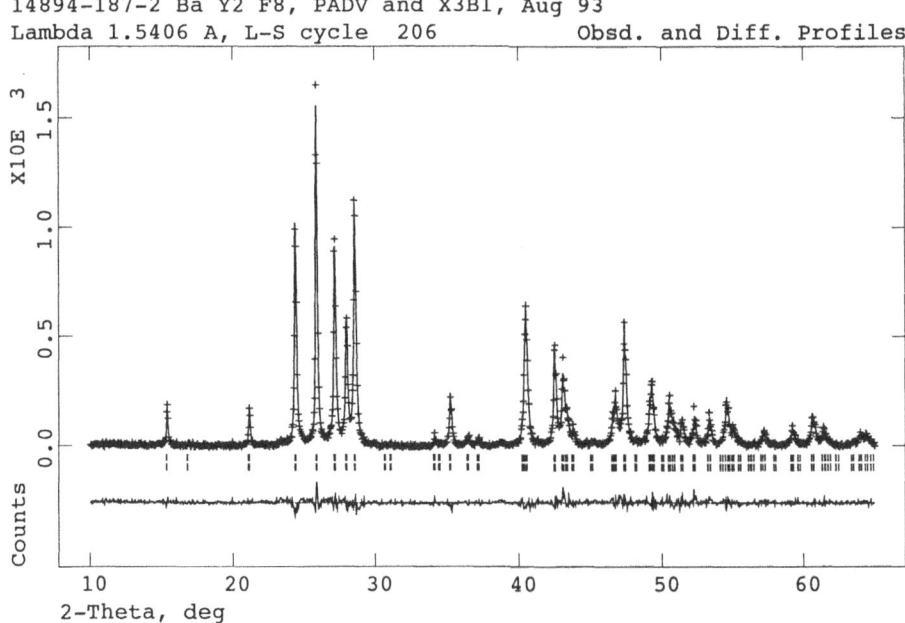

14894-187-2 Ba Y2 F8, PADV and X3B1, Aug 93
Lambda 1.5406 A, L-S cycle 206   Obsd. and Diff. Profiles

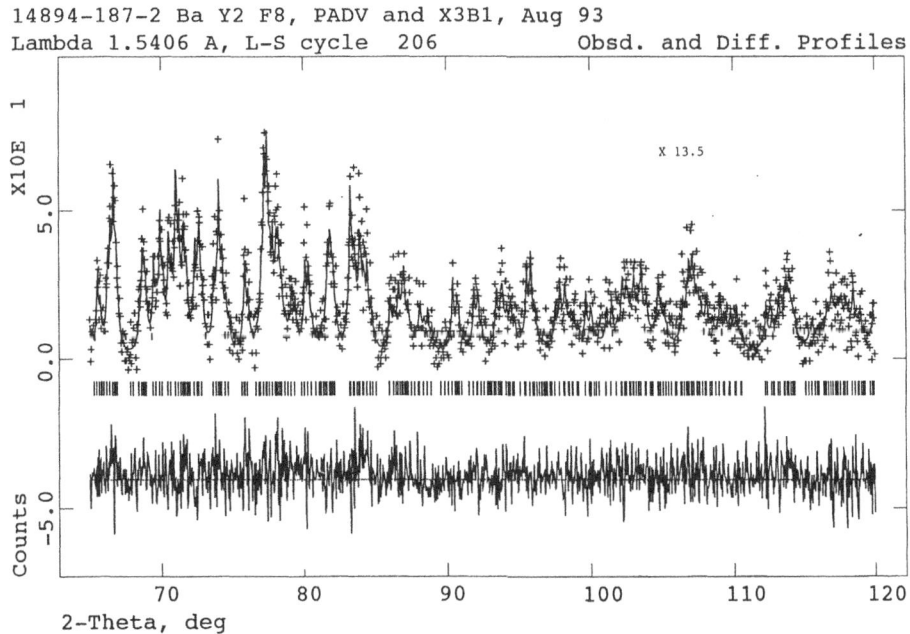

14894-187-2 Ba Y2 F8, PADV and X3B1, Aug 93
Lambda 1.5406 A, L-S cycle 206   Obsd. and Diff. Profiles

Figure 8. (continued)

97

distances to describe the background. The distances which arose from this treatment were 2.305(7) and 4.209(6) Å, which correspond to the first- and second-nearest-neighbor distances around the Y/Yb site. We suspected that these "diffuse background features", if real, could reflect nonrandomness in the site occupancy.

To test the reality of these diffuse features, time-of-flight neutron powder diffraction patterns were measured on the General Purpose Powder Diffractometer (GPPD) at the Intense Pulsed Neutron Source (IPNS) at Argonne National Laboratory. Patterns from the 90° and 148° 2θ detector banks were incorporated into the refinement. The neutron profiles were described by the TOF type 1 profile function[23]; only the σ_1 and σ_2 terms were refined. The neutron backgrounds were fit much better by a 6-term cosine Fourier series than by a pair correlation function. This difference suggests that the "diffuse features" of the synchrotron patterns were artifacts of the overlap of peak tails. No structural parameter changed significantly as the treatment of backgrounds was varied.

For the final refinement, all datasets were used: three X-ray powder patterns, two neutron powder patterns, and the X-ray single crystal data. The final refinement of 97 variables using 41674 observations yielded the overall powder residuals wRp = 0.1199 and Rp = 0.0407; the reduced χ^2 was 4.34. The single crystal residuals were wR(F) = 0.013 and R(F) = 0.061 for the reflections for which $I > 3\sigma(I)$. The refined structural parameters are reported in Table 4.

Results and Discussion

BaY_2F_8 and $BaYYbF_8$ are isostructural. The structure consists of 12-coordinate Ba and 8-coordinate Y, joined through M-F-M bridges. The Ba resides at the origin, a site of 2/m

Table 4
Refined Structural Parameters of BaYYbF$_8$

Space Group C2/m, $a = 6.94534(3)$, $b = 10.47071(5)$, $c = 4.24667(2)$ Å, $\beta = 99.5550(3)°$

Atom	x	y	z	Uiso, Å^2
Ba	0	0	0	0.0136
Y, frac = 1/2	1/2	0.17608(2)	1/2	0.0102
Yb, frac = 1/2	1/2	0.17608(2)	1/2	0.0102
F1	0.1861(2)	0.1410(1)	0.5581(3)	0.0161
F2	0.3943(3)	0	0.2299(5)	0.0164
F3	1/2	0.2386(2)	0	0.0177

Anisotropic thermal parameters $T = \exp(h^2a^{*2}U_{11} + ... + 2hka^*b^*U_{12} + ...)$

Atom	U$_{11}$	U$_{22}$	U$_{33}$	U$_{12}$	U$_{13}$	U$_{23}$
Ba	0.0157(1)	0.0097(1)	0.0157(1)	0	0.0035(1)	0
Y	0.0121(1)	0.0085(1)	0.0107(1)	0	0.0036(0)	0
Yb	0.0121(1)	0.0085(1)	0.0107(1)	0	0.0036(0)	0
F1	0.0146(5)	0.0115(5)	0.0234(6)	-0.0004(4)	0.0066(5)	0.0012(5)
F2	0.0208(9)	0.0101(7)	0.0195(8)	0	0.0064(7)	0
F3	0.0229(10)	0.0169(9)	0.0139(6)	0	0.0047(7)	0

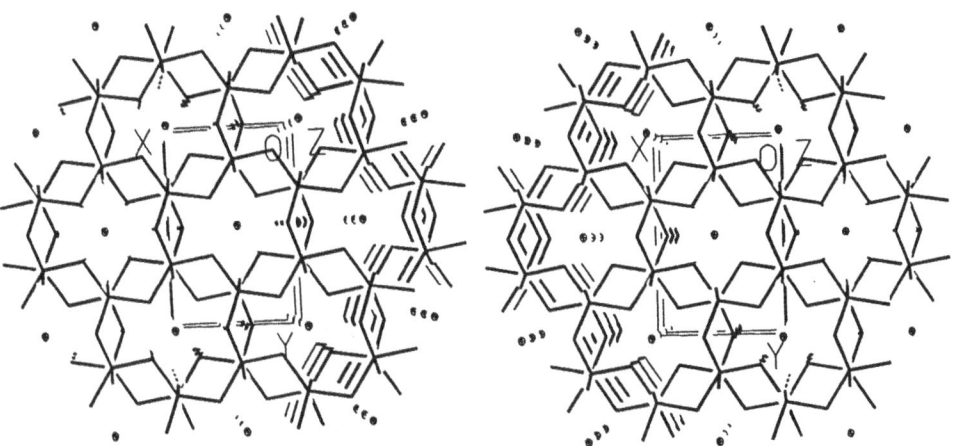

Figure 9. A stereo illustration of the crystal structure of BaY_2F_8, viewed down the monoclinic c-axis. The a-axis points to the left, and the b-axis down. The Y-F bonds are indicated by solid lines. The Ba-F bonds have been omitted for clarity. 50% probability ellipsoids.

symmetry. The Ba coordination is best described as tetracapped square prismatic. Eight F1 form an approximate square prism, and each side face is capped with one of two F2 or F3. The Ba-F distances of 2.8595(23), 2.9092(20), 2.7469(30), and 2.7180(30) Å in BaY_2F_8 fall within the normal range, and differ insignificantly from those in $BaYYbF_8$. The cap distances are significantly shorter than the Ba-F1 distances.

The Y is 8-coordinate, and lies on a 2-fold axis. The coordination is best described as square antiprismatic. The Y-F distances of 2.2916(22), 2.3083(22), 2.2507(18), and 2.2390(10) Å in BaY_2F_8 fall within the normal range, and differ slightly from the distances of 2.2634(14), 2.3008(14), 2.2303(12), and 2.2219(6) Å in $BaYYbF_8$. These small changes in bond distances correspond to a small shift of the Y site within the coordination sphere. The average RMS shift of the F atoms in the two structures is 0.026 Å. F1 is 4-coordinate, and the coordination is tetrahedral. The coordination of F2 and F3 is trigonal planar.

The crystal structure is most easily visualized when viewed down the short c-axis (Figure 9). The Y sites are linked in a pseudohexagonal network perpendicular to c by dual F bridges; the Y coordination polyhedra share edges. The F1 bridges are roughly parallel to a, and the F2 bridges are parallel to b. These layers perpendicular to c are joined into a 3-dimensional network through single Y-F3-Y bridges parallel to c; the Y coordination polyhedra share corners in this direction. The Ba reside in cavities formed by this network.

The lattice parameters (space group $C2/m$) of BaY_2F_8 are significantly larger than those of $BaYYbF_8$. The decrease in lattice dimensions on Yb substitution is consistent with the ionic radii[18]: Y^{3+} = 1.03 Å, and Yb^{3+} = 0.998 Å. The decrease is approximately isotropic.

CATION SITE OCCUPANCIES IN NH$_4$-EXCHANGED Na-FAU

Zeolites are framework aluminosilicates, built from corner-sharing tetrahedral units. The frameworks are open, enclosing cavities and/or channels, which are occupied by cations

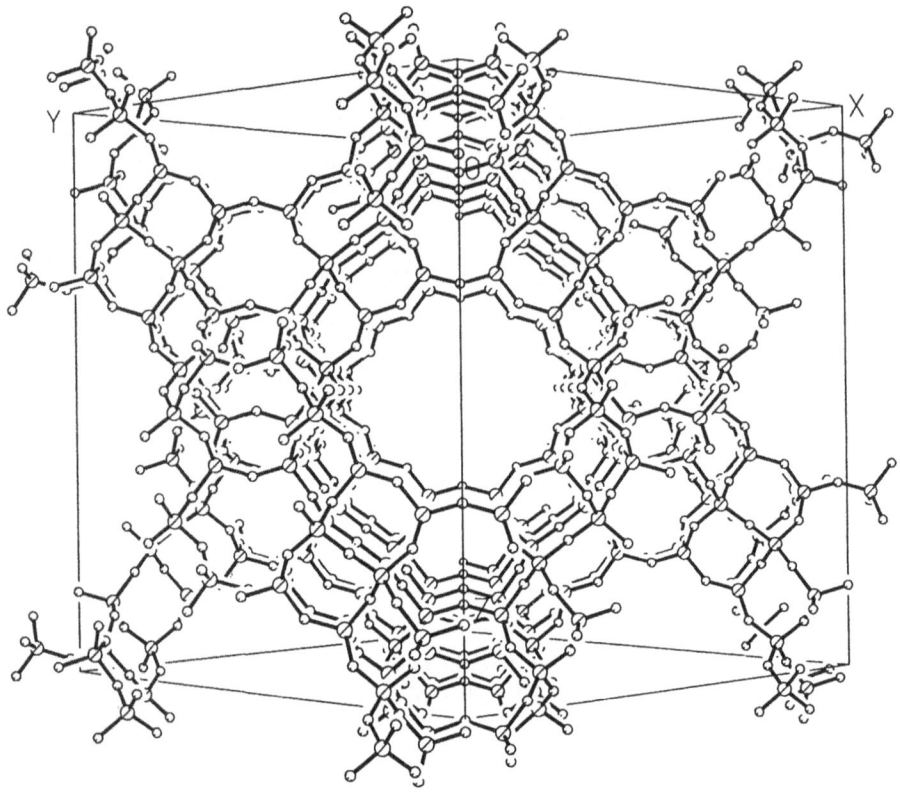

Figure 10. The crystal structure of zeolite Y (structure type FAU), viewed down the cubic [110] direction. The tetrahedral Si and Al (T sites) are indicated by hatched circles, and the oxygens by open circles. The large windows in the structure are formed by circular rings of 12 tetrahedral sites, and are approximately 8 Å in diameter.

and water molecules. Zeolites can be dehydrated reversibly, leaving open space, which can be occupied by other molecules. This porosity makes them very useful in catalytic applications.

The most important catalytic zeolite is zeolite Y (structure type[24] FAU - after the mineral faujasite). The structure (Figure 10) is built from 24-tetrahedra cuboctahedral units (sodalite cages), which are joined through their hexagonal faces (double 6-rings). Pores formed by rings of 12 tetrahedra (approximately 8 Å in diameter) act as windows between large cages (supercages) approximately 12 Å in diameter. If the framework were Al-free, it would be neutral. The presence of Al substitution for Si results in a negative framework charge, which must be balanced by cations.

Faujasite is normally synthesized in the sodium form. Since most commercial zeolite catalysis is acid catalysis, the proton is the desired cation. Faujasite, like most zeolites, cannot be converted from the Na-form to the H-form by direct acid exchange. The normal strategy for preparing H-FAU is to carry out an ammonium ion exchange, and calcine the material at temperatures high enough to release ammonia, while leaving protons behind and the framework intact. To help better understand the process of ion exchange, we have examined the structures of several partially-ammonium-exchanged Na faujasites, three of which have been selected for discussion here.

There are remarkably few structure reports for hydrated faujasites; most attention has centered on the dehydrated forms. Although we can find no references to the crystal structure of hydrated Na/NH$_4$-FAU, the structures of hydrated Na-Y[25,26] and Na-X[27] have been reported.

There is an established nomenclature for the extraframework sites in faujasite-type zeolites[28,29]. Four sites are commonly occupied:

I   at the center of the double 6-rings
I'   in the sodalite cage, adjacent to a hexagonal ring shared by the sodalite cage
          and a double 6-ring
II   in the supercage, adjacent to an unshared hexagonal face of a sodalite cage
II'   in the sodalite cage, adjacent to an unshared hexagonal face

Additional sites are sometimes populated[29]. Site II' is often occupied by small molecules, such as water or ammonia[30]. There are considerable differences in the cation locations between hydrated and dehydrated zeolites[30].

Experimental

The starting material for this study was the commercial Na-Y LZ-Y52, containing 54 Na/cell by chemical analysis (FAU54). Ion exchanges were carried out using aqueous NH_4NO_3 at varying degrees of severity to prepare materials containing different Na concentrations. Two materials, containing 12 Na/cell (FAU12) and 3 Na/cell (FAU03), are discussed here.

The sieves were mixed with known concentrations of micronised α-quartz internal standard. X-ray powder patterns were measured from flat plate samples at beamline X3B1 at NSLS, using a wavelength of 1.14988 Å. Patterns were measured from 4-60° 2θ in 0.01° steps, counting for 4 sec/step. The sample was rocked from ω-1 to ω+1° during each data point.

All data processing was carried out using GSAS[6]. Only the 15-60° 2θ portion of the patterns was included in the refinements. This represents a compromise. The low-angle reflections contain much of the information about the extraframework species, but are hardest to fit. The current GSAS profile function does not describe the asymmetry of the low-angle peaks well. There may also be some systematic distortions of intensity at low angles, the result of surface roughness, microabsorption, and other effects.

The FAU framework structural model was taken from our in-house refinements of LZ-Y52. A soft constraint of 1.63(1) Å was applied to the T-O bond distances. The weight applied to this constraint was gradually decreased through the course of refinement. The Si/Al occupancies of the T site were fixed at 0.734/0.266, derived from the framework Al concentration calculated from the lattice parameter of LZ-Y52 using the Kerr-Dempsey correlation[31]. Only 40% of the calculated shift was applied to any structural parameter in each least-squares cycle.

The quartz structural model was derived from our refinements of the pattern of the pure phase, was refined during the refinement of LZ-Y52, and fixed for the other patterns. Included in the refinements were scale factors for each phase, and the lattice parameters.

The peak profiles were described by a pseudo-Voigt function. For FAU, the gaussian U, V, and W, as well as the cauchy X and Y, and asymmetry terms were refined. For quartz, these terms were refined, but the Y profile parameter was fixed at zero. The background was described by a 6-term real space pair correlation function, using the two fixed characteristic distances of 4.23 and 3.10 Å. These correspond to the two shortest T-T distances in the faujasite structure, and permit a good description of the small concentration of amorphous material present in these zeolites.

Initially only the faujasite framework was included in the structural model. Extraframework peaks located in difference Fourier maps were added gradually. The isotropic displacement coefficients of all extraframework species were initially fixed at 0.04 Å², and refined only later; occupancies were refined individually. The largest ΔF peaks were observed at reasonable cation positions. It is important to proceed carefully in developing structural models for sieves, since both structural and global parameters are highly-dependent on the number and locations of the extraframework species. For example, in LZ-Y52, initially a strong difference peak was observed at site I, the center of the double 6-rings. Its initial occupancy was about 1/2, but decreased steadily as additional cations were added to the model.

The isotropic displacement coefficients of the framework atoms are dependent on the extraframework species; the coordinates seem relatively insensitive. We have not been able to refine occupancies and thermal parameters simultaneously using in-house data - the better resolution and peak/background of the synchrotron data permits such successful refinement. The correlation coefficients between extraframework site occupancies and Uiso vary between 0.5 and 0.7; these values are not high enough to preclude simultaneous refinement.

Cations were added until the largest difference peaks were of the same magnitude as the largest difference holes - generally about 0.5 electrons. Our experience is that peaks less than 1/2 electron are not refineable using powder data. The largest difference peaks were located near the cation positions. Such peaks may represent the "split" sites occupied by Na and O discussed by Rubio, Soria, and Cano[25]. A trace of zeolite P was detected, but was not included in the refinements.

The final refinements using 4503 observations yielded residuals in the ranges: $0.126 < wRp < 0.138$, $0.096 < Rp < 0.108$, $0.121 < R(F) < 0.130$, and $13.46 < \chi^2 < 32.02$. These residuals are much higher than would normally be considered acceptable, but are typical for refinements of faujasite using powder data. The Bragg R(F) are comparable to literature refinements of hydrated faujasites. The final difference Fourier maps are flat enough to suggest that no more structural information can be extracted from these patterns. The largest deviations between the observed and calculated patterns lie in the Bragg peaks. Some deviations are the result of peak asymmetry and sharp quartz peaks at low angles, but the source of most of the differences probably lies in not accounting for the considerable water content of these sieves. Variation in the Al distribution in small domains of the sieve, which results in different cation distributions, may also be a source of the differences between the observed and calculated patterns. The final atom coordinates and isotropic displacement coefficients are reported in Table 5.

Results and Discussion

Framework Structures. The frameworks of all three compositions are very similar (Table 5). The T-O bond distances in the Na/NH$_4$ sieves are more similar to each other than to those in LZ-Y52. O1 represents the "side" of the double 6-ring, O2 the edge shared by two adjacent 6-rings of the sodalite cage, O3 the edge a hexagonal face of a double 6-ring shares with the 4-ring of the sodalite cage, and O4 edge shared by the sodalite cage 4-ring and the 6-ring exposed to the supercage. Only O2 and O3 are coordinated to Na in LZ-Y52.

The largest change in a structural parameter is in the T-O3-T angle, which varies from 143.2(5)° in FAU54 to 149.1(5)° in FAU03. The isotropic displacement coefficients of the framework atoms are approximately constant at 0.02 Å^2. This (reasonable) value is highly-dependent on the extraframework model, and its "reasonableness" lends additional support to the structural models.

Significant electron density was observed at three extraframework sites, I', II', and II. In the initial stages of the refinement of LZ-Y52, density was observed at site I, but the occupancy of this site decreased to zero as additional extraframework sites were included in the structural model. This decrease is consistent with the generally low occupancy of site I in hydrated faujasites.

Table 5
Structural Parameters of Hydrated (Na,NH$_4$)-FAU

		Space Group Fd$\overline{3}$m		
Sieve		FAU54	FAU12	FAU03
	a, Å	24.6773(15)	24.68765(19)	24.73723(21)
Atom	Parameter			
Si/Al	x	0.96364(11)	0.96381(12)	0.96412(12)
	y	0.87555(11)	0.87525(13)	0.87524(11)
	z	0.05399(10)	0.05366(13)	0.05420(12)
	Uiso	0.0243(8)	0.0203(9)	0.0202(9)
O1	x	1	1	1
	y	0.8941(2)	0.8936(2)	0.8927(2)
	z	0.1058(2)	0.1063(2)	0.1072(2)
	Uiso	0.0237(10)	0.0226(12)	0.0202(12)
O2	x	1.0040(2)	1.0045(2)	1.0050(2)
	y	0.8577(2)	0.8579(3)	0.8586(3)
	z	0.0040(2)	0.0045(2)	0.0050(2)
	Uiso	0.0237(10)	0.0226(12)	0.0202(12)
O3	x	0.9244(2)	0.9232(2)	0.9220(2)
	y	0.9244(2)	0.9232(2)	0.9220(2)
	z	0.0344(3)	0.0355(3)	0.0371(3)
	Uiso	0.0237(10)	0.0226(12)	0.0202(12)
O4	x	0.9263(2)	0.9262(2)	0.9279(2)
	y	0.8236(2)	0.8237(2)	0.8220(2)
	z	0.0714(3)	0.0706(3)	0.0697(3)
	Uiso	0.0237(10)	0.0226(12)	0.0202(12)
Na1/N1	x = y = z	-0.0692(5)	-0.0735(7)	-0.0834(5)
site I'	frac	0.36(2)	0.49(3)	0.73(3)
	Uiso	0.044(6)	0.024(7)	0.041(7)
Na2	x = y = z	0	0	0
site I	frac	0.03(2)	0	0
	Uiso	0.044(6)	-	-
Na3	x = y	-0.0823(3)	-0.0821(5)	-0.0803(16)
site II'	z	-0.1677(3)	-0.1679(5)	-0.1697(16)
	frac	0.56(2)	0.41(2)	0.14(2)
	Uiso	0.044(6)	0.024(7)	0.041(7)
Na4/N3	x = y	0.0080(5)	0.0161(5)	0.0142(4)
site II	z	-0.2580(5)	-0.2661(5)	-0.2642(4)
	frac	0.33(1)	0.73(2)	0.84(2)
	Uiso	0.044(6)	0.024(7)	0.041(7)

LZ-Y52. The occupancy of site I' is 0.36(2); this corresponds to 12 Na/cell, in good agreement with Marti, Soria and Cano[25,26]. The occupancy of site II corresponds to 11 Na/cell, and is much higher than that previously reported for hydrated Na-Y. The occupancy of site II' is higher than previously reported[26]. This site is generally described as occupied by water molecules. For reasons described below, we believe that there is significant occupancy of this site by Na.

Site I' is 2.57 Å from 3 framework O3. It is 2.47 Å from 3 site II'; this is the source of the usual identification of site II' as water oxygen. The identification as Na or O cannot be confirmed from only a single diffraction experiment, though slightly better residuals were obtained when modelling the site as Na rather than O. Even modelling as Na, not all of the cations have been located in LZ-Y52. A significant fraction of the Na cations must be disordered among supercage sites.

Ammonium-Exchanged FAU. The NH_4 ion is larger than Na, and has significantly different bonding properties. Partial replacement of Na cations with NH_4 would be expected to change the extraframework site occupancies significantly.

In the ammonium-exchanged sieves, the site I' moves significantly toward the center of the sodalite cage, and away from the framework atoms. The occupancy increases with the degree of NH_4 exchange. Both of these observations suggest the replacement of Na by NH_4 at this site. Likewise, the occupancy of site II and the site-framework distance increase with increasing ammonium exchange. This site is not favored by Na, but seems favorable for NH_4.

The position of site II' remains constant with ion exchange. The occupancy drops when the Na content is reduced from 54 to 12/cell, but decreases even more when the Na content is lowered from 12 to 3/cell. This last decrease accounts completely for the overall change in Na content between FAU12 and FAU03. The changes in occupancy and the constancy of position suggest that this is the last Na site to be exchanged. It may very well be partially-occupied by oxygen in LZ-Y52, but seems to represent Na in the exchanged materials.

The changes in the extraframework site occupancies as a function of ammonium exchange are summarized in Table 6:

Table 6
Cation Site Occupancies in (Na,NH_4)-FAU

Site / Sieve	FAU54	FAU12	FAU03
I'	12 Na	16 NH_4	23 NH_4
II'	18 Na (?), or Na + O	13 Na	4 Na
II	11 Na	23 NH_4	27 NH_4
cations/cell	41	52	54

Sites I' and II become more highly populated as the ammonium content increases, and site II' becomes depopulated. These changes in occupancy and the changes in cation position suggest the identification as NH_4 or Na. The refined occupancies suggest that only about half of the ammonium ions reside in the supercages.

ACKNOWLEDGMENTS

We thank A. Bhattacharyya, S.C. Jevne, L.C. Satek, A.P. Wilkinson, R.H. Jarman, J.T. Miller, and P.D. Hopkins for supplying the materials. Y.-M. Chen, M.N. Kaminsky, and S.M.

Short provided technical support. M. Nguyen and B.L. Meyers performed the TGA analysis of hydrated sodium aluminate, and C.H. Cullen and G.J. Ray carried out the NMR studies, which provided information essential to the structure solution.

The powder diffractometer at X3B1 was built by P.W. Stephens and Dengfa Liu of SUNY Stony Brook, with the support of the X3 PRT. We also thank Robert Dinnebier for assistance in data collection. D.E. Cox and J.A. Hriljac provided assistance during data collection on X7A, and we thank J.A. Hriljac for supplying single crystal data on copper aluminum borate in advance of publication. We acknowledge the support of J.W. Richardson Jr., F.J. Rotella, and R. Hitterman during data collection at IPNS.

This work represents research carried out in part at the National Synchrotron Light Source at Brookhaven National Laboratory, which is supported by the U.S. Department of Energy, Division of Materials Sciences, and Division of Chemical Sciences. The SUNY X3 beamline at NSLS is supported by the Division of Basic Energy Sciences of the U.S. Department of Energy (DE-FG02-86ER45231).

This work has benefited from the use of the Intense Pulsed Neutron Source at Argonne National Laboratory. This facility is funded by the U.S. Department of Energy, BES-Materials Science, under contract W-31-109-Eng-38.

REFERENCES

1. C. Misra, *Industrial Alumina Chemicals*. ACS Monograph 184; American Chemical Society: Washington, D.C., 1986; pp. 151-155 and references therein.

2. A. G. Elliot and R. A. Huggins, *J. Amer. Ceram. Soc.*, **58**, 497(1975).

3. W. Gessner, M. Weinberger, D. Müller, L.P. Ni, and O.B. Chaljapina, *Z. Anorg. Allg. Chem.*, **547**, 27-44 (1987).

4. L.I. Ryskina, M.V. Zakharova, N.V. Kirchanovaa, and Yu.F. Klyuchnikov, *Tr. Inst. Met. Obogashch. Akad. Nauk Kaz. SSR*, **47**, 25-28(1972).

5. J.W. Visser, *J. Applied Cryst.*, **2**, 89-95(1969); Version 12.

6. A.C. Larson and R.B. Von Dreele, *GSAS, The General Structure Analysis System*, Los Alamos National Laboratory, February 1993.

7. *SHELXTL Plus*, Version 3.4, Nicolet Instrument Corporation, 1988.

8. R. Gramlich-Meier and W.M. Meier, *J. Solid State Chem.*, **44**, 41-49(1982).

9. I.D. Brown and D. Altermatt, *Acta Cryst.*, **B41**, 244-247(1985).

10. A.D. Mighell and V.L. Himes, *Acta Cryst.*, **A42**, 101-105(1986); 1992 version of the Crystal Data database (182666 entries).

11. R.X. Fischer, *J. Applied Cryst.*, **18**, 258-262(1985).

12. L.C. Satek, J.A. Kaduk, and P.E. McMahon, Characterization, structure, and active site hypothesis for novel copper aluminum borate dehydrogenation and dehydrocyclization catalysts, in: "Catalysis of Organic Reactions", W.E. Pascoe, ed., Marcel Dekker, New York (1992).

13. L. Richter, "Synthese und Strukturuntersuchungen von Eisen- und Kupfer-Aluminum-Boraten", Technische Hochschule, Aachen (1977).

14. L.C. Satek, J.A. Hriljac, R.D. Brown, J.A Kaduk, and A.K. Cheetham, unpublished results.

15. P. Coppens, "Synchrotron Radiation Crystallography", Academic Press, London (1992).

16. V. Petricek, Y. Gao, P. Lee, and P. Coppens, *Phys. Rev.*, **B42**, 387 (1990).

17. M. Martinez-Ripoli, S. Martinez-Carrerra, S. Garcia-Blanco, *Acta Cryst.*, **B27**, 677-681 (1971).

18. *a.* G. Bergerhoff, R. Hundt, R. Sievers, and I.D. Brown, *J. Chem. Inform. Comput. Sci.*, **23**, 66-69 (1983). *b.* G. Bergerhoff, R. Sievers, and R. Hundt, The Inorganic Crystal Structure Database, in: "Crystallographic Databases", F.H. Allen, G. Bergerhoff, and R. Sievers, eds., International Union of Crystallography, Chester (1987).

19. G.W. Zajac, J. Faber, and S. Pei, unpublished results.

20. O.E. Izotova and V.B. Aleksandrov, *Dok. Akad. Nauk SSSR*, **192**, 1037-1039 (1970).

21. D.K. Smith, *Advances in X-ray Analysis*, **35A**, 1-15 (1992).

22. H.P. Klug and L.E. Alexander, "X-ray Diffraction Procedures", Wiley, New York (1974), p.360.

23. *a.* J.D. Jorgensen, D.H. Johnson, M.H. Mueller, T.G. Worlton, and R.B. Von Dreele, *Proc. Conf. on Diffraction Profile Analysis*, Krakow, 14-15 August 1978, p.20-22. *b.* R.B. Von Dreele, J.D. Jorgensen, and C.G. Windsor, *J. Applied Cryst.*, **15**, 581-589 (1982).

24. W.M. Meier and D.H. Olson, "Atlas of Zeolite Structure Types", International Zeolite Association, London (1992).

25. J.A. Rubio, J. Soria, and F.H. Cano, *J. Colloid Interface Sci.*, **73**(2), 312-323 (1980).

26. J. Marti, J. Soria, and F.H. Cano, *J. Colloid Interface Sci.*, **60**(1), 82-86 (1977).

27. D.H. Olson, *J. Phys. Chem.*, **74**(14), 2758-2764 (1970).

28. J.V. Smith, *Adv. Chem. Series*, **101**, 171-200 (1971).

29. W.J. Mortier, "Compilation of Extra Framework Sites in Zeolites", International Zeolite Commission, London (1982).

30. J.J. Van Dun and W.J. Mortier, *J. Phys. Chem.*, **92**, 6740-6746 (1988).

31. G.T. Kerr, *Zeolites*, **9**, 350-351 (1989).

SURFACE PHYSICS WITH SYNCHROTRON RADIATION

F.J. Himpsel[1]*, D.A. Lapiano-Smith[1], H, Akatsu[1], J.A. Carlisle[2], E.A. Hudson[2], L.J. Terminello[2], T.A. Calcott[3], J.J. Jia[3], M.G. Samant[4], J. Stöhr[4], D.L. Ederer[5], R.C.C. Perera[6], and D. K. Shuh [6]

[1] IBM Research Divsion, Thomas J. Watson Research Center, Yorktown Heights, NY 10598
[2] Lawrence Livermore National Laboratory, Livermore, CA 94551
[3] University of Tennessee, Knoxville, TN 37996
[4] IBM Research Division, Almaden Research Center, 650 Harry Road, San Jose, CA 95120-6099
[5] Tulane University, New Orleans, LA 70118
[6] Lawrence Berkeley National Laboratory, Berkeley, CA 94720
* Present Address: Department of Physics, University of Wisconsin, Madison, Madison, WI 53706

This article illustrates the use of synchrotron radiation for illuminating the electronic structure of surfaces. After a brief introduction into the most common spectroscopies, their capabilities are exposed by a set of examples from the authors' area of expertise. It comprises the surface and interface chemistry of semiconductors and magnetic materials, but leaves out a large body of work on surface chemistry of heterogeneous catalysis.

Synchrotron-Based Techniques

Synchrotron radiation offers a variety of spectroscopies that take advantage of its special properties, such as tunability and polarization (For an overview see Reference 1). Figure 1 shows how the three most common spectroscopies operate, using a schematic potential diagram of electronic states at surfaces as template. In photoemission (Figure 1a) a bound electron is ionized, and its binding energy determined by subtracting the photon energy from the kinetic energy of the photoelectron. Energy shifts of the substrate and adsorbate levels are a measure of the charge transfer and chemical bonding at surface atom. The main advantage of synchrotron radiation with respect to traditional light sources, such as resonance lamps and x-ray tubes, is the possibility to tune the kinetic energy of the photoelectrons by tuning the photon energy. Thereby one can reach the escape depth minimum of the electrons, typically in the 30–70 eV kinetic energy range. At the minimum one has a probing depth of a few atomic layers, making this technique ideal for probing surfaces and adsorbates.

Absorption spectroscopy (Figure 1b) determines the pattern of unoccupied orbitals by exciting them optically from a core-level. Instead of measuring transmission or reflectivity, the absorption coefficient is determined in surface experiments by collecting secondary products, such as photoelectrons, Auger electrons, and adsorbate fluorescence. The short escape depth of electrons compared to photons provides surface sensitivity. Absorption spectroscopy is virtually the exclusive domain of synchrotron radiation, since it requires continuous tunability over a wide photon energy range. Only high energy electron energy loss spectroscopy has such properties, but lacks surface sensitivity and is prone to radiation damage effects.

Synchrotron Radiation Techniques in Industrial, Chemical, and Materials Science
Edited by D'Amico *et al.*, Plenum Press, New York, 1996

107

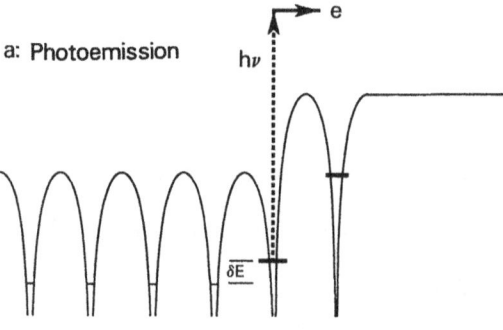

a: Photoemission

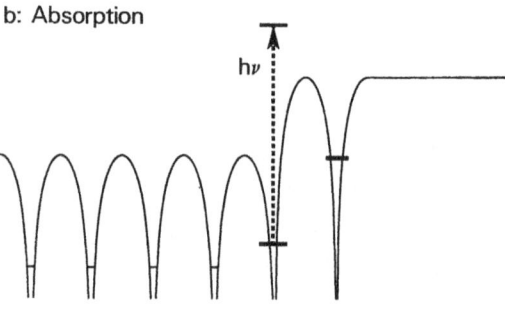

b: Absorption

c: Fluorescence

Figure 1   Schematic of the most common surface spectroscopies using synchrotron radiation, using a potential diagram of an adsorbate atom (right) and the outermost substrate atoms.

Fluorescence spectroscopy (Figure 1c) can be viewed as the reverse of absorption spectroscopy. Here, the valence orbital structure is obtained from the spectral distribution of the fluorescent light emitted during the recombination of a core hole with a valence electron. As a purely optical spectroscopy, fluorescence is not surface sensitive per se, but it can be made sensitive enough to see monolayers by the brute force of third generation synchrotron light sources (see below). The rationale for this tour de force is the detection of buried interfaces, which are inaccessible to techniques with a short probing depth. Another area, where monolayer sensitivity together with deep penetration offers interesting new opportunities, is the study of chemical reactions at surfaces in the presence of ambient gas.

An important consideration in these spectroscopies is access to the core-levels of all elements. If one considers the sharpest core-level of each element (Figure 2), one finds that photon energies up to 1 keV suffice to excite them all. For absorption and fluorescence spectroscopy, one is typically interested in the first 10–20 eV above threshold, for photoemission about 30–70 eV, to achieve optimum surface sensitivity. In fact, the current third generation of ultraviolet synchrotron light sources such as the Advanced Light Source (ALS), is designed around these requirements. At the center of the desirable energy range, i.e., 300–500 eV, lie the C, N, O, 1s levels, which are so important for organic surface chemistry.

Semiconductor Surface Chemistry
Semiconductor processing has been shifting continuously from wet to dry processes, such as reactive ion etching (RIE) and chemical vapor deposition (CVD). These techniques offer a variety of benefits, such as a reduction in toxic waste material and faster cycling from one process stage to he next. While these methods have become already quite sophisticated by trial and error, there is ample room for improvement by systematic study of the surface chemical reactions. They determine the bottleneck in many cases, and they are crucial for substrate-selective deposition and etching, a rapidly-growing field. In the etching area it took until the first synchrotron radiation photoemission studies to find out that the bottleneck of the removal of silicon by a fluorocarbon plasma was the removal of SiF_3 species from the surface, not SiF_2, as previously believed (2). In CVD of silicon there has been a development of low temperature, low pressure techniques that require ultrahigh vacuum cleanliness standards, but produce devices with excellent electrical properties, which are beating the speed records for silicon-based electronic devices (3). A critical step in this method is the cleaning of the starting surface. The native oxide layer is etched away by dipping the wafer into an ultrapure 10% HF solution. The explanation, why this surface treatment worked so well compared to a high temperature heating cycle, became obvious with core-level spectroscopy. The Si 2p photoemission spectra showed that the surface became hydrogen-terminated, with only a fraction of a monolayer of sub-oxides and no SiO_2 buildup whatsoever (4). Sub-oxides are reduced during the CVD process, while SiO_2 remains at the interface as a defect.

An interesting development in semiconductor surface chemistry is the quest for monolayer control during etching and growth, i.e., "digital" etching and growth. Processes are sought which are self-limited and stop etching or growth after each layer, and then are activated again in a second step. Atomic layer epitaxy (ALE) is already being used commercially with II–VI semiconductors (5), while exploratory work is going on with III–V and Group IV semiconductors. A particularly fascinating material is SiC, which has many phases that correspond to different stacking sequences. They exhibit different band gaps, and thus become possible targets for designing modulation-doped heterojunctions of the type that produce the fastest GaAs/GaAlAs and Si/SiGe transistors today. With SiC these devices could be produced in the same materials with perfect lattice match. Figure 3 shows some preliminary work in that direction, where Si- and C-containing precursors are tested on a Si(100) surface by using photoemission from the C 1s core-level and the C 2s valence state (6, 7). Traditional deposition of SiC involves cracking of hydrocarbons and silanes at high temperatures (900–1600°C), and is not amenable to layer by layer control. In the search for C-containing compounds that can achieve self-limiting growth one encounters several obstacles. Traditional precursors in CVD of SiC, such as trichloromethylsilane, are found to stick poorly to the surface, probably because of the inert nature of the chloro and methyl groups. Hydrogen splits off from Si much easier. It seems to be a prerequisite for obtaining a sticking probability that is high enough to prevent contamination of the surface at the lower temperatures that are required to achieve layer-by-layer control and prevent interdiffusion. Alkyl groups, such as shown in Figure 3 for dimethylsilane and diethylsilane, provide the desired saturation of the surface after a monolayer is adsorbed. From the bonding/anti-bonding splitting between the two C 2s valence states in diethylsilane the alkyl group saturating the surface can be readily identified. In general, an alkyl chain of n atoms

The Sharpest Core Levels and their Binding Energies (eV)

H 1s 14																	He 1s 25
Li 1s 55	Be 1s 112											B 1s 189	C 1s 284	N 1s 410	O 1s 543	F 1s 698	Ne 1s 870 2p 22
Na 2p 31	Mg 2p 49											Al 2p 73	Si 2p 100	P 2p 135	S 2p 163	Cl 2p 200	Ar 2p 248 3p 16
K 2p 295 3p 18	Ca 2p 346 3p 25	Sc 2p 399	Ti 2p 454	V 2p 512	Cr 2p 574	Mn 2p 639	Fe 2p 707	Co 2p 778	Ni 2p 853	Cu 2p 933	Zn 3d 10	Ga 3d 19	Ge 3d 29	As 3d 42	Se 3d 55	Br 3d 69	Kr 3d 94 4p 14
Rb 3d 112 4p 15	Sr 3d 134 4p 20	Y 3d 156	Zr 3d 179	Nb 3d 202	Mo 3d 228	Tc 3d 253	Ru 3d 280	Rh 3d 307	Pd 3d 335	Ag 3d 368	Cd 4d 11	In 4d 16	Sn 4d 24	Sb 4d 32	Te 4d 40	I 4d 50	Xe 4d 68 5p 12
Cs 4d 78 5p 12	Ba 4d 90 5p 15	La 4d 103	Hf 4f 14	Ta 4f 22	W 4f 31	Re 4f 41	Os 4f 51	Ir 4f 61	Pt 4f 71	Au 4f 84	Hg 5d 8	Tl 5d 13	Pb 5d 18	Bi 5d 24	Po 5d 31	At 5d 40	Rn 5d 48 6p 11
Fr 5d 58 6p 15	Ra 5d 68 6p 19	Ac 5d 80															

Ce 4d 109 4f 1	Pr 4f 3	Nd 4f 5	Pm 4f 5	Sm 4f 5	Eu 4f 2	Gd 4f 8	Tb 4f 2	Dy 4f 4	Ho 4f 5	Er 4f 5	Tm 4f 5	Yb 4f 1	Lu 4f 7
Th 5d 85	Pa 5d 94	U 5f 1	Np 5f	Pu 5f 2	Am 5f	Cm 5f	Bk 5f	Cf 5f	Es 5f	Fm 5f	Md 5f	No 5f	Lw 5f

Figure 2   Periodic table with the sharpest core-level of each elements and its binding energy.

gives rise to n C 2s levels (8), making the identification of these species at the surface a straightforward task. Alkyl ligands are commonly-used in chemical vapor deposition of many other materials, e.g. III–V semiconductors. They have problems for the deposition of SiC, though. When trying to activate the surface thermally for the deposition of the next layer, the alkyl groups desorb as a whole and leave very little carbon at the surface. This observation leads to the next class of compounds, i.e., organosilanes where C is bound to more than one Si, e.g., bis(trimethylsilyl)methane. Current work indicates that it deposits twice as much carbon per cycle as the alkylsilanes and produces exclusively the carbidic configuration, not the graphitic inclusions seen with alkylsilanes (7). This chain of experiments shows how one might approach a difficult problem, such as layer-by-layer growth of SiC, in a systematic fashion using photoelectron spectroscopy.

Bonding and Electronic Structure of Interfaces

By far the most significant interface system in the area of semiconductors is the Si/SiO_2 interface, particularly on Si(100), which corresponds to the orientation in Si devices. The high

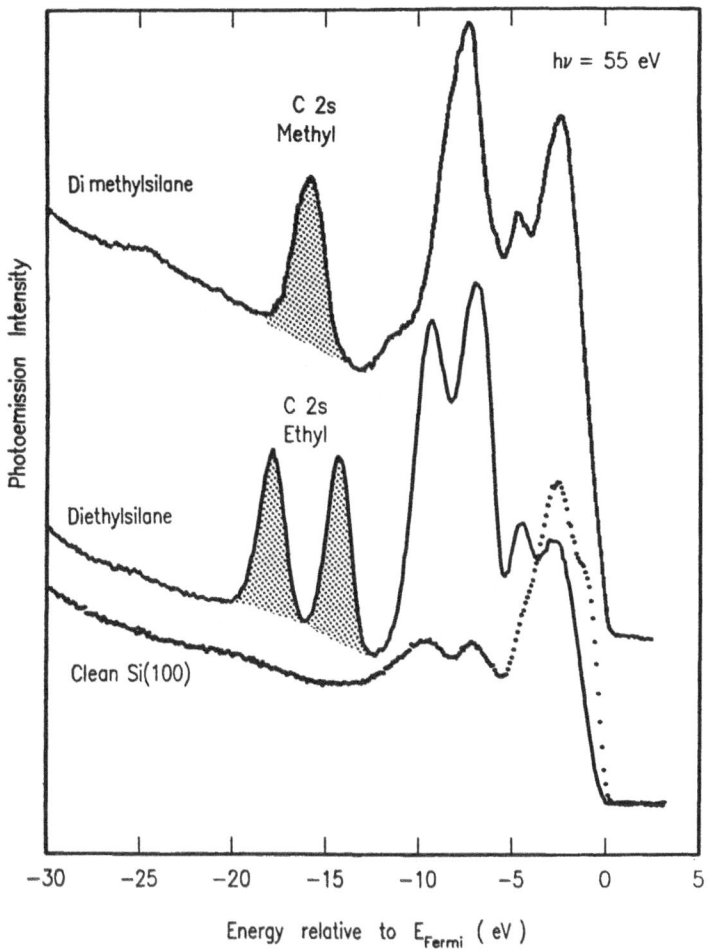

Figure 3   Identification of ethyl and methyl groups at the Si(100) surface using the split C2s
level. The number of C2s components gives directly the number of C atoms in the
alkyl group (7).

electrical quality of this interface with less than 10^{-5} active defects per interface atom is one of
the primary reasons for today's dominant role of silicon technology. Despite an extensive history
of measurements at this interface, its local atomic and electronic structure are not yet resolved
completely. A big obstacle to a clear solution is the amorphous nature of the SiO_2 overlayer,
which allows only average quantities to be determined.   Core-level spectroscopy with
synchrotron radiation has been able to shed light onto the bond topology at the interface by
clearly resolving the different oxidation states of Si at the interface and the interface and their
relative abundance (Figure 4, 9). Apart from the Si 2p levels of the substrate and the SiO_2
overlayer one can clearly recognize the three remaining oxidation states of Si in between. They
correspond to Si atoms bonding to 1, 2, and 3 oxygen atoms, respectively. In the bulk these
bonding configurations are unstable and disproportionate into Si and SiO_2. At the interface they
are enforced by geometric constraints: Si interface atoms have to bond to Si in the substrate and
to O in the overlayer. The total intensity and distribution of intermediate oxidation states is a
good measure of the bond topology at the interface and provides a critical test of interface
models. Both of these measures depend on the preparation of the oxide overlayer. Therefore,

one can try to link the interface bond geometry to its electrical properties, despite the fact that the actual number of electrically-active interface defects is much too low to be seen directly in such an experiment. The top spectrum in Figure 4 is representative of the interface grown in thermal equilibrium, i.e., at low oxygen pressure and elevated temperature. The area under the intermediate oxidation states 1, 2, and 3 corresponds to about two Si layers, suggesting an extended interface, where the Si atom density mismatch between Si and SiO_2 is taken up by a gradual increase in the coordination of Si by oxygens. Such a bond topology agrees with strain minimization calculations. The interface shown at the bottom of Figure 4 represents the opposite case, i.e., an interface grown far from thermal equilibrium at room temperature and atmospheric pressure. Another feature of this interface is the use of molecular beam epitaxy to obtain a particularly flat starting surface. This interface comes close to the atomically-abrupt limit, with only a single layer of Si atoms in intermediate oxidation states, most of them bonding to a single oxygen. The bond density mismatch is taken up by a pairing of Si atoms in this case, which reduces the number of bonds across the interface by a factor of two. Interfaces of this type have been found to exhibit superior electrical behavior. Properties, such as increased breakdown stability will become increasingly-important for the next generations of MOS devices, which push the thickness of the gate oxide layer down to the physical limits.

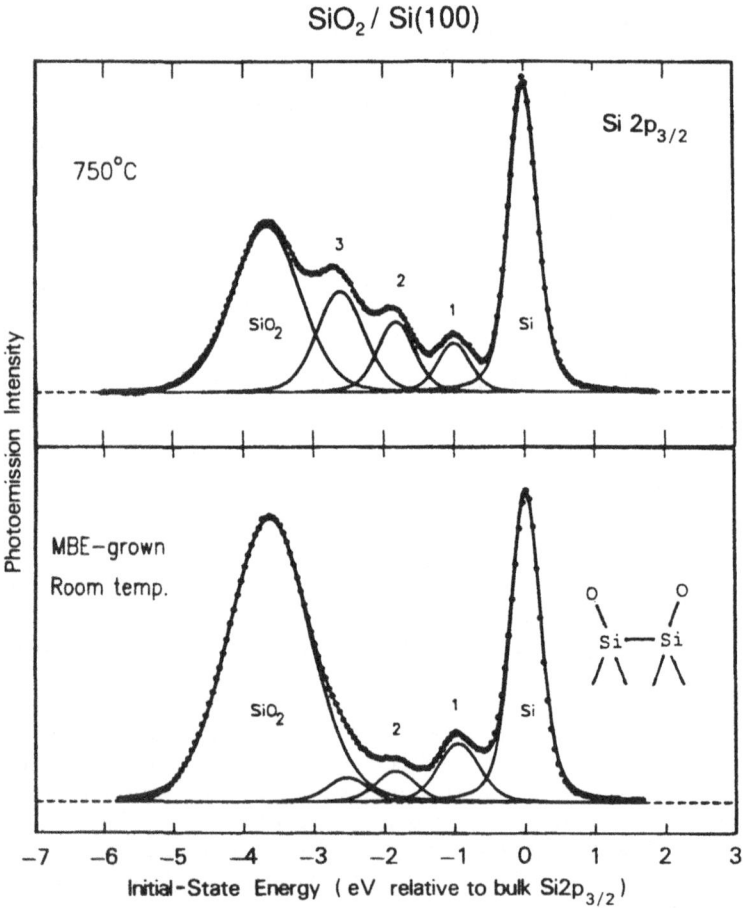

Figure 4   Resolving the oxidation states of Si at the SiO_2/Si(100) interface using core-level photoelectron spectroscopy. Apart from Si and SiO_2 one observes immediate oxidation states, corresponding to Si atoms bonding to 1, 2, and 3 oxygen atoms. Their distribution reflects the bond topology at the interface, which changes with the growth conditions (9).

Several ways have been tested for extending the envelope of gate oxide performance by adding constituents to SiO_2, for example in oxynitrides. Nitrogen increases the lifetime of MOS devices with respect to charges injected into the oxide and becoming trapped. Figure 5 shows absorption spectra of oxynitrides grown in N_2O (10). Using the absorption induced at the N 1s and O 1s absorption edges as a measure of the N and O coverage (left and right side of Figure 5, respectively), a dramatic reversal in stoichiometry is observed with increasing film thickness. While the first layer is dominated by a nitrogen termination of the Si bonds, a thicker film of 42 Å is almost pure oxide. This measurement was performed at a state-of-the-art undulator beam line at the ALS, but with a simple current meter. By normalizing to the background absorption below the edges it is possible to obtain a coverage calibration. A monolayer corresponds to a 7% change in absorption. The dynamic range of the range of the measurement ranges from tenth of a monolayer to 100 layers (note the scale change from bottom to top in Figure 5). It is limited by beam stability at the low end and by saturation of the signal due to the finite photoelectron escape depth at the high end.

By using fluorescence photons instead of photoelectrons for monitoring N 1s and O 1s absorption, such measurements should be possible over a much larger dynamic range since the pre-edge background can be discriminated out by its lower fluorescence photon energy, and the escape depth of fluorescence photons is an order of magnitude longer than for electrons.

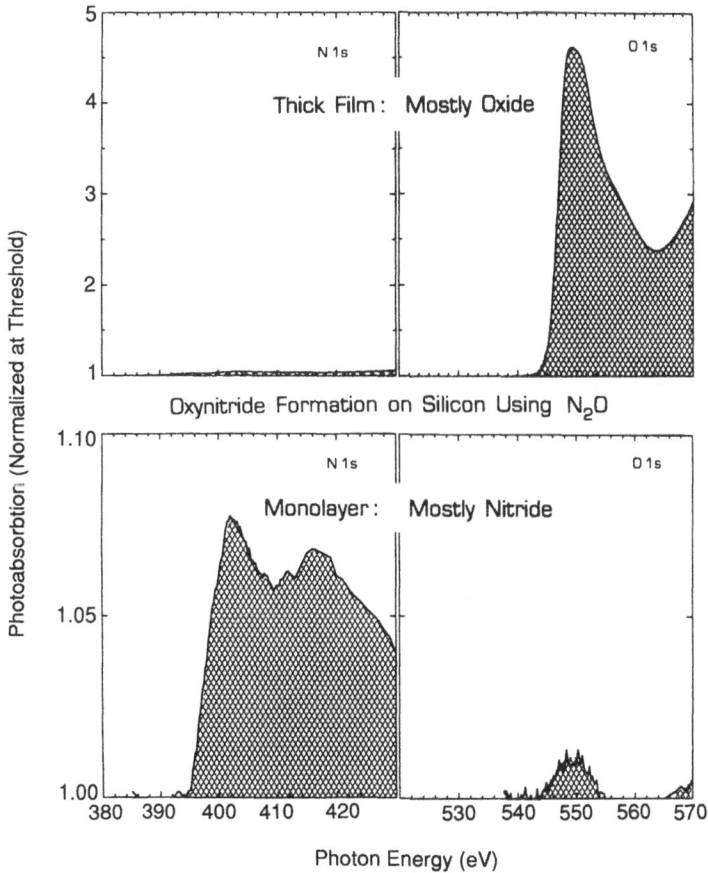

Figure 5   Reversal of the stoichiometry with film thickness in oxynitride films grown on SiO_2, determined by absorption spectroscopy at the N 1s and O 1s edges (10).

Particularly appealing is the prospect of performing such an experiment while the oxidation takes place, penetrating the ambient gas with fluorescence photons. Layer-by-layer growth and etching should be observable in real time with such a setup. The only handicap is the low yield of fluorescence photons, which typically lies three orders of magnitude below that of photoelectrons for the sharp core-levels listed in Figure 2. It can be overcome at third generation light sources, as shown in the following paragraph. Prototype experiments of this kind have already been performed (11), and the fluorescence technique has quickly gained notoriety at the new light sources.

Most interfaces in actual devices are buried too deeply to keep them accessible to probes with a short escape depth, such as photoelectrons. Here again, fluorescence comes to the rescue. Figure 6 shows that it is feasible with a third generation undulator light source to detect a monolayer of BN, buried under 50 Å of carbon (12). Here, a special trick helps to enhance the

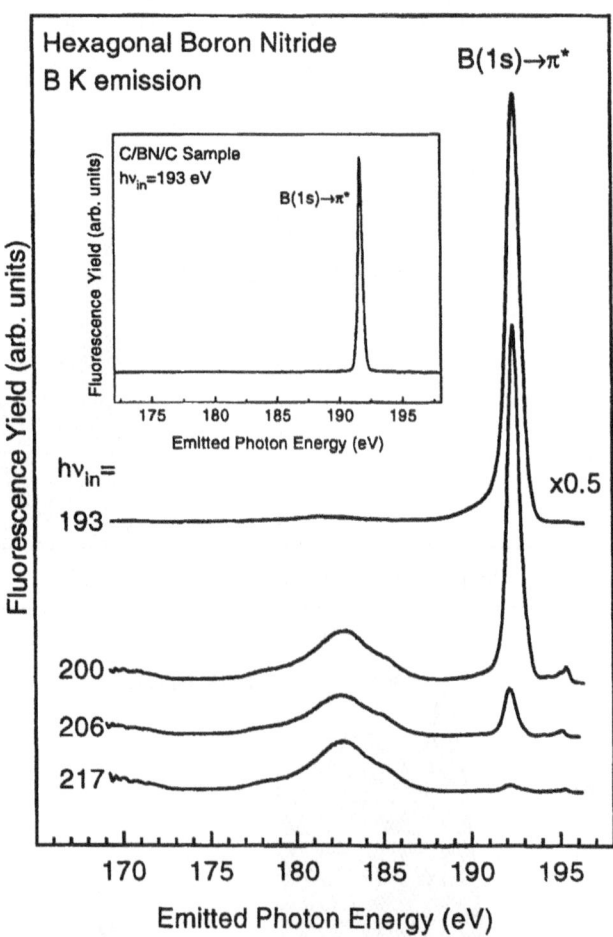

Figure 6   Fluorescence spectra from the B 1s core-level in BN, both for a bulk BN sample and a buried BN monolayer (inset). Using a resonant enhancement of the fluorescence at the C1s-to-π^* transition (hν = 193 eV) one can achieve sub-monolayer sensitivity, combined with deep penetration (12).

sensitivity. By tuning the excitation energy to the lowest B 1s absorption feature one finds a resonantly-enhanced fluorescence excited at higher energies. It is due to B 1s electrons excited into the lowest empty orbital of B, which recombine with the hole. The presence of such an orbital also gives clues about the electronic configuration of boron. For example, it is absent in cubic BN, where no equivalent to the long-lived π^* orbital of hexagonal BN exists. Likewise, electrically-active boron dopants have been detected by their unoccupied doping level.

Quantum Effects in Magnetic Multilayers

Magnetic multilayers have become a very popular topic since the discovery of "giant" magnetoresistance and oscillatory magnetic coupling in these structures. Their resistivity can change by a factor of two in a relatively low external field, which makes them attractive materials for reading heads of magnetic data storage disks, where state-of-the-art heads use the 2% magnetoresistance of permalloy. The magnetic coupling between two magnetic layers (e.g., Co) oscillates between parallel and anti-parallel, depending on the thickness of a noble metal spacer (e.g., Cu). These phenomena pose intellectual challenges for surface physics. How can the resistivity change so much in an external field? How can a non-magnetic noble metal

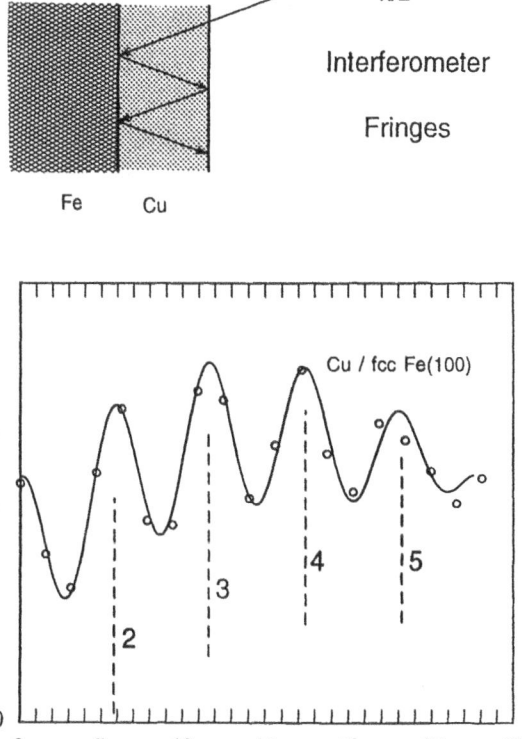

Figure 7   Quantum well states in a noble metal overlayer on a magnetic substrate, giving rise to oscillations in the density of states at the Fermi level (13). They have the same periodicity as the oscillations in magnetoresistance and magnetic coupling observed in the corresponding multilayers.

transmit magnetic information? Searching for the electronic states that cause thickness-dependent oscillations in the magnetic coupling of multilayers one has to understand electronic states of thin films first. In general, the continuum of bulk bands becomes discrete along the direction perpendicular to the interface when electrons are confined to a thin film. The number of discrete states per band corresponds to the number of atomic layers in the film. These states are often labeled quantum well states, since they are particularly pronounced when they are confined to the thin film by a potential well. This segmentation can be seen for thin noble metal films on magnetic substrates (13). A particularly elegant way of visualizing the discrete states is to consider the noble metal overlayer as Fabry-Perot interferometer for electrons, as shown in Figure 7. Quantum well states correspond to standing waves in the Cu film, reflected back and forth between the two interfaces. Looking at the density of states versus overlayer thickness one can see these regularly-spaced interference fringes with photoemission or inverse photoemission (13). They are spaced by half the wavelength of the electrons in a quantum well state. The

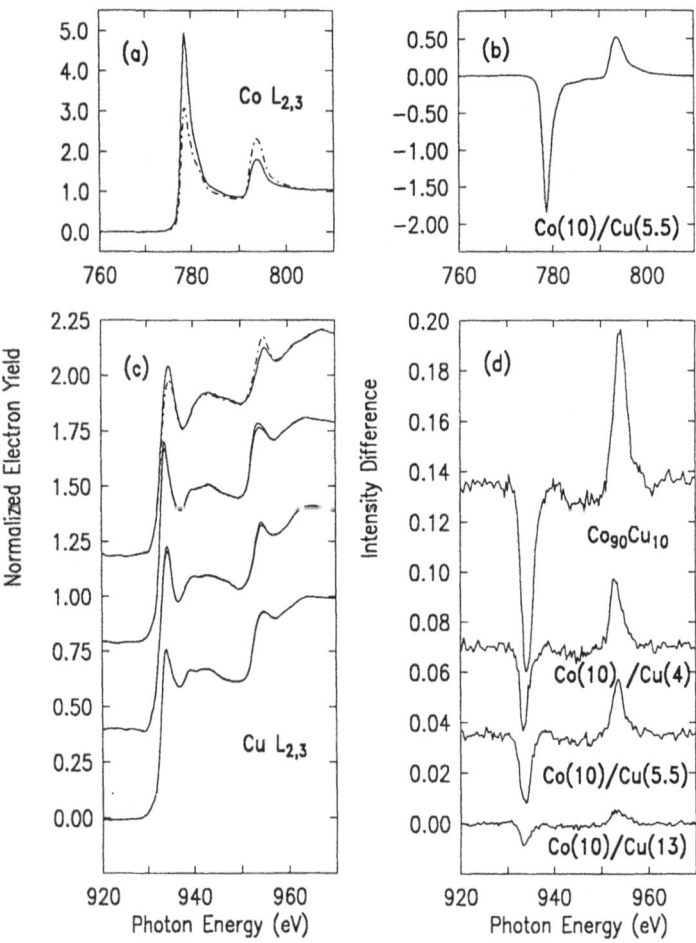

Figure 8 Co and Cu 2p absorption spectra of Cu/Co multilayers recorded with circularly-polarized light (solid lines for parallel alignment of photon spin and sample magnetization, dashed for anti-parallel, differences on the right hand side). The change of absorption with helicity (magnetic circular dichroism (MCD)) shows that Cu has become spin-polarized in the neighborhood of a magnetic Co substrate (16).

periodicity of 6 atomic layers agrees with the oscillations in the magnetic coupling. Measurements of such interference fringes with spin-polarized photoemission at various synchrotron light sources have shown that the spin-polarization also oscillates with the same period (14, 15). Hence, the noble metal has produced spin-polarized states and become magnetized. Since spin detection involves a loss of 3–4 orders of magnitude intensity and since tunability is required to enhance the transition from the s,p-band, this type of experiment is a prime application of undulator-based synchrotron light sources. In order to visualize the process or magnetic coupling it would be interesting to have a picture of the spin density distribution in the Cu spacer layer. Calculations readily provide this information, but one has also some experimental access to it by using magnetic circular dichroism, i.e., a change in the core-level absorption from left- to right-handed, circularly-polarized light (see Figure 8).(16, 17) By selecting the Co 2p and Cu 2p core-levels and using sum rules involving the angular and spin magnetic moment it is possible to obtain the average moments in the Co and Cu layers. Varying the Cu layer thickness provides depth information. Such circular dichroism experiments on a weakly-polarized noble metal have to detect small absorption changes of a few percent. They can only be done with intense, tunable, and polarized synchrotron light sources.

Multilayers represent just one of many conceivable quantum structures. Considering more general concepts one arrives at lateral superlattices, quantum wires, and quantum dots. At this point in time one can only speculate whether or not such structures will provide interesting and useful phenomena. The task at hand is to fabricate them with atomic precision. Traditional methods, such as lithography, fail at this level. Moving atoms with a scanning tunneling microscope is tricky and tedious. Borrowing from the experience with semiconductor quantum structures (18) one can try to use the regular array of steps at a vicinal surface as template for lateral magnetic superlattices and metallic quantum wires. The steps may be decorated by evaporating a metal that seeks out high binding energy site at a step, and if the surface energetics are right, additional rows of adatoms will also incorporate at the step edges and produce a step flow growth mode. Stopping growth at submonolayer coverage and switching to a different material will then provide the desired striped structure. First experiments on metal substrates are under way to test the feasibility of such an approach (19). To explore the electronic and magnetic structure of such "quantum wires" and "quantum stripes" one needs an atom-specific, background free probe that suppresses the substrate signal. Core-level fluorescence has these properties, and it can be driven to sub-monolayer detection limits at third generation synchrotron light sources, as demonstrated in Figure 6. Using fluorescence to detect core-level absorption and dichroism should do the trick. Such experiments promise to become an interesting part in the arsenal of techniques available at the new, third generation synchrotron light sources

Acknowledgment
This work was supported in part by the Office of Naval Research contract number N00014-91-C-0080. Part of it was performed under the auspices of the U.S. Department of Energy by Lawrence Livermore National Laboratory under contract number W-7405-ENG-48. The synchrotron radiation results were obtained at NSLS, SSRL, and ALS, which are supported by the Department of Energy.

References
1. F.J. Himpsel and I. Lindau, "Photoemission and Photoelectron Spectra," Encyclopedia of Applied Physics, ed. by G.L. Trigg and E.H. Immergut, VCH publishers, New York, to be published.

2. F.R. McFeely, J.F. Morar, and F.J. Himpsel, Surf. Sci. 165, 277 (1986).

3. B.S. Meyerson, Scientific American, 270, No. 3, p. 42 (1994).

4. B.S. Meyerson, F.J. Himpsel, and K.J. Uram, Appl. Phys. Lett., 57, 1034 (1990).

5. T. Suntola, Thin Solid Films, 225, 96 (1993).

6. D.A. Lapiano-Smith, F.J. Himpsel, and L.J. Terminello, J. Appl. Phys., 74, 5842 (1993).

7. D.A. Lapiano-Smith, F.J. Himpsel, and L.J. Terminello, Surf. Sci., submitted.

8. J.J. Pireaux, S. Svensson, E. Basilier, P.A. Malmqvist, U. Gelius, R. Caudano, K. Siegbahn, Phys. Rev. A, 14, 2133 (1976).

9. F.J. Himpsel, F.R. McFeely, A. Taleb-Ibrahimi, J.A. Yarmoff, and G. Hollinger, Phys. Rev. B, 38, 6084 (1982); F.J. Himpsel, D.A. Lapiano-Smith, J.F. Morar, and J. Bevk, Proceedings of the Second Symposium on "The Physics and Chemistry of SiO2 and the Si-SiO2 Interface 2," ed. by C.R. Helms and B.E. Deal, Plenum Press, New York, (1993), p. 237.

10. F.J. Himpsel, H. Akatsu, J.A. Carlisle, L.J. Terminello, E.A. Hudson, J.J. Jia, T.A. Calcott, R.C.C. Perera, M.G. Samant, J. Stöhr, and D.L. Ederer, Proceedings of the Symposium on "Silicon Nitride and Silicon Dioxide Thin Insulating Films," Electrochemical Society, submitted.

11. M. Georgson, G. Bray, Y. Claesson, J. Nordgren, C.G. Ribbing, and N. Wassdahl, J. Vac. Sci. Tech., A9, 638 (1991).

12. J.A. Carlisle, L.J. Terminello, E.A. Hudson, R.C.C. Perera, J.H. Underwood, T.A. Calcott, J.J. Jia, and F.J. Himpsel, to be published.

13. J.E. Ortega and F.J. Himpsel, Phys. Rev. Lett., 69, 844 (1992); J.E. Ortega, F.J. Himpsel, G.J. Mankey, and R.F. Willis, Phys. Rev. B, 47, 1540 (1993).

14. K. Garrison, Y. Chang, and P.D. Johnson, Phys. Rev. Lett., 71, 2801 (1993).

15. C. Carbone, E. Vescovo, O. Rader, W. Gudat, and W. Eberhardt, Phys. Rev. Lett., 71, 2805 (1993).

16. M.G. Samant, J. Stohr, S.S.P. Paring, G.A. Held, B.D. Hermsmeier, and F. Herman, Phys. Rev. Lett., 72, 1112 (1994).

17. S. Pizzini, C. Giorgetti, A. Fontaine, E. Dartyge, G. Krill, J.F. Bobo, and M. Piecuch, Mat. Res. Soc. Symp. Proc., 313, 625 (1993).

18. P.M. Petroff, A.C. Gossard, and W. Wiegmann, Appl. Phys. Lett. 45, 620 (1984); R. Nötzel, N.N. Ledentsov, L. Däweritz, M. Hohenstein, and K. Ploog, Phys. Rev. Lett., 67, 3812 (1991).

19. F.J. Himpsel and J.E. Ortega, Phys. Rev. B, 50, Aug. 15 (1994); Y.W. Mo and F.J. Himpsel, Phys. Rev. B, 50, Sept.15 (1994).

CHEMICAL REACTION DYNAMICS USING THE ADVANCED LIGHT SOURCE

X. Yang,[1,2] D. A. Blank,[1,2] J. Lin,[1,2] P. A. Heimann,[3] A. M. Wodtke,[1,2] Y. T. Lee,[1,2,4] and A. G. Suits[1,2]

[1]Dept. of Chemistry, University of California, Berkeley, CA 94720
[2]Chemical Sciences Division, Lawrence Berkeley National Laboratory
[3]Accelerator and Fusion Research Division, Lawrence Berkeley
National Laboratory, Berkeley CA 94720
[4]Present address: Institute of Atomic and Molecular Sciences,
Academia Sinica, Taipei, Taiwan, ROC

INTRODUCTION

The studies of bimolecular and unimolecular reactions using molecular beam techniques have greatly improved our understanding of chemical reactions. The angular resolved time of flight (TOF) method for reactive scattering studies gives us the dynamical information that is needed to understand the detailed dynamics of reactive processes.[1] Currently used universal TOF detection techniques for crossed beam reactive scattering studies are mostly associated with electron bombardment ionization. While electron bombardment ionization is the standard technique for mass spectrometric detection, it does have certain disadvantages, such as: extensive dissociative ionization, limited TOF resolution because of its finite ionization region, high background for certain masses, and certainly no selective ionization of different species.

Light, from the early flash lamp source to the modern lasers, has been widely used in the studies of chemical processes. Tunable VUV coherent light as the ionization detection technique in TOF measurements, instead of electrons, has tremendous advantages over the widely used electron bombardment ionization method. Firstly, tunable VUV ionization can be species selective since the ionization potentials for different species are different in nature. As a result it is also possible to eliminate the high background at mass 15(CH_3^+), 16(CH_4^+), 18(H_2O^+), 28(CO^+). This is particularly important when detecting radical species with IP much lower (in many cases) than those of the background molecules.

Synchrotron Radiation Techniques in Industrial, Chemical, and Materials Science
Edited by D'Amico *et al.*, Plenum Press, New York, 1996

Tunable VUV light also allows us to obtain information about the internal energy of the reactive species due to the different characteristics of energy dependent ionization cross sections between "cold" and "hot" species. Secondly, soft (low energy) VUV ionization is expected to cause much less fragmentation than electron bombardment ionization, making the detection and analysis of multiple channel reaction processes simpler and more straightforward. Thirdly, since undulator VUV light is focusable, the ionization region can be much smaller than a comparable electron bombardment ionizer, making both the TOF and angular resolution much higher. Since VUV ionization is focusable with a relatively small associated heat load, the ionization region can also be easily cooled to very low temperature, making the background much smaller in the detector. However, in order to make the VUV ionization a standard universal detection technique for chemical reactions, one has to meet the following two requirements: high VUV photon flux (high enough sensitivity) and wide range tunability, which are hard to meet even by the modern state-of-art laser techniques. The recently developed third generation synchrotron radiation source at Berkeley, namely the Advanced Light Source, provides us such a light source. An undulator based VUV beamline has been constructed at the ALS, providing $\sim 10^{16}$ photon/s (2% bandwidth) VUV light, tunable from 5 to 30 eV.[2] In addition to the full undulator light, high resolution VUV light is also available for the beamline, which can be used for high resolution studies on photoionization processes in the VUV region.

In this paper we will briefly describe the ALS chemical dynamics beamline and endstation one (rotating source cross beam machine), then present our first results using synchrotron radiation to study photodissociation of methylamine and ozone using endstation one.

EXPERIMENTAL SET UP

The ALS Chemical Dynamics Beamline

The Advanced Light Source is one of the first third generation, low emittance, synchrotron radiation facilities.[3] The ALS is designed based on insertion devices (undulators,wigglers) which produce tunable VUV to soft X-ray radiations with much higher photon flux than bending magnets. A U8 undulator, which will be replaced soon with a U10 undulator, is currently being used as the light source of the chemical dynamics beamline. The lowest energy achievable for the U8 undulator is 18 eV for a 1.5 GeV storage ring energy, and 8 eV for a 1.0 GeV storage ring energy. The future U10 undulator can produce VUV light as low as 5 eV for a standard 1.5 GeV storage ring energy.

Figure 1 shows the layout for the chemical dynamics beamline. Light produced from the undulator is directed through horizontal and vertical apertures, whose openings are matched to the central cone size. The M1 and M2 mirrors absorb unwanted light at short

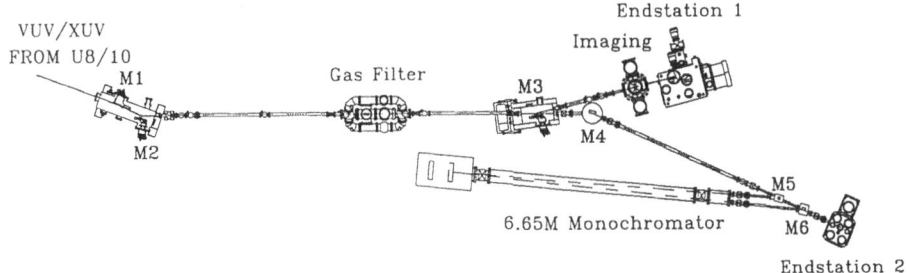

Figure 1. The chemical dynamics beamline layout.

wavelengths and focus the light at the center of the harmonic filter, which is developed to suppress the higher harmonics of undulator radiation. A doubly differentially pumped scheme is used in the harmonic filter to fulfill the requirement of high number density (~30 torr) for the gas cell in order to achieve effective suppression while preserving the beamline high vacuum to $< 5 \times 10^{-9}$ torr. It has achieved greater than 10^4 suppression of the higher harmonics with no measurable (<5%) attenuation of the fundamental. A detailed description of the instrument can be found in Ref. 4. After the harmonic suppressor, the VUV light is reflected by the retractable M3 mirror to endstation one, which is the rotating source crossed beam machine. M3 also focuses the light at the detection region of the crossed beam machine. An imaging endstation will be installed in front of endstation one in the near future, primarily for complementary crossed beam studies using the recently developed imaging technique.

If M3 is removed from the beam path, the VUV light will be reflected by M4 towards M5, which focuses the light onto the entrance slit of the 6.55 meter eagle monochromator, manufactured by McPherson, Corp. The achievable resolution of the monochromator will be 10^5 or better, providing endstation two with an extremely high resolution VUV light source. After the monochromator, the light is then refocused by the M6 mirror into endstation two, primarily for various photoionization studies.

The Rotating Source Crossed Beam Endstation

The newly built endstation one on the chemical dynamics beamline features the world's first crossed beam machine using synchrotron radiation as the detection method. Figure 2 shows the schematic of the crossed molecular beam apparatus. It has three main chambers: the rotating source, the main chamber and the detector. The rotating source has two main source chambers, each of which has a differentially pumped region. The two sources fixed at 90 degrees, and the whole source chamber is rotatable from -20 to 110 degrees (total 130 degrees rotation). The entire machine is pumped by oil-free magnetic bearing turbomolecular pumps. Each of the two main source region for the two sources is

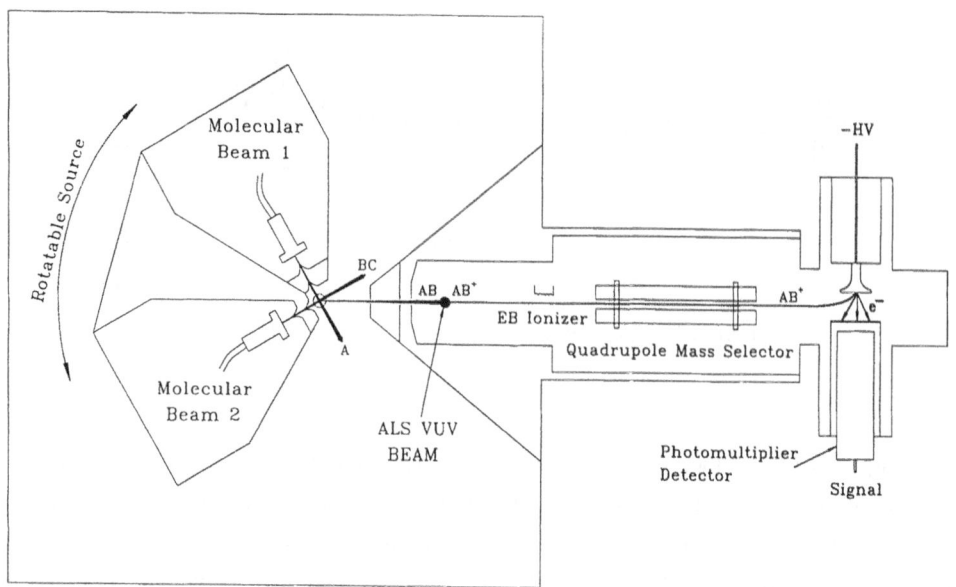

Figure 2. Schematic of the rotating source crossed molecular beam endstation (endstation one).

pumped by a high-throughput 2000 l/s turbopump (Seiko-Seiki STPH2000C), and each of the two differential pumping regions is pumped by a 300 or 400 l/s turbopump (Seiko-Seiki STP 300/400). The main chamber, which serves as the region where chemical reactions take place, is pumped by a 2000 l/s turbopump (Seiko-Seiki STP2000C). A base plate cooled by liquid nitrogen is also available for cryopumping of the main chamber. There are four regions in the detector. Region I, pumped by a 300 l/s turbopump, is the first differential region. Region II, also pumped by a 300 l/s pump, is the second differential region which also connects with the advanced light source. Region III is the ionization region, which includes a VUV ionizer, an electron bombardment ionizer and ion optics to transport ions. It is pumped by a 600 l/s pump. Quadrupole mass selection (Extrel) then follows in region IV which also houses a standard Daly ion detector. Time of flight chopper wheels are available in front of the detector for velocity and cross-correlation measurements.

Three types of experiments are possible at endstation one. a) Crossed molecular beam reactions, detected either by VUV ionization with mass selection, or electron bombardment ionization; b) Single molecular beam laser photodissociation experiments using the detection schemes same as a; c) Photodissociation by VUV synchrotron radiation experiments using electron bombardment ionization detection.

The first set of experiments carried out on the endstation one are focused on photodissociation of molecules by an excimer laser using tunable VUV ionization for detection. Only one molecular beam source is used in these experiments. An excimer laser is focused at the axis of rotation of the molecular beam source. Photodissociation products are then detected as a function of flight time and scattering angle. Results will be shown in

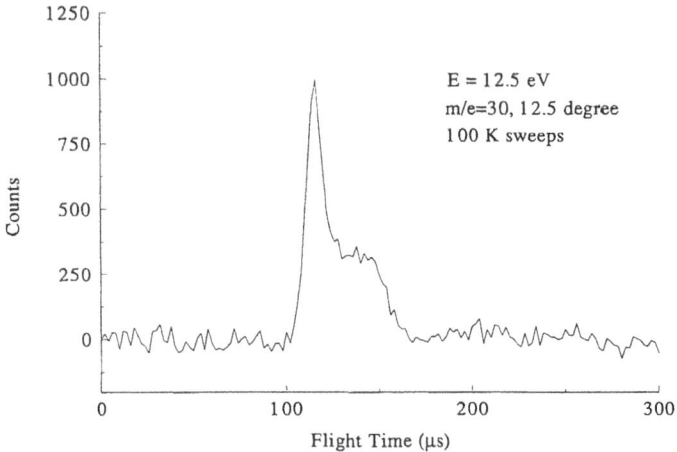

E = 12.5 eV
m/e=30, 12.5 degree
100 K sweeps

Figure 3. Time of flight spectrum of methylamine photodissociation at 193 nm detected at m/e=30, θ=12.5 degrees.

the following section for methylamine photodissociation at 193nm and ozone photodissociation at 193 nm and 248 nm.

RESULTS

Photodissociation of Methylamine

Photodissociation study of hydrocarbons is an important topic since it helps us to understand the energetics and the dynamics of how these molecules break apart. It can also be used to measure the bond energies. In this experiment, the photochemistry of methylamine at 193 nm was studied using the angular resolved time of flight technique. Tunable VUV light of 8.5 eV or higher was used in this experiment with a storage ring running at 1.0 GeV. Figure 3 shows the TOF spectrum of m/e=30 / CH_3NH^+, θ=12.5 degree (0 is the angle between the molecular beam and the detector). This shows the hydrogen atom loss channel

$$CH_3NH_2 + h\nu \rightarrow CH_3NH + H$$

for methylamine at 193 nm photodissociation. Signal at m/e=29 was also observed, which was the strongest signal observed; however its TOF profile is distinctively different from that of m/e=30. The origin of the m/e=29 signal is not yet clear. There are three possibilities: a) the m/e=29 signal is from secondary dissociation of m/e=30 because of the unstable nature of m/e=30; b) it is from a concerted H_2 loss process; or c) both. Further analysis is needed in order to clarify the picture. Experimental measurements were also made to determine the relative ionization cross section for the m/e=29 product by tuning

123

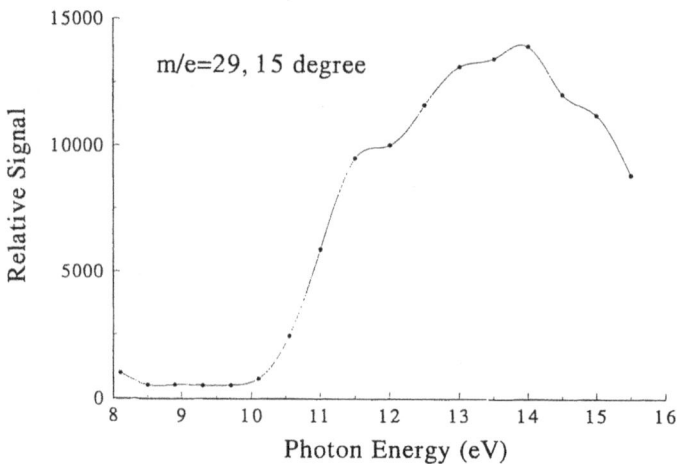

Figure 4. VUV ionization energy dependence of total ionization signal of the m/e=29 pro
duct at θ=12.5 degree from photodissociation of methylamine at 193 nm.

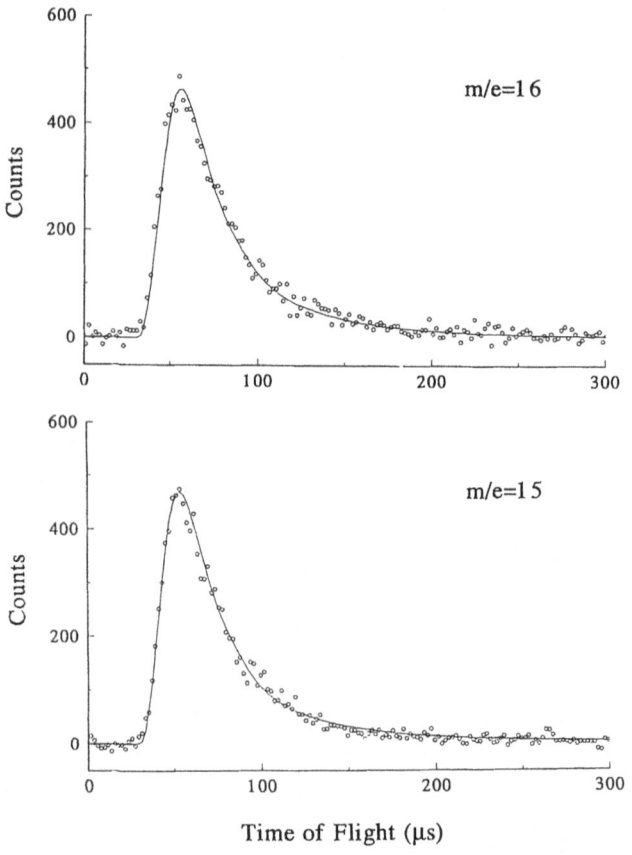

Time of Flight (μs)

Figure 5. Time of flight spectra of m/e=15, 16 at θ=25 degrees, photon energy 13.0 eV.
The open circles represent the experimental data. The solid line shows the fit to both
spectra using a single translational energy probability distribution.

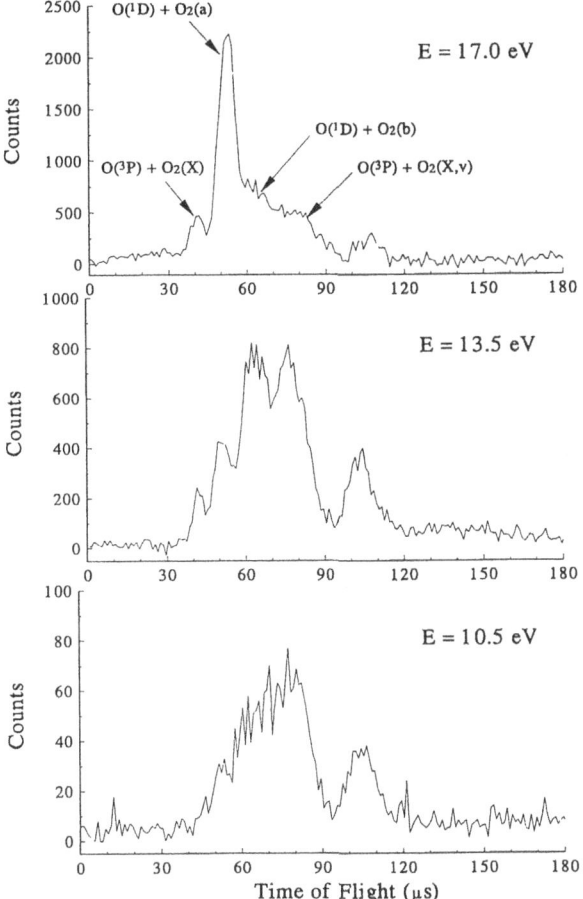

Figure 6. Time of flight spectra of m/e=32 from ozone photodissociation at 193 nm detected at θ=20 degrees. Upper: ionization energy 17 eV, middle: 13.5 eV, lower: 10.5 eV. TOF peaks are labeled consistent with results from Ref. 5.

the undulator gap, shown in Figure 4. This type of data contains information on the internal excitation of the photofragment. However, in order to extract such information, one need to know the relative ionization cross section near the IP for the same species at low temperature. Accurate theoretical calculations are definitely required in this case.

Signal at m/e=15, 16 are also observed from methylamine photodissociation at 193nm. Figure 5 shows the time of flight spectra of m/e=15 at θ=25 degrees, and m/e=16 at the same angle. Eventhough this is a very minor channel (a few percent of the total dissociation), good signal to noise can still be obtained due to the extremely low background at both m/e=15 and 16. By tuning the VUV energy lower than the IP of the background gas, the background can be almost eliminated. Clearly, the m/e=15, 16 signals are corresponding to the methyl group loss channel,

$$CH_3NH_2 + h\nu \rightarrow CH_3 + NH_2$$

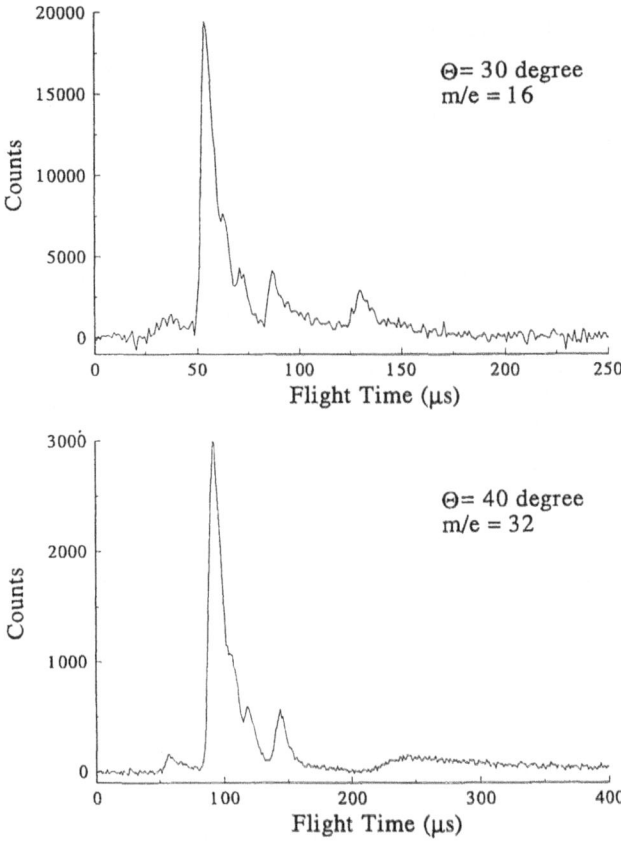

Figure 7. Time of flight spectra of m/e=16, 32 from ozone photodissociation at 248 nm.

Figure 5 shows the momentum matched fits to both m/e=15 and m/e=16 at 25 degrees using a single translational energy probability distribution.

In conclusion, at least two dissociative channels have been observed for methylamine photodissociation at 193 nm. These experimental results also show some of the advantages of VUV ionization over the filament ionization: much lower background and less dissociative ionization.

Photodissociation of Ozone

UV photodissociation of ozone is an important process in the upper atmosphere since it is the very process that protects living things from solar UV radiation. Recent studies show that a newly identified channel in the UV photodissociation of ozone might be responsible for a new ozone creation channel operating in the stratosphere, in addition to the established Chapman mechanism involving O_2 photodissociation.[5,6]

In order to better understand the ozone problem, a series of experiments was carried out using endstation one with the storage ring running at 1.0 GeV. Figure 6 shows a series of TOF spectra, measured at m/e=16 and θ = 30 degree, with 193 nm as the photolysis source. In this experiment, the VUV light source is tunable from 8 eV. Therefore, the dependence of each photodissociation channel of product O_2 on the ionization energy can be mapped out easily. One can see that individual peaks in the TOF spectrum change very differently as the ionization energy increases. This set of data contains information on how O_2 is excited: vibrationally, electronically or both, since each type of excitation should show its unique characteristics of cross-section dependence on ionization energy. As all the electronic states of O_2 below 8 eV are known, it is possible through simulations to determine the state specific distribution of product O_2. Similar experiments can be done for product O atom to identify the product state distribution of the O atoms. This can be much simpler because atoms do not have rotational and vibrational states. Detail analyses and further experiments are underway in this direction.

Photodissociation of ozone at 248 nm has also been carried out using the full undulator light (without harmonic suppression) as the detection light source, with the fundamental at 18 eV. In this experiment, a storage ring running at 1.5 GeV is used with the U8 undulator. The lowest energy achievable is 18 eV, therefore tunable VUV selective ionization is not possible. Figure 7 shows the time of flight spectra of O_2 and O when ozone was photolysed at 248 nm. The data is not yet fully analyzed. Nevertheless this figure shows one of the main advantages of the new endstation: even at 18 eV ionization energy there is little dissociative ionization in comparison with the previous results of photodissociation at 193nm using electron bombardment ionization where dissociation ionization is extensive (Figure 4 in Ref. 5). Direct peak-to-peak momentum matching of the products can be seen, with little or no fragmentation of O_2 to give mass 16 signal. Vibrationally resolved structures in the TOF spectra have been observed.

CONCLUSION

A unique chemical dynamics beamline has been established for reaction dynamics studies at the Advanced Light Source. The new crossed beam endstation is now in full operation. Preliminary results are reported for photodissociation studies of methylamine and ozone. Experimental studies show that the new endstation using ALS VUV light as the ionization detection scheme has tremendous advantages over similar machines using electron bombardment ionization as the detection technique.

ACKNOWLEDGMENTS

This work was supported by the Director, Office of Energy Research, Office of Basic Energy Sciences, Chemical Sciences Division, of the U.S. Department of Energy under contract no. DE-AC03-76SF00098. We also appreciate the excellent jobs done on the beamline by the LBNL engineering staff, and all the support from the ALS staff.

REFERENCES

1. Y. T. Lee, Science **236**, 793 (1987)

2. M. Koike, P. A. Heimann, A. H. Kung, T. Namioka, R. DiGennaro, B. Gee, and N. Yu, Nucl. Instrum. and Meth. A **347**, 282 (1994).

3. A. S. Schachter, New Directions in Research With Third-Generation Soft X-ray Synchrotron Radiation Sources, pp.1-22, NATO ASI Series E: Applied Sciences - Vol. 254, edited by A. S. Schlachter and F. J. Wuilleumier, Kluwer Academic Publishers, 1992.

4. A. G. Suits, P. A. Heimann, X. Yang, M. Evans, C. -W. Hsu, K. -T. Lu, A. H. Kung, and Y. T. Lee, Rev. Sci. Instrum., in press.

5. Domenico Stranges, Xueming Yang, James D. Chesko, and Arthur G. Suits. J. Chem. Phys. **102**(15) 6067 (1995).

6. R. L. Miller, A. G. Suits, P. L. Houston, R. Toumi, J. A. Mack, and A. M. Wodtke, Science **265**, 1831 (1994).

HIGH RESOLUTION PHOTOIONIZATION AND EXCITATION USING THIRD GENERATION SYNCHROTRON RADIATION SOURCES

N. Berrah, B. Langer, A. Farhat, and A. Happelman*

Physics Department
Western Michigan University
Kalamazoo, MI 49008

INTRODUCTION

Atomic and molecular structure and dynamics can now be explored with unprecedented resolution with third generation synchrotron radiation sources. Insertion devices, like undulators and wigglers, coupled with well designed monochromators, make synchrotron radiation a powerful tool since it can offer tunability, intensity, polarization and time structure.

The objective of the reported work is to probe fundamental aspects of atomic structure and the dynamics of the interaction of high energy photons with 2-electron atoms and multi-electron atoms. The principal focus is the investigation of many body effects due to electron-electron interaction. The emphasis of the reported work is on photoionization and excitation of atoms and on near threshold effects following either outer-shell or inner-shell photoexcitation and photoionization. In this paper, we present recent developments in four areas of this field.

PHOTOIONIZATION NEAR THE HE DOUBLE-IONIZATION THRESHOLD

Photo-double-ionization is one of the fundamental processes of physics because it requires a solution of the three-body Coulomb problem where the boundary conditions for the two continuum electrons must be included. There has been much interest in the study of photo-double-ionization of He because it is a system that is dominated by electron-electron correlations. Because the independent electron model failed to provide adequate agreement with measurements, new theoretical approaches have had to be developed. Since the classic work of Wannier,[1] numerous theoretical studies have been made on near-threshold ionization.[2-4] The various theories yield predictions for three different observable

Synchrotron Radiation Techniques in Industrial, Chemical, and Materials Science
Edited by D'Amico *et al.*, Plenum Press, New York, 1996

129

situations: (a) the energy dependence of the cross section (b) the energy sharing of the two outgoing electrons and © the angular correlation of these electrons. Wannier theory predicts, in the energy range just above threshold, that $\sigma^{++}=\sigma_0 E^{\alpha}_{exc}$. Kossmann et al[5] made an extensive study of the threshold law for the cross section of double ionization in helium. Ion analysis was carried out with a pulsed-field time-of-flight (TOF) e/m analyzer. Their results provide quantitative information about the Wannier exponent, $\alpha=1.05(2)$ which agrees with the theoretical prediction of 1.056; a threshold value $\sigma^0=1.02(4)\text{x}10^{-21}$ cm^2 and $E_{th}=79.013(10)$ eV. Furthermore, their experimental results find the range of validity of the cross-section threshold law to be approximately 2-eV excess energy above threshold. Lablanquie et al[6] used coincidence measurements between low energy electrons and doubly charged ions to study the dynamics of double photoionization and confirmed the range of validity of the Wannier theory. They found that the energy distribution of the two outgoing electrons is flat, within 20%, in agreement with the theoretical prediction, but in a 15 eV energy range above threshold. Photoionization phenomena near the double-ionization threshold has also been extensively studied by Hall et al[7]. Using a photoelectron/photoion coincidence technique they find the value of the exponent α to be consistent with the Wannier prediction. They also investigate the behavior of the asymmetry parameter, ß, near threshold and obtain a nearly constant value close to -0.4. Their result is in disagreement with the prediction of the Wannier theory which appears to underestimate the angular correlation between the two electrons[8]. Dawber et al[9]. have exploited the photoelectron-photoelectron coincidence technique to measure the triple differential cross section (TDCS) at very low excess energies E (0.6 eV $<$E$<$ 2 eV) for both equal and unequal energy sharing between the two outgoing electrons. The measured data are compared with the Wannier(1) predictions and also with recent ab initio calculations[10-12] that are not based on a Wannier-like treatment. Their measurements suggest a departure from the predictions of the Wannier model at the largest excess energy studied, E=2 eV. Lablanquie et al[13] have also very recently studied the effect of electron energy sharing near the double photoionization threshold. In their energy and angle resolved measurements, they observed that although the angular distributions do not depend much

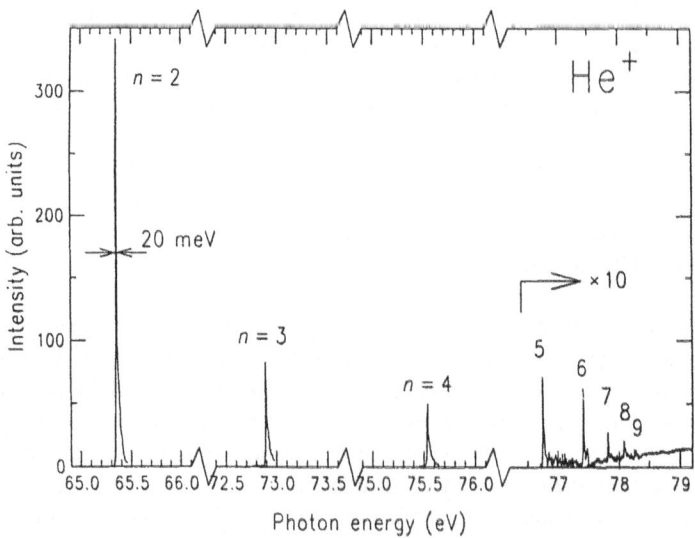

Fig. 1. Threshold ($E_e=0$ eV) photoelectron spectrum obtained with a monochromator bandpass of 6 meV and photon energy increment of 5 meV.

on the energy sharing of the two electrons at 4 eV above threshold, a strong effect is measured at E=18.6 eV.

We have used a zero-volt spectrometer built in Becker's group[14] to study with higher resolution than 50 meV[15] photoionization phenomena near the double-ionization threshold. Fig. 1 shows a preliminary spectrum taken with photons from an undulator beamline coupled with a spherical-grating monochromator of the Advanced Light Source at Lawrence Berkeley Laboratory. The 1.5 GeV storage ring was filled to 40 mA in the two bunch mode at injection. The monochromator bandpass was 6 meV near 79 eV with 31 μm entrance and 27 μm exit slit widths. The scan in Fig.1 shows eight satellite lines which are the result of electron correlations. These satellite lines originate from an ionization process with additional excitation leaving the ion in a He$^+$ nl (n>1) state. The line widths are about 20 meV, an improvement by a factor of 2.5 over previous measurements.[15]

ENERGY DEPENDENCE OF THE 4P^{4}ND (^{2}S$_{1/2}$) (N=4-9)
ELECTRON CORRELATION SATELLITES IN Kr
FROM 68.5-250 eV

Since the discovery of photoelectron satellites lines nearly 30 years ago,[16,17] there has been a continuing interest in understanding their behavior. This interest is attributed to the fact that such satellites result entirely from electron-electron correlation. According to the configuration interaction picture of valence shell photoionization, these mp^4 (^{1}D)nd(^{2}S$_{1/2}$) (n≥m) satellites are expected to be the dominant kind in the rare gases because their ionic-state configurations mix very strongly with the primary configuration consisting of the ms(^{2}S$_{1/2}$) hole. A correct description of the behavior of these satellites in the heavier rare gases, Kr and Xe, posses a serious challenge to theory. The calculation would need to be carried out in an intermediate coupling scheme and inclusion of relativistic and interchannel effects might be essential.

High resolution photoelectron spectra of the nd Rydberg satellites in Kr have been measured for the first time for six different angles and for energies between 68.5 and 250 eV.[18] Figure 2 shows spectra recorded at 0° for three different energies. The experiment was performed at the Hamburger Synchrotronstrahlungslabor (HASYLAB) on the new undulator beamline (BW3) utilizing a high resolution SX-700 monochromator. The measurements were made during single bunch and double bunch operation of the electron storage ring. Spectra were recorded using two time-of-flight (TOF) electron spectrometers.[19] The high flux, high resolution combination of this beamline allowed us to record the highest resolution spectra yet of the Kr valence satellite region using monochromatized synchrotron radiation.

All spectra have been corrected for transmission effects, differences in detector efficiencies and source volume anisotropies. The satellites are designated according to the scheme used by Krause et al.[20] The assignments of these lines are based on the semi-empirical calculations of Hansen[21] whose binding energies were derived from optical data. We notice from figure 2 that the nd(^{2}S$_{1/2}$) satellite intensities decrease with increasing energies. This behavior is similar to the 4s main line.[22] Analysis of the measurements reveals that the relative intensities of these satellites with respect to the 4s and 4p main lines are in very good accord with the measurements at 68.5 eV.[20] The results also confirm that present theoretical predictions for the relative intensities of these satellites are too high. From our data we also extracted the angular distribution parameters ß as a function of energy. These results will be published in a forthcoming paper.[23]

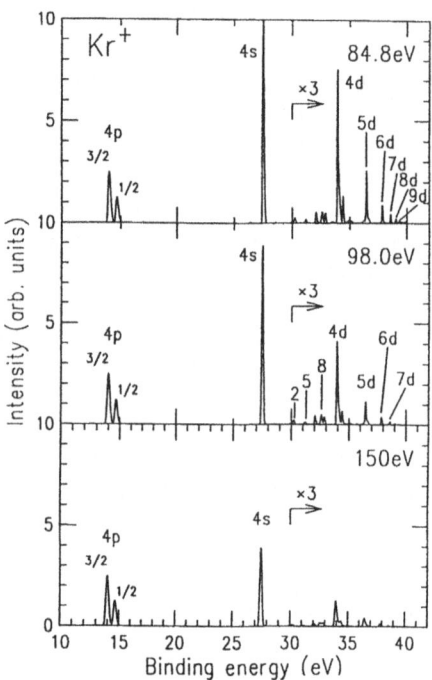

Fig. 2. High resolution photoelectron spectrum of the Kr correlation satellites recorded at three different energies for a 0° angle.

HIGH-RESOLUTION ANGLE-RESOLVED PHOTO-ELECTRON SPECTROSCOPY: THE Ar 3S⁻¹→ NP (N=4-16) RESONANCES

The discovery of a series of autoionizing states in He[24] and its explanation[25] were crucial for the understanding of electron correlation effects. Theoretical work has shown that electron angular distributions[26] and the shape of autoionization resonances[27] are essential to understanding electron-electron correlations. Theoretical analysis[28] has shown that the angular distributions of photoelectrons in resonance regions are significantly different from those observed in non-resonance regions. The angular distribution can vary rapidly over an energy range on the order of a resonance width.

Accurate studies of electron angular-distributions parameters (ß) are a sensitive probe[29] of atomic wave functions. According to Starace,[30] dramatic changes in the parameter ß is an indication of strong effects due to e-e correlations. The experimental determination of the anisotropy parameter ß in the resonance region is an important source of information on the dynamics of resonance photoionization which cannot be obtained by absorption or ion mass spectrometry. Electron angular distributions can give a nearly complete picture of the photoionization process (apart from the spin polarization of the emitted electron) because information on the angular momenta involved, the partial-wave matrix elements and the phase shifts between partial waves of the outgoing electrons can be obtained.

Autoionization resonances studied in this work result from the decay of the excited discrete states Ar* 3s3p⁶ np into the continuum state Ar⁺ 3s²3p⁵ + e⁻ (ks,kd). Because the continuum also can be reached by direct photoionization, the two paths give rise to interference effect that produce the characteristic Beutler-Fano line shape.[31] Detailed measurements of the shape of this autoionization series were conducted previously using absorption techniques[32] and ion mass spectrometry.[33] Also, angular distributions for the first three autoionization resonances have been measured using photoelectron spectrometry.[34] These autoionization resonances have been calculated using many-body perturbation theory,[35] multichannel quantum-defect theory,[36] the eigenchannel R-matrix method,[37] the K-matrix procedure,[38] and the random-phase approximation with exchange.[39]

We have measured detailed high-precision angle-resolved electron-spectrometry measurements of the Ar $3s^23p^6 \rightarrow 3s3p^6$ np (n=4-16) autoionization resonances in the energy range between 26 and 29.3 eV. The aim of this work was to provide a critical test of present calculations and a testing ground for further theoretical advances. We have taken advantage of the high resolution and high brightness of an undulator beamline at the Advanced Light Source, coupled with our electron time-of-flight spectrometers, to obtain for the first time (1) accurate measurements of the photoelectron angular-distribution parameters for all members of the series and (2) first observation from n=8 to n=16 of the Ar $3s^23p^6 \rightarrow 3s3p^6$ Rydberg series. We have fit the angular-distribution data using a model function derived by Kabachnik and Sazhina[40] and the comparison of our results with the R-matrix calculations of Taylor[26] is found to be excellent. We also have analyzed the cross-section shape of each resonance, using a model function originally given by Fano,[41] and have determined the values of the shape parameter q, the correlation parameter ρ^2 and the resonance width Γ of the Beutler-Fano profiles which best describe them. Our results confirm previous measurements[32,33] for the lower 6 resonances.

Experimental Procedure

The experiment was performed at the Advanced Light Source at Lawrence Berkeley Laboratory. The 1.5 GeV storage ring was filled to 40 mA in two bunch mode at injection. Two components of instrumentation were essential in these high-resolution gas-phase photoemission measurements: (1) a monochromator with a resolving power E/ΔE of at least 10,000, and (2) an angle-resolved technique to measure angular distributions.

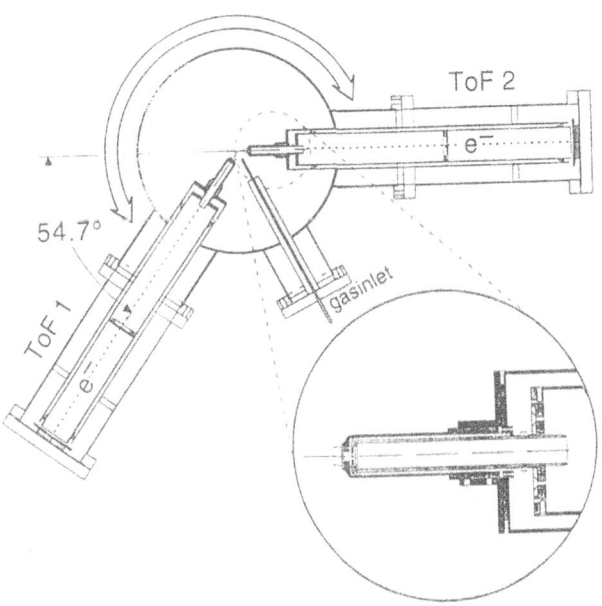

Fig. 3. Schematic diagram of the apparatus showing the interaction region in the rotating chamber that houses two TOF spectrometers.

Our measurements were carried out using the 8-cm-55-period U8 undulator and ALS beamline 9.0.1 which utilizes a spherical-grating monochromator. For the reported measurements the monochromator was used with the 380 l/mm grating. The photon beam had a spot size of about 0.1mm x 1mm in the interaction region of our experimental chamber. We measured a photon beam linear polarization of better than 99%. Near 30 eV, a calibrated diode[42] determined a flux of about 1.10^3 ph/s, for 400 mA of stored current in the ring, with 33 μm entrance and 33 μm exit slit widths. The experiment was carried out with our newly built apparatus, similar in design to previous ones,[43] but equipped with upgraded and advanced TOFs. Briefly, it consists of a rotating vacuum chamber that houses two advanced time-of-flight (TOF) spectrometers[44] to record spectra simultaneously and to allow *both* partial-cross section and angular-distribution measurements. The two TOFs are mounted perpendicularly to the incoming photon beam, as shown in Fig. 3. Details of the experimental apparatus will be described in another work.[44]

Argon gas was leaked into the chamber interaction region through a glass capillary array to a pressure of about 5×10^{-5} Torr, while the background pressure was about 1×10^{-8} Torr. Photoemitted electrons were detected by the two spectrometers and recorded simultaneously while the photon beam was scanned from 26 to 29.3 eV. Figure 4 shows one of our scans of the electron yield measured at an angle of 0° with respect to the electric field vector of the linearly polarized synchrotron beam. The monochromator bandpass was 6 meV at 27 eV with 40 μm entrance and 60 μm exit slit widths and the scan was normalized to the incident photon flux. Figure 4 shows the Ar $3s^2 3p^6 \rightarrow 3s3p^6$ np Rydberg series of *window* type resonances from n=4 to n= 16.

Theoretical Background

Experimental advances toward a highly differentiated analysis have been achieved by determining the electron angular distribution. In the dipole approximation, valid for low energy photons, hv, the differential photoionization cross-section, $d\sigma_{if}$ (hv)$/d\Omega$, and the photoelectron angular-distribution parameter, ß, resulting from photoionization of state $|I>$ by linearly polarized photons leaving the ions in state $|f>$ is given by[45]:

$$\frac{d\sigma_{if}}{d\Omega} = \frac{\sigma_{if}}{4\pi} [1 + \beta_{if} P_2 (\cos \theta)] \tag{1}$$

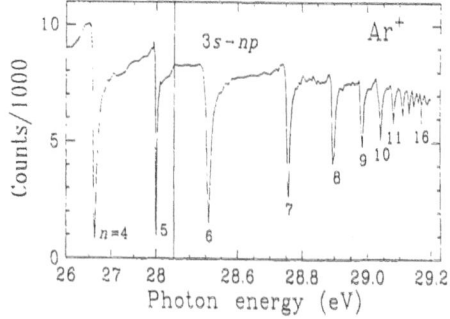

Fig. 4. Photoelectron yield scan of the Ar $3s^2 3p^6 \rightarrow 3s3p^6$ np autoionization resonances at 0° with respect to the polarization of the synchrotron light.

where σ_{if} is the total photoionization cross section for producing state $|f\rangle$ of the ion, θ is the angle between the photon's polarization vector and the photoelectron's momentum direction, $P_2(x)=(3x^2-1)/2$, and β_f is the electron angular-distribution parameter. The physical range of ß is between -1 and 2 since otherwise the partial cross section in equation 1 would be negative. From this, one notes that the angular distribution of any photoionization process (in the dipole approximation) can be described by a single parameter, ß, which in general is dependent on the photon energy.

As mentioned in the introduction, the presence of a discrete state embedded in one or more continua causes an interference in the photon absorption process because of the indistinguishability of direct ionization and autoionization to the same final state. In the case of an isolated discrete state interacting with one or more continua, we have used a parameterization of the variation in ß over autoionizing resonances introduced by Kabachnik and Sazhina.[40] It is based on the Fano[41] parameterization for the total photoionization cross section, σ_{if}, given by[26]:

$$\sigma = \sigma_a \frac{(q + \epsilon)^2}{1 + \epsilon^2} + \sigma_b \qquad (2)$$

where q is the line-profile index and ϵ is given by:

$$\epsilon = \frac{E - E_r}{\frac{1}{2}\Gamma} \qquad \text{and} \qquad \rho^2 = \frac{\sigma a}{\sigma a + \sigma b}$$

where E_r is the position of the resonance and Γ its width. The cross sections σ_a and σ_b are slowly varying background cross sections. Also derived[40,26] is the expression for ß given by:

$$\beta = -2 \frac{X\epsilon^2 + Y\epsilon + Z}{A\epsilon^2 + B\epsilon + C}$$

with:

$$A = \frac{\sigma_a + \sigma_b}{4\pi} \qquad B = \frac{1}{4\pi} 2q\sigma_a \qquad C = \frac{1}{4\pi} (\sigma_a q^2 + \sigma_b)$$

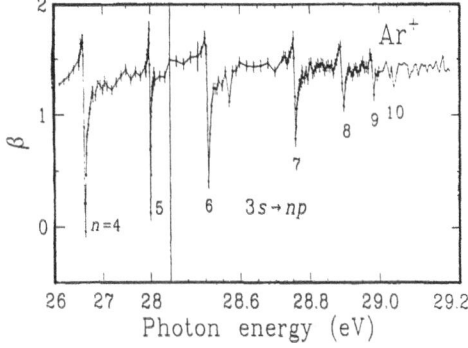

Fig. 5. Angular-distribution anisotropy-parameter ß measurements of the Ar $3s^23p^6 \rightarrow 3s3p^6$ np autoionization resonances.

with X, Y, and Z considered as free parameters in the fit to the data.

Results and Discussion

The variation of ß as a function of photon energy is shown in Fig. 5. This spectrum was obtained using two simultaneously recorded electron-yield signals at $\Theta = 0°$ and at the magic angle, $\Theta = 54.7°$. We have fit the above expressions to both the angular distribution and to total photoionization cross section data. The angular-distribution data compared with the fit based on the parameterization[40] gave a very good accord.

The angular-distribution data was also compared to the untested theory of Taylor[26] based on the R-matrix method. Taylor[26] used in the first approximation a single-configuration (SC) wavefunction to represent the $^2P^0$ and $^2S^e$ ionic states of the direct and indirect photoionization paths. He also made a more sophisticated approximation[46] (CI) where multiconfigurational wavefunctions are used in the representation of these states. Both calculations were done in the length and velocity form. The CI calculations improved the agreement between the length and velocity results for the only two calculated resonances $n=4$ and $n=5$ by Taylor[26]; the length and velocity results were coincident. Comparison of the $n=4,5$ resonances with the CI calculation of Taylor[26] is shown in Fig. 6 (a),(b). Figure 6 © shows the comparison of the $n=7$ resonance with the R-matrix calculation of Gorczyca.[47] In all cases, the dotted lines are deconvoluted fits (6 meV). As can be seen, agreement between the data and both theories is excellent. The parameters X, Y, and Z produced by our fit were compared for $n=4$ and $n=5$ with the results from Taylor.[26] The CI calculation was indeed a better model for the data because, in the case of $n=4$, the SC calculation did not even predict the correct sign of the Y parameter. Excellent agreement between the data and the CI calculations was found.

We also fit the total photoionization cross-section data using the above expressions and extracted the fitting parameters, the width Γ, the line profile q and the correlation coefficient ρ^2, for the photoionization cross section in the region of the first six resonances. Our

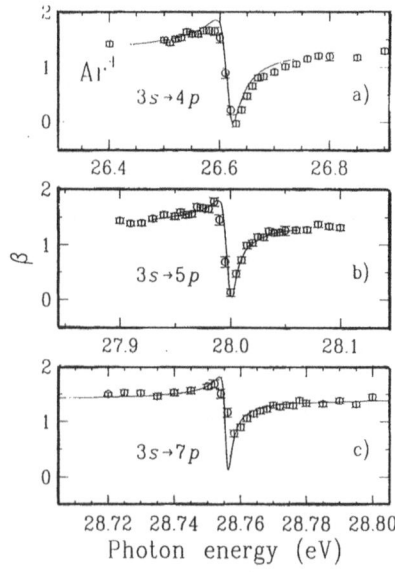

Fig. 6. Angular distribution of the Ar $3s^2 3p^6 \rightarrow 3s 3p^6$ np autoionization resonances compared to the calculation of Taylor[3] for $n=4$ and $n=5$ and to the calculations of Gorczyca[24] for $n=7$.

results were in good agreement with previous quantitative measurements[32,33] and calculations by Burke and Taylor.[46] We also found that there is excellent agreement with Burke and Taylor's[46] CI theory. In particular, in the case of n=4, one observes that the CI theory agrees very well, while the SC calculation and multiconfiguration Dirac-Fock (MCDF) calculation Tulkki's[33] theory agree with each other but not with the data.

ANGLE RESOLVED RESONANT RAMAN SPECTROSCOPY

In order to understand the importance of electron correlation effects in detail, it is important to experimentally resolve the individual fine structure transition. Such partial cross sections offer a more stringent test of theoretical approximation methods than do the unresolved cross section. There has been extensive theoretical efforts[48] to disentangle geometrical effects from dynamical effects.

The sensitivity of the branching ratios of resonant Auger transitions to electron correlation is largely hidden if data have to be averaged over the various states belonging to a recorded peak with non-resolved transitions. The resonant Auger Raman spectroscopy provides a unique method to eliminate the lifetime broadening in resonant Auger spectra. This method can be therefore used to resolve experimentally the fine structure of particular transitions. By comparing the observed intensity distribution, one deduces the strength of the electron correlation effects since multiconfiguration Dirac-Fock predictions[49] are available in some cases.

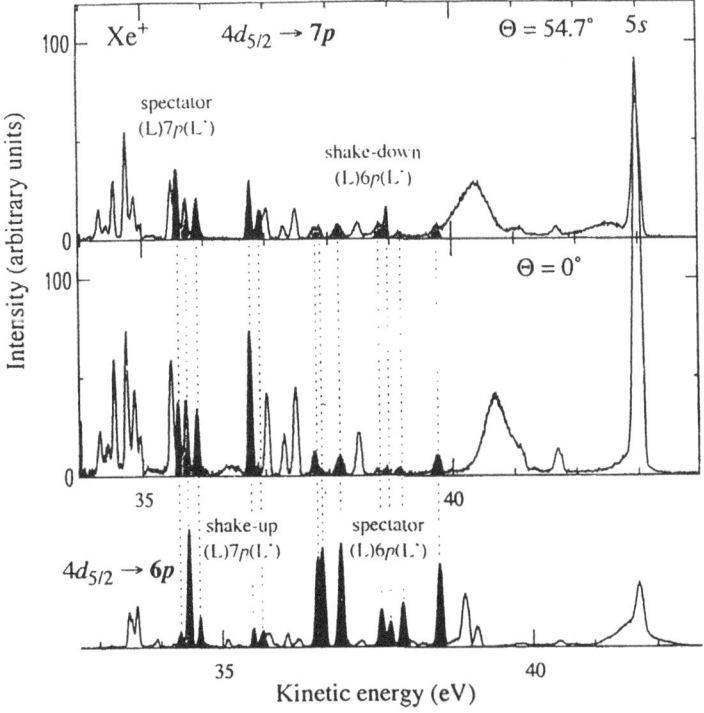

Fig. 7. Photoelectron spectrum of the Xe $4d^{-1}_{5/2} \rightarrow 5p^{-2}7p$ and $4d^{-1}_{5/2} \rightarrow 5p^{-2}$ 6p resonant Auger transitions using the resonant Raman effect.

Resonant Raman Auger effects have been pioneered by Craseman et al.[50] in the x-ray regime in the early 1980's. They found that in the vicinity of core-level energy thresholds, atomic photoexcitation followed by x-ray or Auger emission can occur as a single second-order quantum process, called the resonant raman effect. In this case, because the intermediate state is virtual, and there is no relaxation, the width of the x-ray or Auger line reflects only that of *the incident radiation and the spectrometer resolution*, but **not that of the natural line width.** Therefore, the width of the line can be *narrower* than the *natural lifetime* of the initial hole state. This is very different from the normal Auger process. By eliminating the lifetime broadening, the Auger resonant Raman spectroscopy offers a unique experimental method of determining the intensity distribution of the spectator Auger transitions with a far greater accuracy that has been previously possible. Very recently, this method[51] has been used in the uv domain without measuring the angular distribution. Figure 7 shows one of our very recent preliminary spectra obatained at the ALS using the same apparatus as in section III. This data measured at the magic angle and at 0^0 with respect to the electric field, shows that the angular distribution varies greatly. One can also observes the probabitlity for shake-up and shake-down processes. Analysis of the data is underway.

CONCLUSIONS

We have reported on detailed angle-resolved high-resolution measurements using a third generation synchrotron radiation source, the Advanced Light Source This was achieved by using a photon resolution better than 10,000 from an undulator beamline in conjunction with a new apparatus that allowed the observation of new resonances and dynamic effects. These measurements have already tested the most recent, refined theories and have for goal to keep pushing the theoretical and experimental limits for a better understanding of electron correlation processes.

* Permanent address: Fritz Haber Institut der Max Planck Gesellshaft, Berlin, Germany.

ACKNOWLEDGMENTS

We would like to acknowledge U. Becker, J. Bozek, O. Hemmer, J. Viefhaus, R. Wehlitz, S. Whitfield and T. Gorczyca. We are indebted to the DOE support of the ALS and the ALS staff.

This work was supported by the U.S. DOE, Office of Basic Energy Science, Division of Chemical Science under contract No. DE-FG02-92ER14299.

REFERENCES

1. G.H. Wannier, *Phys. Rev.* 90:817 (1953).
2. A.R.P. Rau, *Phys. Rev. A* 4:207 (1971).
3. H. Klar and W. Schlecht, *J. Phys. B* 9:1699 (1976).
4. C.H. Greene and A.R.P. Rau, *Phys. Rev. Lett.* 48:533 (1982) and J. *Phys. B* 16:99 (1983).
5. H. Kossmann, V. Schmidt, and T. Andersen, *Phys. Rev. Lett.* 60:1266 (1988).
6. P. Lablanquie, K. Ito, P. Morin, I. Nenner, and J.H.D. Eland, *Z. Phys. D* 16:77 (1990).
7. R.I. Hall, A.G. McConkey, L. Avaldi, K. Ellis, M.A. MacDonald, G. Dawber, and G.C. King, *J.Phys. B: At. Mol. Opt. Phys.* 25: 1195 (1992).

8. G. Dawber, R.I. Hall, A.G. McConkey, M.A. MacDonald, and G.C. King, *J. Phys. B: At. Mol. Opt. Phys.* 27:L341 (1994).

9. G. Dawber, L. Avaldi, A.G. McConkey, H. Rojas, M.A. MacDonald, and G.C. King, *J. Phys. B: At. Mol. Opt. Phys.* 28:L271 (1995).

10. A. Huetz, P. Selles, D. Waymel, and J. Mazeau, *J. Phys. B: At. Mol. Opt. Phys.* 24:1917 (1991).

11. A. Kazansky and V.N. Ostrovski, *J. Phys. B: At. Mol. Opt. Phys.* 27:447 (1994).

12. F. Maulbetsch and J.S. Briggs, *Phys. Rev. Lett.* 68:2004 (1994).

13. P. Lablanquie, J. Mazeau, L. Andric, P. Selles, and A. Huetz, *Phys. Rev. Lett.* 74:2192 (1995).

14. F. Heiser, U. Hergenhahn, J. Viefhaus, K. Wieliczek and U. Becker, *J. Electron Spectrosc. Relat. Phenom.* 60: 337 (1992).

15. R.I. Hall, L. Avaldi, G. Dawber, M. Zubek, K. Ellis, and G.C. King, *J. Phys. B: At. Mol. Opt. Phys.* 24: 115 (1991).

16. T.A. Carlson, *Phys. Rev. A* 156:142 (1967).

17. M.O. Krause, T.A. Carlson, and R.D. Dismukes, *Phys. Rev.* 170:37 (1968).

18. N. Berrah, B. Langer, J. Viefhaus, S.B. Whitfield, F. Heiser, and U. Becker, "XVIII International Conference on the Physics of Electronics and Atomic Collisions (ICPEAC)," p.18, July 1993.

19. U. Becker, R. Holzel, H.G. Kerkhoff, B. Langer, D. Szostak, and R. Wehlitz, *Phys. Rev. Lett.* 56:1120 (1986).

20. M.O. Krause, et al. *J. Electron. Spectros. Relat. Phenom.* 58:79 (1992).

21. J.E. Hansen and W. Persson, *Phys. Scr.* 36:602 (1987) and references therein.

22. J. Tulkki, et al. *Phys. Rev. A* 45:4640 (1992).

23. N. Berrah, B. Langer, A. Farhat, R. Wehlitz, J. Viefhaus, S.B. Whitfield, and U. Becker (to be submitted to Phys. Rev. A).

24. R.P. Madden and K. Codling, *Phys. Rev. Lett.* 516 (1963).

25. J.W. Cooper, U. Fano, and F. Prats, *Phys. Rev. Lett.* 10:518 (1963).

26. K.T. Taylor, *J. Phys. B: At. Mol. Opt.* 10:L699 (1977).

27. H. Feshbach, *Ann. Phys.* 19:287 (1962).

28. V.V. Balashov, N.M. Kabachnik, I.P. Sazhina, "Proc. 5th Sov. Conf. on Physics of Electronic and Atomic Collisions," Uzhgorod, Nauka, p.118.

29. M.Ya. Amusia and A.S. Kheifets, *Phys. Lett.* 82A:407 (1981).

30. A.F. Starace, *Phys. Rev. A* 16:231 (1977).

31. H. Beutler, *Z. Phys.* 93:177 (1935).

32. R. P. Madden, D. L. Ederer, and K. Codling, *Phys. Rev.* 177:136 (1969); M.A. Baig and M. Ohno, *Z. Phys. D* 3:369 (1986).

33. S.L. Sorensen, T. Aberg, J. Tulkki, E. Rachlew-Källne, G. Sundström, and M. Kirm, *Phys. Rev. A* 50:1218 (1994).

34. A. Svensson, M.O. Krause, and T.A. Carlson, J. Phys. B: At. Mol. Phys. 20:L271 (1987); K. Codling, et al. *J. Phys. B: At. Mol. Phys.* 13:L693 (1980).

35. H.P. Kelly, in X-ray and Inner-Shell Processes, "AIP Conference Proceedings No. 215," T.A. Carlson, M.O. Krause, and S.T. Manson, eds., AIP, New York (1990), p. 292.

36. M. Seaton, *Rep. Prog. Phys.* 46:167 (1983).

37. C.H. Greene and L. Kim, *Phys. Rev. A* 38:5953 (1988).

38. J.P. Connerade and A.M. Lane, *Rep. Prog. Phys.* 51:1439 (1988).

39. M.Ya. Amusia, "Atomic Photoeffect," Plenum, New York (1990).

40. N.M. Kabachnik and I.P. Sazhina, *J. Phys. B: At. Mol. Opt. At.* 9:1681 (1976).

41. U. Fano, *Phys. Rev.* 124:1866 (1961).

42. P. Heimann (private communication).

43. R.Z. Bachrach, F.C. Brown, and S.B. Hagström, *J. Vac. Sci. Technol.* 12:309 (1975); M.G. White, R.A. Rosenberg, G. Gabor, E.D. Poliakoff, G. Thornton, S.H. Southworth, and D.A. Shirley, *Rev. Sci. Instrum.* 45:494 (1974); U. Becker, R. Hölzel, H.G. Kerkoff, B. Langer, D. Szostak, and R. Wehlitz, Phys. Rev. Lett. 56:1120 (1986).

44. B. Langer, N. Berrah, O. Hemmers, and J. Bozek (to be submitted).

45. T. Manson and A. Starace, *Rev. Mod. Phys.* 54:389 (1982).

46. P.G. Burke and K.T. Taylor, *J. Phys. B: Atom. Molec. Phys.* 8:2620 (1975).

47. T. Gorczyca (private communication).

48. A.R P. Rau and U. Fano, *Phys. Rev. A* 4:1751 (1971); A.R.P. Rau, in "Electron and Photon Interaction with Atoms," H. Kleinpoppen and M.R.C. McDowell, eds., Plenum, New York (1976), pp.141-148.

49. J. Tulki, H. Aksela, and N.M. Kabachnik, *Phys. Rev. A* 50:2366 (1994).

50. G.S. Brown, M.H. Chen, B. Crasemann, and G. Ice, *Phys. Rev. Lett.* 45:1937 (1980).

51. H. Aksela, et al. *Phys. Rev. A* 49:R4269 (1994).

RECENT ADVANCES TOWARD A STRUCTURAL MODEL FOR THE PHOTOSYNTHETIC OXYGEN-EVOLVING MANGANESE CLUSTER

Matthew J. Latimer,[1,2] Holger Dau[3], Wenchuan Liang,[1,2] Joy C. Andrews,[1,2] Theo A. Roelofs,[1] Roehl M. Cinco,[1,2] Annette Rompel,[1,2] Kenneth Sauer,[1,2] Vittal K. Yachandra,[1] and Melvin P. Klein.[1]

[1]Structural Biology Division, Lawrence Berkeley National Laboratory and [2]Department of Chemistry, University of California, Berkeley, CA 94720 and [3]FB Biologie/Botanik, Phillips-Universität Marburg, D-35032, Germany.

INTRODUCTION

Photosynthetic water oxidation occurs in the oxygen evolving complex (OEC) of photosystem II (PS II). One-electron photo-oxidations in the reaction center of PS II are coupled to the four-electron oxidation of water in the OEC. As the PS II reaction center sequentially extracts electrons, the OEC cycles through five intermediate oxidation states (S_0-S_4) where S_0 is the least oxidized and S_4 is a transient state which decays to S_0 with the release of a dioxygen molecule. A complex of four manganese atoms has been shown to function in charge accumulation and is thought to form the catalytic site for the water oxidation reaction. The structure of this manganese complex and the mechanism of water oxidation have been the subject of vigorous inquiry by a great many research groups (Sauer et al., 1992).

We have proposed a model (Yachandra et al., 1993) for the Mn complex based largely upon extended X-ray absorption fine structure (EXAFS) spectroscopy of Mn in PS II preparations. The core feature of this model is two di-μ-oxo bridged Mn pairs yielding two Mn-Mn interactions of ~2.7 Å. The Mn pairs are connected by a single bridging oxygen which results in another Mn-Mn distance of ~3.3 Å. These distances and structural interpretations arose from a comparison of the Mn-EXAFS fitting results of PS II compared to known structural motifs from a large number of inorganic Mn complexes (DeRose et al., 1994). Experiments on partially oriented preparations indicated that the scattering vectors giving rise to the ~2.7Å interaction are oriented close to the plane of the membrane (averaging 60°±7° from the membrane normal), while the vectors contributing to the ~3.3Å interaction are aligned closer to the membrane normal (43°±10°) (Mukerji et al., 1994). This angular information was included in the model to indicate the position of the complex within the membrane. A closely bound Ca was added to the model, based on EXAFS from Sr-substituted preparations, and a directly-ligated chloride was also included, based on a large body of indirect evidence for a chloride binding site in PS II. Finally, a histidine ligand to Mn was included based on electron spin-echo envelope modulation experiments.

In this paper we describe a revised structural model for the manganese cluster in PS II based on the results of recent studies on PS II preparations subjected to various treatments

Synchrotron Radiation Techniques in Industrial, Chemical, and Materials Science
Edited by D'Amico *et al.*, Plenum Press, New York, 1996

141

(Sr-substitution, ammonia or fluoride inhibition, or low temperature illumination to produce the S_2-state exhibiting the g=4.1 EPR signal). The information gained from these studies has helped confirm and extend our structural model. We also present the results of Mn X-ray absorption K-edge spectra from PS II samples advanced through the S-states with saturating flashes of light.

RESULTS AND DISCUSSION

The Mn-Mn Interaction at ~2.7Å

The second Fourier peak in the EXAFS spectrum of PS II is generally agreed to arise from 2-to-3 di-μ-oxo bridged Mn pairs. Although there is certainly more than one Mn-Mn interaction, spectra from unmodified PS II preparations in the S_1-state, and in the S_2-state where the multiline EPR signal is exhibited, show only a single resolved distance (DeRose et al., 1994). In recent work we have shown that this distance degeneracy is lifted in the S_2-state where the g=4.1 EPR signal is exhibited (Liang et al., 1994) and in preparations which have been inhibited with ammonia (Dau et al., 1995) or with fluoride (DeRose et al., 1995). The Fourier transforms of the EXAFS from these preparations are shown in Figure 1. In each of these preparations, the second Fourier peak (peak II) is observed to have a smaller amplitude than in control samples. Also, in each case, curve-fitting results have indicated increased disorder and that the data can be fit with two Mn-Mn interactions, one at approximately 2.69-2.73 Å and a second interaction at ~2.78-2.87 Å.

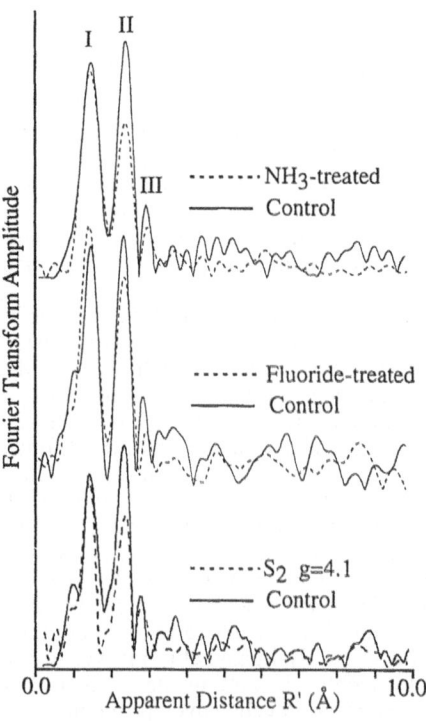

Figure 1. Fourier transforms of the EXAFS from ammonia or fluoride treated PS II samples in the S_2-state, and from an untreated sample illuminated at 140K to generate the S_2-state exhibiting the g=4.1 EPR signal. For comparison each trace is plotted with a control sample in the S_2-state exhibiting the multiline EPR signal.

A variety of models can be employed to explain the increased distance heterogeneity including direct binding of ammonia to the Mn complex, ligand exchange reactions which are affected by temperature, protonation of oxo-bridges, or changes in oxidation state of the individual dimers dependent upon ligand environment. Fundamentally, however, the resolution of two different ~2.7Å distances (one relatively unchanged and the other lengthened from that found in untreated/unilluminated preparations) implies two relatively independent Mn-Mn dimeric units.

EXAFS spectra from partially oriented ammonia-treated membranes has yielded further information about the relative angles of the individual manganese vectors (Dau et al., 1995). EXAFS curve fitting revealed Mn-Mn interactions at 2.72 and 2.87Å, and changes with angle of the apparent coordination number of each vector allowed determination of angles of 55° (2.72Å interaction) and 67° (2.87Å interaction) relative to the membrane normal. Angles determined for the ammonia-treated complex are likely to be similar to the angles found in the untreated complex. In Figure 2 we present the individual EXAFS partial waves determined from ammonia treated preparations, along with polar plots which show the dependence of coordination number on angle.

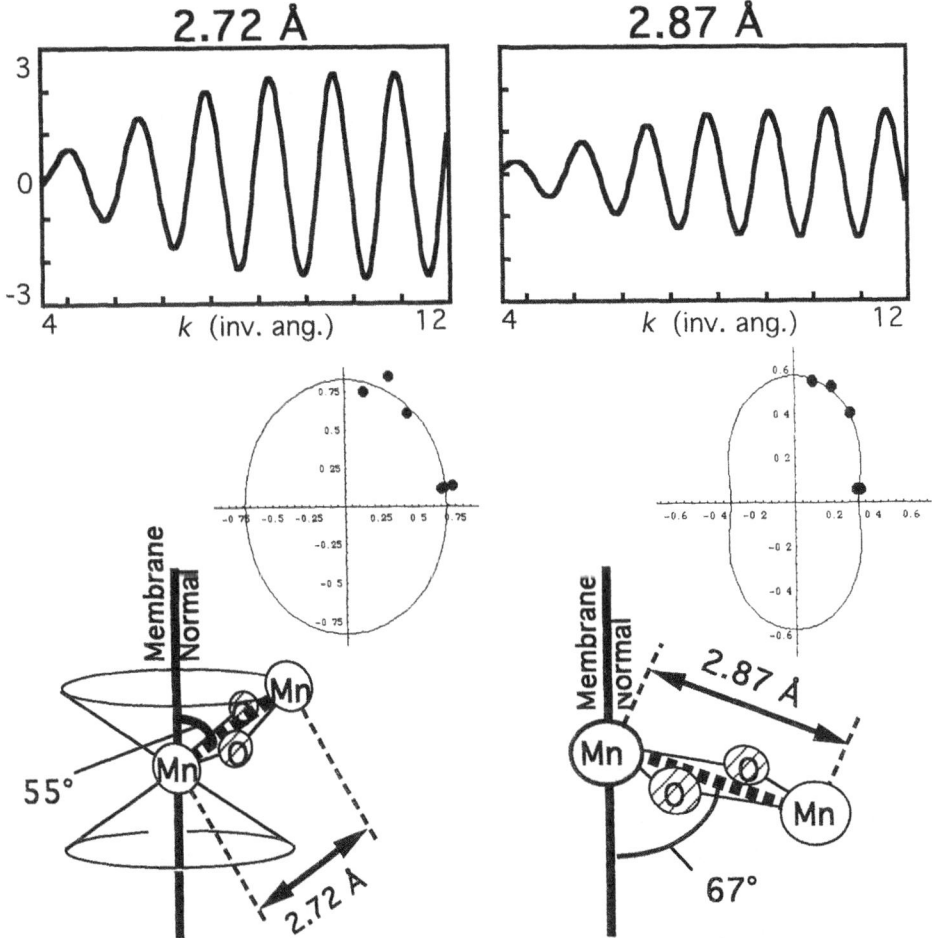

Figure 2. EXAFS of oriented ammonia-treated annealed S_2 reveals significant heterogeneity of the two Mn-Mn di-µ-oxo bridged moieties. Polar plots of the apparent coordination number vs. angle θ, indicate vectors with angles of 55° (2.72Å) and 67° (2.87Å) with respect to the membrane normal.

143

Ca/Sr Experiments

Ca is an essential co-factor for oxygen evolution and a great deal of effort in recent years has centered on elucidating its role in PS II. We have shown, through EXAFS experiments on PS II preparations which have had Sr substituted for Ca that the amplitude of the third Fourier peak is enhanced relative to Ca-containing preparations (Fig. 3) (Yachandra et al., 1993; Latimer et al., 1995). Detailed fitting of the EXAFS data from these preparations (Latimer et al., 1995) has indicated that this change in amplitude can be explained by an exchange of strontium for calcium in a binding site ~3.5Å from a manganese atom (see Figure 4). By comparison with compounds of known structure which contain calcium-metal distances, we proposed that calcium and manganese may be bridged by one or more single oxygen atoms derived from carboxylate or phenolate protein residues, protein backbone carbonyls, or possibly water or hydroxide. Although the data can be explained with just calcium contributing to peak III, a careful consideration of chemically reasonable structural models indicates that the most likely model is that both Mn-Mn and Mn-Ca interactions contribute to peak III.

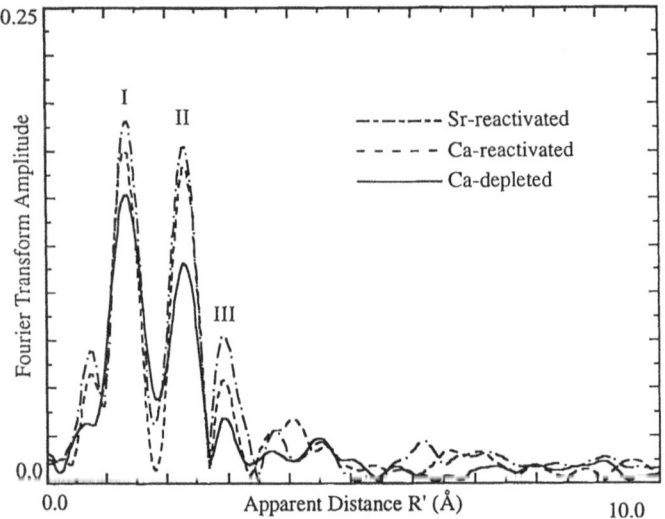

Figure 3. Fourier transforms of EXAFS spectra of Ca-depleted preparations in the S_1-state which have been reactivated with Sr, Ca, or with no reactivation. The major difference between Ca- and Sr-reactivated preparations is seen in the amplitude of peak III. In the Ca-depleted spectrum, changes are observed in amplitude of all three peaks relative to the samples containing Ca and Sr. Calcium depletion of PS II preparations was achieved by use of a low pH/citrate method (Ono and Inoue, 1988). Calcium and strontium reactivation procedures and sample characterization are described in Latimer et al. (1995) and Latimer (1995).

Spectra from Ca-depleted preparations appear to be further modified in the amplitude of all three Fourier peaks. Peak III is definitely decreased in these preparations and is more affected by the noise level of the data. There also appears to be increased heterogeneity in peak II. These structural modifications produced by the calcium depletion procedure, however, may arise from possible effects of the chelators used in the process, as well as removal of bound calcium. Thus, the relevance of the structure of the Ca-depleted preparations to that in untreated preparations is unclear.

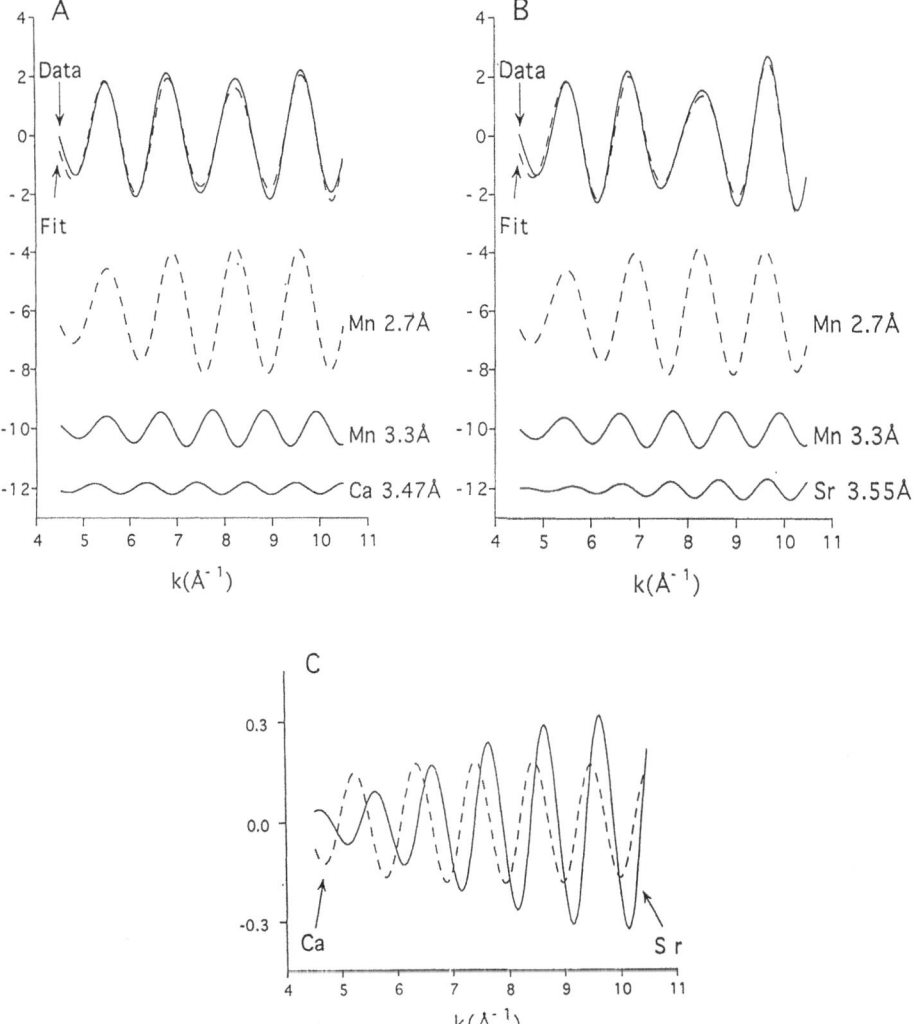

Figure 4. Fits to the EXAFS of Ca- (A) and Sr-reactivated PS II (B). Both fits are to Fourier isolates of peaks II and III. Fit parameters are shown in Table 1. The partial waves from individual scattering contributions are shown as well as the overall fit (top). The individual Ca and Sr partial waves used in these fits are also shown overplotted on an expanded scale (C) to demonstrate the clear differences in phase and amplitude for these two contributions.

TABLE 1 Restricted fit values for Ca and Sr-reactivated PS II samples.
Fit to peaks II and III using FEFF 5 parameters.

Atom	R (Å)	N	σ^2 (Å^2)	ΔE_0 (eV)
Ca-PS II				
Mn	2.74	1.0	0.002	-11
Mn	3.29	0.5	0.003	
Ca	3.47	0.25	0.003	
Sr-PS II				
Mn	2.73	1.0	0.002	-14
Mn	3.28	0.5	0.003	
Sr	3.55	0.25	0.003	

Coordination numbers (N) were fixed.
Debye-Waller factors (σ^2) were fixed.
Distances and ΔE_0 values were floated.

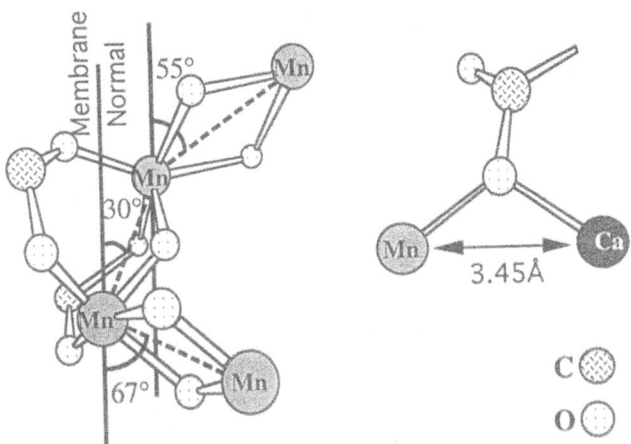

Figure 5. Revised structural model for the Mn cluster of PS II. One possible positioning of the cluster within the membrane is shown (left). Also shown is a proposal for a calcium binding site (right), although which Mn atom of the complex is involved in this site is undetermined at this time.

The Revised Model

Our revised model for the Mn cluster in PS II is presented in Figure 5 (Dau et al., 1995). The histidine and chloride ligands are not shown, although at least one histidine ligand has been proven and a chloride ligand is very likely. The model we propose here for the manganese cluster in PS II is a combination of structural elements and the exact positions of these elements relative to each other is unknown. In addition, we note that there may be significant structural changes that occur on the S_2-to-S_3 transition, thus the structural model presented here is applicable to the S_1 and S_2 states.

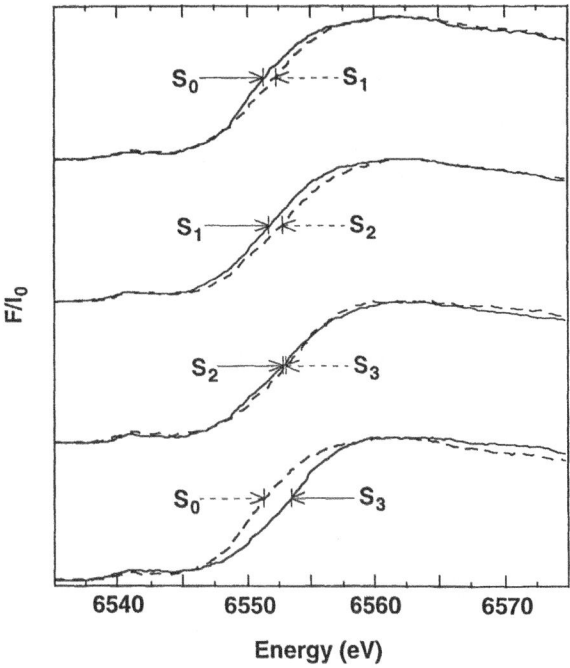

Figure 6 Mn K-edge spectra of PS II samples in the S_0, S_1, S_2, and S_3 states.

The Oxidation States of Mn in the Various S-states

Oxidation state changes of the Mn complex in the S-states have been studied by Mn X-ray K-edge spectroscopy. PS II samples were prepared by flash illumination and the S-state composition was characterized by measuring the oscillation in the amplitude of the multiline EPR signal. Mn K-edge spectra of the flash-induced S-states are shown in Figure 6. Increases in the energy of inflection point (measured as the zero crossing of the second derivative) are observed for the S_0 to S_1 and S_1 to S_2 transition, indicating the oxidation of Mn during these S-state advances. However, the small increase in the inflection point energy for the S_2 to S_3 transition and the similarity of the shape of the second derivative (not shown) in the S_2 and S_3 states indicates that Mn oxidation probably does not occur during the S_2 to S_3 transition.

Acknowledgments

This research was supported by the Director, Office of Basic Energy Sciences, Division of Energy Biosciences of the U.S. Department of Energy, under Contract No. DE-AC03-76SF00098, and by a grant from the National Science Foundation (DMB91-04104).

REFERENCES

Dau, H., Andrews, J. C., Roelofs, T. A., Latimer, M. J., Liang, W., Yachandra, V. K., Sauer, K., and Klein, M. P., 1995, Biochemistry 34, 5274-5287

DeRose, V. J., Latimer, M. J., Zimmermann, J.-L., Mukerji, I., Yachandra, V. K., Sauer, K., and Klein, M. P., 1995, Chem. Phys. 194, 443-459

DeRose, V. J., Mukerji, I., Latimer, M. J., Yachandra, V. K., Sauer, K., and Klein, M. P., 1994, J. Am. Chem. Soc. 116, 5239-5249

Latimer, M. J., 1995, Ph.D. Thesis, University of California, Berkeley, CA

Latimer, M. J., DeRose, V. J., Mukerji, I., Yachandra, V. K., Sauer, K., and Klein, M. P., 1995, Biochemistry in press.

Liang, W., Latimer, M. J., Dau, H., Roelofs, T. A., Yachandra, V. K., Sauer, K., and Klein, M. P., 1995, Biochemistry 34, 5274-5287

Mukerji, I., Andrews, J. C., DeRose, V. J., Latimer, M. J., Yachandra, V. K., Sauer, K., and Klein, M. P., 1994, Biochemistry 33, 9712-9721.

Ono, T. and Inoue, Y., 1988, FEBS Lett. 227, 147-152

Sauer, K., Yachandra, V. K., Britt, R. D., and Klein, M. P., 1992, in Manganese Redox Enzymes (Pecoraro, V. L., ed.) pp. 141-175, VCH Publishers, New York

Yachandra, V. K., DeRose, V. J., Latimer, M. J., Mukerji, I., Sauer, K., and Klein, M. P., 1993, Science 260. 675-679

CESIUM XAFS STUDIES OF SOLUTION PHASE CS-IONOPHORE COMPLEXATION

K. M. Kemner[1]*, D.B. Hunter[2], W. T. Elam[1], and P.M. Bertsch[2]

[1]US Naval Research Laboratory, Code 6685, 4555 Overlook Ave., Washington, D.C., 20375
[2]Division of Biogeochemistry, Savannah River Ecology Laboratory, University of Georgia, P.O. Drawer E, Aiken SC 29803

ABSTRACT

X-ray absorption fine structure (XAFS) measurements have been made at the Cs LIII absorption edge on 1) 0.04 M concentrations of CsBr in 0.04 M Dibenzo-18-Crown-6 ethers (D18C6) in acetonitrile solution, and 2) crystalline CsBr, CsCl and CsF powders that were used as standards. Due to the many difficulties associated with obtaining high quality XAFS data on these systems, a custom-manufactured 5 mil thick Sc foil was used in conjunction with a Soller slit assembly to improve the XAFS signal to noise ratio by almost a factor of 6. XAFS analyses of the Cs-D18C6 solution show the presence of a 1:1 Cs-D18C6 complex with the Br counter-ion still in contact with the Cs atom. These results are consistent with previous wet chemical studies. The choice of a heavy backscattering Br anion reduces the error in determining the presence of a single backscattering counter ion among the lighter backscattering oxygen and carbon atoms of the crown ether complex. The ability of XAFS to directly probe the anion contact pairing to Cs complexed to a macrocycle opens many exciting avenues for improving extraction methodologies.

Introduction

There is an increasing awareness for the need to understand the speciation and interactions of elements in many fields of research, particularly, the development of separations procedures for contaminant elements through the design of reagents capable of highly specific interactions with those elements. [137]Cesium is a common radio-contaminant in nuclear waste. An understanding of its structural binding properties with ligands is important in designing highly specific chelates in waste treatment processes. Since X-ray Absorption Fine Structure (XAFS) spectroscopy provides average local environment information about specific atomic species, it is potentially well suited to the study of these highly specific interactions and is capable of very detailed molecular structure determinations in chemical systems.

In the field of developing macrocyclic compounds with high binding efficiencies for specific alkali cations, macrocyclic polyether (crown ether) compounds have received much study and several excellent review articles have been written[1]. The size of the intramolecular hole in the crown ether can discriminate in binding affinity to different cations. This high binding efficiency or macrocyclic effect is also dependent on the physical properties of the cation being investigated (concentrations, solvation, number of donor atoms, size etc.)[2,3]. Previous studies on the complexation of [133]Cs+ with 18-crown-6 ether by potentiometric, calorimetric and [133]Cs NMR techniques, provided a useful data base for an XAFS analysis of solution complexation of Cs in 18-crown-6 ether[4,5,6,7]. They also represent an

Synchrotron Radiation Techniques in Industrial, Chemical, and Materials Science
Edited by D'Amico *et al.*, Plenum Press, New York, 1996

149

experimental system suitable for XAFS structural studies by supposedly presenting well defined first and second shell environments about the bound Cs.

As one of the heaviest of the alkali metals, Cs possesses some unique and subtle differences in chemical behavior over the lighter alkali metals. The large atomic radius coupled with its low charge to surface ratio makes the Cs^+ ion poorly solvated with a very low solvation energy. This low solvation energy results in an effectively smaller solvated radius than Li^+, Na^+, K^+, or Rb^+ thus permitting accessibilty of Cs^+ to size restrictive binding sites [8]. Conversely, in solvents less polar than water, Cs^+ is more liable to form contact ion pairs with the corresponding anion than the lighter alkali metals as evidenced by extremely large chemical shifts in ^{133}Cs nuclear magnetic resonance spectra [9]. Although introduction of the 18-crown-6 ether macrocycle into Cs selective exchange resins will provide methods to remove 137 Cs from aqueous solutions of high level nuclear wastes, we have chosen the fundamental D18C6 unit which has been well characterized in solvents (such as acetonitrile, in which it is highly soluble).

There are, however, many difficulties associated with obtaining high quality XAFS data on these systems. Since these experiments are performed in the solution phase with very dilute Cs concentrations, there is a very large background signal due to the inelastic scattering from the acetonitrile solution. XAFS measurements at the Cs LIII absorption edge also greatly hinder the use of the transmission mode since the energy of the incoming X-ray beam (5 KeV) cannot penetrate through the sample. Therefore, measurements of the Auger electron or fluorescent X-ray signals are required to correctly perform these experiments. Because of the many difficulties associated with performing an electron yield measurement on a liquid, the fluorescent technique was chosen. Another problem, due to performing these experiments in the liquid phase, is that there are very large XAFS Debye-Waller factors coupled with the typical, quickly-dampened, k-dependent backscattering amplitude of oxygen and carbon shells. Secondly, the branching ratio of Auger electron to fluorescent X-ray emission strongly favors Auger electron production near the Cs LIII absorption edge energy[10]. Thirdly, the radial distances of the oxygen and carbon backscattering shells are such that their backscattering frequencies destructively interfere with each other throughout a large portion of the XAFS data. Finally, the range of useful data is limited to only 350 eV above the LIII absorption edge because of the presence of the Cs LII absorption edge.

Even though all these difficulties exists, the ability of XAFS to accurately elucidate the structural environment of the crown ether about a Cs ion will permit an extension of this form of analysis to many other systems in which the Cs is selectively bound within a cage of electronegative oxygen donors. Such examples would include humic and other organic substances, colloidal clay minerals and ion selective exchange resins used to treat nuclear wastes.

Previous XAFS studies of ions complexed to crown ethers have been performed[11,12,13], to the authors' knowledge, however, this is the first Cs XAFS study undertaken in solution phase. XAFS analyses of CsBr dissolved in acetonitrile solutions of dibenzo-18-crown-6 (D18C6) ether show the presence of a 1:1 Cs-dibenzo-18-crown-6 ether complex with the Br counter-ion still in contact with the Cs atom. The choice of a heavy backscattering Br anion greatly reduces the error in determining the presence of a single backscattering counter ion from the signal arising from 6 oxygen atoms and 12 carbon atoms of the D18C6 ether chelator. Potential errors in other XAFS analyses arising from inattention to the counter ion will be discussed.

EXPERIMENTAL

A dibenzo-18-crown-6 ether complex was prepared by dissolving 0.04 M CsBr in 0.04 M dibenzo-18-crown-6 (D18C6) ether in acetonitrile. Previous nuclear magnetic resonance (^{133}Cs NMR) data have demonstrated that a Cs:Crown ether ratio of 1:1 exists at this molar ratio.[14] Similarly, stability constants have shown the ratio of CsBr in solution to Cs bound to crown-ether to be $1:10000$.[15] A 2 cm^3 volume of the solution was transferred into a polypropylene bag with 5 μm (0.2 mil) thick sides.

XAFS experiments at the Cs LIII edge were performed on beamline X23B[16] at the National Synchrotron Light Source with an electron beam energy of 2.54 GeV and stored currents between 130 and 250 mA. (The optical and X-ray properties of this beamline are presented in Ref. 16.) One helpful characteristic of the X23B beamline is its ability to focus the X-rays. Focusing the X-rays, as compared to masking a nonfocused beam, can sometimes increase the monochromatic X-ray beam's intensity by up to two orders of magnitude.

Measurements on the above mentioned solution were made at room temperature in the fluorescence mode utilizing a filter-slit combination.[17,18]. Transmission measurements were also made on 99.7% ultradry (<50 ppm H_2O, OH, oxide) CsF, CsCl and CsBr powders. These powders, while in an Argon-filled atmosphere, were ground with a mortar and pestle, mounted on adhesive tape, and enclosed in an airtight mylar sample cell to reduce any effects of sample hydration. Different sample thicknesses were measured to investigate sample thickness effects and uniformity. No detectable differences were noticed in the amplitude or phase of the normalized data, and the amplitude of the absorption edge scaled appropriately. Calculated spectra using FEFF3[19] were used to simulate first shell backscatterers of F, Cl and Br around Cs and checked with the experimentally obtained transmission data. The S_0^2 and V0 values[19] required for the FEFF3 code to accurately simulate Cs with F backscatterers were used as inputs to simulate oxygen and carbon backscatterers which were later used as standards to analyze the Cs-D18C6 sample. CsF was used in lieu of Cs_2O because of the difficulty in locating a well-defined single shell standard with a Cs to oxygen backscatterer. The S_0^2 and V0 values required to accurately simulate Cs with Br backscatterers were used as inputs to the FEFF3 code to simulate a single bromine backscatterer. Results of theoretical simulations of the XAFS signal due to 6 oxygen and 12 carbon backscatterers are shown in Fig. 1. This depicts the destructive interference of the two signals throughout a major portion of the usable data range.

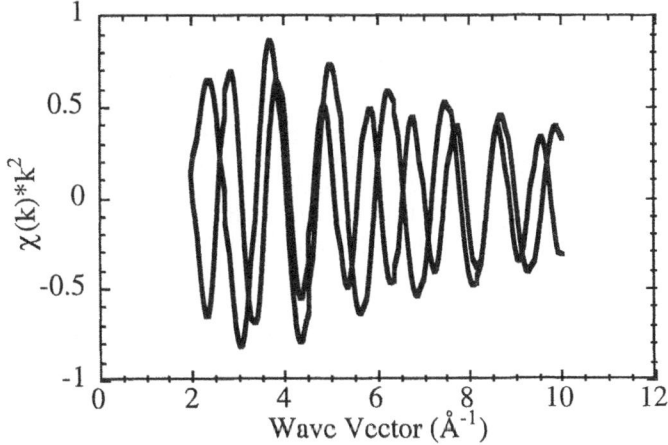

Figure 1. Theoretical simulation of 6 oxygen and 12 carbon backscatterers surrounding a Cs absorber. Note the destructive interference between the two backscattering shells.

A custom manufactured 10 μm (0.4 mil) thick Sc foil filter (Rhone-Poulenc, Phoenix Arizona) combined with Soller slits was used to reduce the large background signal due to scattered radiation from the solution. This improved the signal to background ratio by a factor of 6. The signal to background ratio was chosen to be the ratio of the absorption edge amplitude to the amplitude of the background signal at 30 eV below the absorption edge. The unnormalized raw data acquired with and without the use of the Sc filter and Soller slits are shown in Fig. 2.

The incident X-ray intensity was monitored by an ionization chamber containing 100 percent free flowing N_2. The fluorescent radiation was monitored by an ionization chamber containing 100 percent free flowing Ar. The chamber space surrounding the sample was constantly flushed with free-flowing He gas to further reduce the effects of scattering. Due to the relatively small XAFS signal from the Cs LIII edge, good linearity between the incident and fluorescent ion chambers is of the utmost importance. Linearity checks[20] between the incident signal and the fluorescent signal were performed and exhibited less than a 0.2% difference for a 50% reduction in the incident signal.

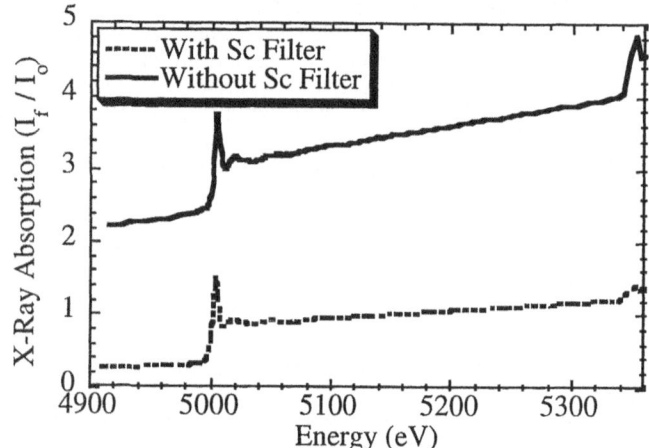

Figure 2. Unnormalized raw XAFS data acquired with and without the use of a Sc filter and soller slits.

Due to the rapid decrease in the oxygen and carbon k-dependent backscattering amplitude, the destructive interference of their backscattering shells due to their radial distance from the Cs, and the low concentration of Cs in the solution, integration time above the absorption edge was set to 30 seconds per point. Due to the small signal to noise ratio, the effect of slight changes in the background signal on the merged, successive raw XAFS scans could not be easily detected. Thus, long integration times were chosen to provide raw data with a good enough quality to enable correct background subtraction, edge step normalization, and conversion to k-space that could then be merged. The total fluorescent signal from the solution was typically 10^8 counts/second. Three successive unnormalized Cs LIII-edge XAFS data from the D18C6 sample are shown in Fig. 3.

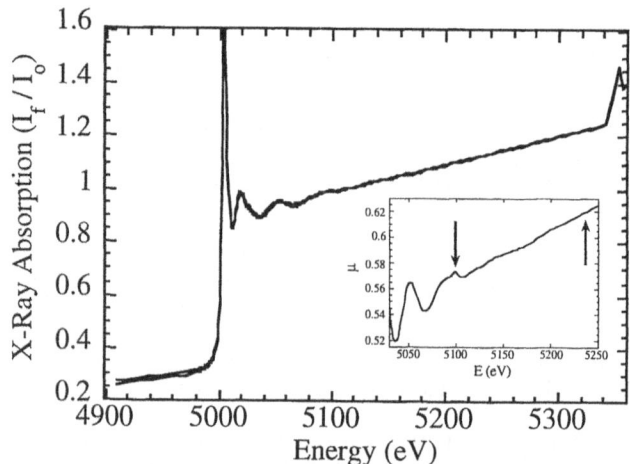

Figure 3. Three successive unnormalized Cs LIII-edge XAFS data from the CsBr/D18C6 solution taken in the Fluorescence mode. The arrows in the inset show the locations of two multi-electron excitation absorption edges.

DATA ANALYSIS AND RESULTS

The inset to Fig. 3 shows the locations of the two multi-electron excitations near 100 eV and 240 eV above the Cs LIII absorption edge. The ratio of these multi-electron excitation intensities to the Cs LIII absorption edge intensity has been calculated elsewhere to be 1.4 and approximately 0.4 percent respectively[21]. A contour of these absorption edges with their amplitudes calibrated to the Cs LIII absorption edge amplitude was subtracted from the raw data before performing subtraction.

Data were analyzed using the University of Washington/Naval Research Labs/Notre Dame analysis package which follows the Standards and Criteria for an XAFS experiment presented elsewhere.[22] The pre-edge background absorption was determined from a linear fit to the data roughly 100 - 50 eV below the edge energy and then extrapolated over the entire range of the spectrum. The LIII edge absorption signal was then isolated by subtracting the extrapolated pre-edge from the XAFS signal. The post-edge data were then fitted with a cubic-spline function. The spectra were then step-normalized by dividing by the amplitude at the absorption edge. The location of the absorption edge, E_0, was assigned to be the first inflection point after the onset of absorption. The normalized XAFS $\chi^2(k)$ data for eight successive scans were merged and the resulting data along are shown in Fig. 4. The data between 2.7 Å^{-1} and 9.1 Å^{-1} was Fourier transformed with a k^2 weighting and a Hanning window modification function of 0.25 Å^{-1} to reduce truncation ripple. This transform is shown in Fig. 5. In order to obtain quantitative information about the Cs atoms in solution, the r-space transformed data were inverse-transformed over a spectral range of 1.6 Å to 4.0 Å. In this instance, a Hanning window modification function of 0.20 Å was used to suppress truncation ripple.

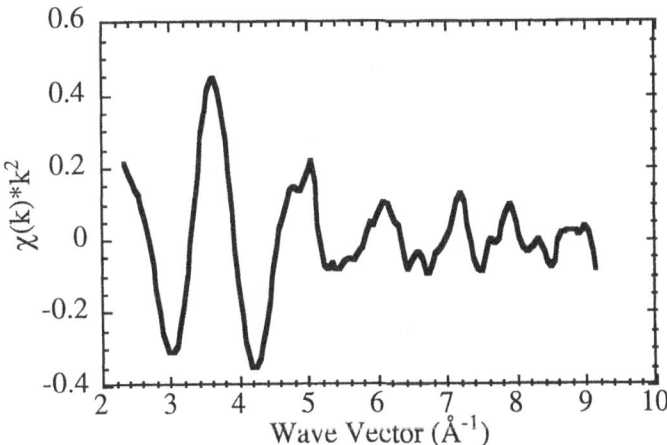

Figure 4. Normalized XAFS $\chi(k)*k^2$ data for eight successive merged scans of the CsBr/D18C6 solution.

Whenever the possibility exists that there are more than one atomic species contained in one coordination shell, nonlinear least-squares fitting with a combination of multiple backscattering shells is necessary. Many different combinations of backscattering shells were investigated during the fitting analysis of the Fourier filtered data. Table I shows the main classes of backscattering shells used during this fitting analysis.

For the transform range (2.7 Å^{-1} to 9.1 Å^{-1}) and inverse-transform range (1.6 Å to 4.0 Å) used here, the maximum number of fit parameters allowed to vary at one time is 8.[23] While fitting the XAFS data, three parameters for each of the backscattering shells were allowed to vary: coordination number (N), coordination distance (R) and mean-square disorder (σ^2). When two backscattering shells with the same atomic species were used in the fit, the sum of backscatterers in both shells was constrained to be a constant. In this way, when three distinct backscattering shells were used in the fit, the total number of independent floating parameters was 8.

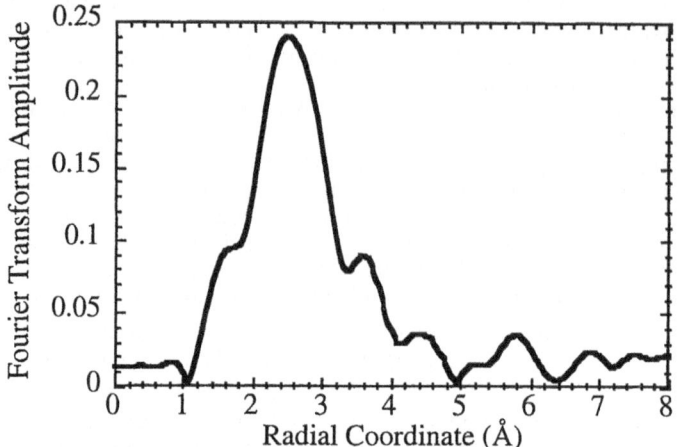

Figure 5. Result of Fourier transformation of the data in Figure 4 between 2.7 Å^{-1} and 9.1 Å^{-1} with a Hanning modification window of 0.25 Å^{-1} to reduce truncation ripple.

Table 1. Main classes of combinations of backscattering shells used during fitting analysis of CsBr/C18D6 solution with corresponding Residual value (χ^2) for each class' best fit.

Scenario #	# of oxygen atoms	# of carbon atoms	# of Bromine atoms	χ^2 value for best fit (note text)
1	1 shell of 6	1 shell of 12	0	0.22
2	1 shell of 6	2 shells with sum of 12	0	0.15
3	1 shell of 12	1 shell of 24	0	0.22
4	2 shells with sum of 6	1 shell of 12	0	0.14
5	1 shell of 6	1 shell of 12	1	0.08

The best fitting parameters are obtained by minimizing the normalized χ^2 value. This value is defined to be the difference in phase and amplitude of the filtered data of the unknown and computer-generated combination of backscattering shells. The minimum χ^2 values for each combination of coordination shells for a fitting range of 3.3-9.0 Å^{-1} are listed in Table I. Throughout the analysis, fitting solutions with Debye-Waller factors in excess of 0.2 Å^2 were considered unrealistic and disregarded.

Using the above mentioned criteria, the best fit to the Fourier filtered data from the acetonitrile solution with CsBr was found for one shell of six oxygen atoms at 3.04±0.03 Å, one shell of 12 carbon atoms at 3.96±0.05 Å and 1 bromine atom at 3.60±0.05 Å. Filtered data and the generated fit with this configuration of backscatterers are shown in Fig. 6. The χ^2 value for this combination of backscattering shells (Note Table I) was almost a factor of two smaller than the χ^2 value for any of the other 4 combinations of backscattering shells previously mentioned. For comparison, the best fit for one shell of 6 oxygen atoms and one shell of 12 carbon atoms is shown in Fig. 7.

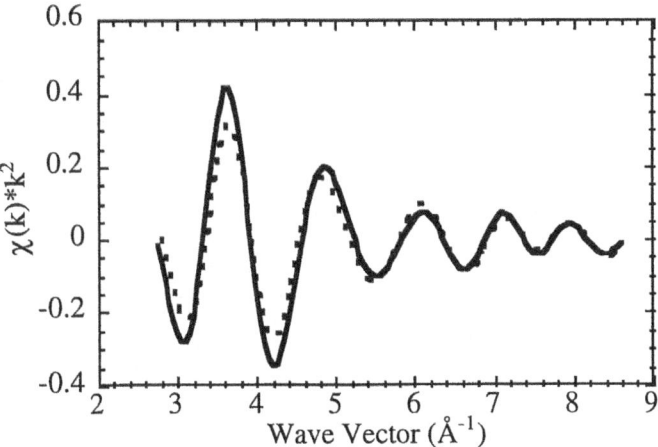

Figure 6. Best Fit (dashed line) to filtered data (solid line) using 6 oxygen atoms, 12 carbon atoms, and 1 bromine atom. (see text)

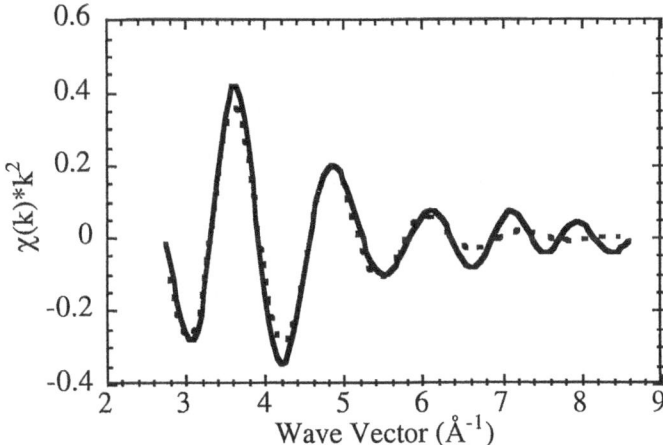

Figure 7. Best fit (dashed line) to the filtered data (solid line) without the inclusion of a bromine backscattering atom. (see text)

Errors for the radial distances represent a goodness of fit and do not include any experimental errors. Estimates from the reproducibility of normalized $\chi^2(k)$ data indicate that the experimental error is much smaller than the fitting errors and do not appreciably effect the quantitative results. The goodness of fit errors were determined by allowing the remaining variables to "float" and monitoring the χ^2 value as a function of the fixed variable. Error bar limits were set to be the value of the variable in question when χ^2 increased by 50 percent. Using this criterion, the fitting errors for scenario #5 are less than 25% for the Cs-O, Cs-C and Cs-Br coordination numbers and less than 50% for the Cs-O, Cs-C and Cs-Br XAFS Debye-Waller factors.

DISCUSSION

Results from this Cs XAFS study illustrate the importance of many different considerations required to accurately obtain and analyze dilute concentrations of Cs in a crown-ether in the liquid phase. First, since the sample is dilute Cs in a liquid phase, the low energy of the incoming X-ray beam (5 keV) cannot penetrate through the sample. Therefore, the fluorescence technique is required. Secondly, the best possible quality of data is of utmost importance since the range of usable data is limited by the onset of the Cs LII absorption edge. Third, the amplitude of the signal is not only reduced because of the small backscattering amplitudes of oxygen and carbon, but also because of the destructive interference of the backscattering oxygen and carbon shells at their respective radial distances. Fourth, as shown in Fig. 3, the effect of multi-electron excitations must be understood to insure that the XAFS results are not significantly distorted. And finally, as shown in Fig. 2, the use of a correct filter (in this case, a Sc foil) in combination with Soller slits increases the signal to background ratio.

The multi-electron transitions have been attributed to a 2p-4d two electron ionization threshold[21,24]. Earlier XAFS studies have neglected to address the contribution of these transitions to the normalized absorption data. Since their presence is on the same scale as the XAFS oscillations, especially in this instance with the weak backscattering amplitude of the oxygen and carbon in the first and second shell, they could have potentially large influences on the results from the data analysis (i.e. the inability to accurately distinguish the presence of one Br backscattering anion).

In another Cs XAFS study[12], the presence of a counter ion contact pair was not detected in transmission measurements conducted on D18C6 that had been incorporated into an ion selective polymeric film. Cs was exchanged onto the films as $CsNO_3$. Results from this study were interpreted as Cs surrounded by two subshells of oxygen (whose sum of O atoms was 6), and one highly disordered carbon shell. Due to the similarity of backscattering amplitude and phase shift between oxygen and nitrogen, it is highly unlikely that the experiment, as designed, would have enabled the detection of the Cs-NO_3 contact pair nor its distinction from the Cs coordinated to the oxygen in the crown ether. Also, sample preparation in this experiment entailed drying the film to perform transmission measurements. Had microcrystals of $CsNO_3$ been formed on the film, the possibility for two different Cs-O bondlengths would exist. Introducing Cs into the crown-ether complex via CsBr, however, enables the detection of the Cs-Br contact pair because of the larger difference in backscattering amplitude and phase shift between bromine and the lighter oxygen and carbon.

The interpretation of our XAFS analyses for the presence of Br as a contact-pair with the Cs-D18C6 complex are consistent with wet chemical studies; potentiometric, calorimetric and ^{133}Cs NMR techniques[6,7]. In addressing the complex equilibria between Cs^+, 18-crown-6, and anion (X^-), the presence of the triple ion complex was inferred from exhaustive studies on the ^{133}Cs NMR chemical shifts as influenced by component concentrations and temperature in order to derive the thermodynamic parameters of formation constants, enthalpy and entropy. The authors raised a question on which of two kinds of ion pairing might have been occurring: contact (i.e., direct Cs-Br) or non-contact (i.e., solvent-separated or ligand-separated) ion pairs. Although it was not possible to separate the effects of these two types of ion pairing by NMR methods, they proposed non-contact triple ion pair formation based on competitive ion pair (Cs^+X^-) formation versus complex formation (Cs^+-D18C6) for a variety of anions (X^-) despite conflicting interpretations from the thermodynamic parameters.

Direct confirmation between the two kinds of ion-pairing can be made from the XAFS analyses. Traditional ionic radii for 7 coordinate Cs^+ would be approximately 1.7 Å and 1.96 Å for Br^- [25]. Hence, an estimate of a minimum direct contact ion pair distance between Cs and Br would be the sum of the two ionic radii, 3.66 Å. If the Cs^+-Br^- pair were solvent-separated, the Cs^+-Br^- distance would be much larger and Br^- would not be as easily detectable as a backscatterer. The best fit of the XAFS data yielded a shell containing a Br atom at 3.60±0.05 Å, a distance that is in good agreement with direct contact-pairing occurring.

The significance of mechanisms underlying the nature of contact pairing has many implications for applying crown ethers to selective ion binding and extraction technologies. Competition for binding between the anion and the crown ether can lower the efficiency of the extraction process[6]. In contrast, formation of triple ions can further enhance removal of the anion as well as the cation[26]. Finally, the potential for ternary interactions between the Cs-crown ether complex can also have synergistic effects for the extraction of other cations[27]. The ability for XAFS to directly probe the anion contact pairing to Cs complexed to a macrocycle opens many exciting avenues for improving extraction methodologies.

CONCLUSIONS

There are few methods to probe the local environments of Cs binding sites of ion selective ligands. However, despite the many difficulties in obtaining statistically clean XAFS data, an unprecedented level of local chemical environment information for Cs has been derived from this liquid phase XAFS study. Results from this study show the presence of a 1:1 Cs-D18C6 complex with the Br counter-ion still in contact with the Cs atom. Due to the large size of Cs, and its low charge to radius value, Cs has a poorly defined coordination geometry and low solvation energy. Despite its large atomic radius, Cs can readily shed its solvation shell and effectively fit into size occluded binding sites that are inaccessible to cations with higher solvation energies and larger effective solvated radii. These XAFS results should permit an extension of this type of experiment to study many other systems in which the Cs is selectively bound within a cage of electronegative oxygen donors. A few examples of these systems include humic and other organic substances, colloidal clay minerals and ion selective exchange resins used to treat nuclear wastes.

ACKNOWLEDGEMENTS:
This research was funded by contract DE-AC09-76SR00819 between the University of Georgia and the U.S. Department of Energy. This research was carried out in part at the National Synchrotron Light Source (Brookhaven National Laboratory, Upton, NY), which is sponsored by the United States Department of Energy (Division of Materials Science and Division of Chemical Science of the Office of Basic Energy Sciences). This research was performed while K. M. K. held a National Research Council - Naval Research Laboratory Research Associateship.

[1] B. J. Dietrich, *Chem. Ed.* 62:954 (1985).

[2] M. Shamsipur, G. Rougaghi and A.I. Popov, *J. Sol. Chem.* 9:701 (1980).

[3] M. Shamsipur and A. I. Povpov, *J. Sol. Chem.* 101:4051 (1979).

[4] Jean-Marie Lehn, *Angew. Chem. Int. Ed. Engl.* 27:89 (1988) .

[5] B. J. Dietrich, *J. of Chem Ed.* 62:954 (1985).

[6] Khazaeli, S., A.I. Popov, and J.L. Dye, *J. Phys. Chem.* 86:5018 (1982).

[7] Khazaeli, S., A.I. Popov, and J.L. Dye, *J. Phys. Chem.* 87:1830 (1983).

[8] F. A. Cotton and G. Wilkinson *in:* "Advanced Inorganic Chemistry," John Wiley, New York, (1980).

[9] W. J. DeWitte, R.C. Schoening and A.I. Popov, *Inorg. Nucl. Chem. Lett.*, 12:251 (1976).

[10] S. M. Heald *in:* "X-Ray Absorption, Principles, Applications, Techniques of XAFS, SEXAFS and XANES," edited by D. C. Koningsberger and R. Prins, Wiley, New York, p. 92, (1988).

[11] F. Beniere, N. Bertru, C. R. A. Catlow, M. Cole, J. Simonet and L. Angely, *J. Phys. Chem. Solids* 53:449 (1992).

[12] C. R. A. Catlow, A. V. Chadwick, G. N. Greaves, L. M. Moroney and M. R. Worboys, *Solid State Ionics* 9 & 10:1107 (1983).

[13] A. Ishida, S. Emura, M. Takahashi, Y. Ito, H. Kananaru, S. Takamuku, *Physica B* 208 & 209:711 (1995).

[14] W. J. DeWitte, R. C. Schoening and A. I. Popov, *Inorg. Nucl. Chem. Lett.* 12:251 (1976).

[15] S. Khazaell, A. I. Popov, and J. L. Dye, *J. of Phys. Chem.* 86-25:5018 (1982).

[16]R. A. Neiser, J. P. Kirkland, W. T. Elam, and S. Sampath, *Nucl. Instrm. Methods Phys. Res. Sect.* A266:220, (1988).

[17]E. A. Stern and S. M. Heald, *Rev. Sci. Instrum.* 50:1579 (1979).

[18]E. A. Stern, B. A. Bunker, and S. M. Heald, *Phys. Rev. B*, 21:5521 (1980).

[19]J. J. Rehr, J. Mustre de Leon, S. I. Zabinsky and R. C. Albers, *J. Am. Chem. Soc.* 113:5135 (1991).

[20]K. M. Kemner, J. Kropf, B. A. Bunker, *Rev. Sci. Instrum.*, 65 (12):3667 (1994).

[21]A. Kodre, I. Arcon, M. Hribar, M. Stuhec, F. Villain, W. Drube, L. Troger, *Physica B* 208 & 209:379 (1995).

[22]D. E. Sayers and B. A. Bunker *in:* "X-Ray Absorption, Principles, Applications, Techniques of EXAFS, SEXAFS and XANES," edited by D. C. Koningsberger and R. Prins Wiley, New York, p. 211, (1988).

[23]Report on the "International Workshops on Standards and Criteria in XAFS," *in:* "X-ray Absorption Fine Structure," edited by S. Samar Hasain, Ellis Horwood Limited, England, p. 751 (1991).

[24]P. Strobel, J. Durr, M-H. Tuilier and J.-C. Charenton, *J. Mat. Chem.* 3:453, (1993).

[25]R. D. Shannon, *Acta Crystallogr.*, A32:751 (1976).

[26]I. M. Kolthoff, M.K. Chantooni, W.J. Wang, *J. Chem. Eng. Data*, 38:556 (1993).

[27]B. A. Rusdiarso, B. A. Messaoudi and J.P. Brunette, *Talanta*, 40:805 (1993).

X-RAY SPECTROSCOPIC STUDIES OF MICROBIAL

TRANSFORMATIONS OF URANIUM

Cleveland J. Dodge[1], Arokiasamy J. Francis[1], and Clive R. Clayton[2]

[1]Brookhaven National Laboratory
Department of Applied Science
Upton, NY 11973
[2]SUNY at Stony Brook
Department of Materials Science and Engineering
Stony Brook, NY 11794

INTRODUCTION

The presence of radionuclide and toxic metal contaminants at Department of Energy (DOE) sites is a major environmental concern. Uranium, in particular is of interest due to its presence in soils, sediments, and wastes resulting from nuclear weapons production. Characterization of uranium in these materials is necessary so that appropriate treatment strategies can be developed. Information on X-ray spectroscopic analysis of uranium has been limited to oxides[1], intermetallic compounds[2], and encapsulated forms[3,4]. In this study a systematic approach was undertaken to characterize various uranium compounds in order to elucidate the basic mechanisms involved in remediation of contaminated sites. We used X-ray photoelectron spectroscopy (XPS) and X-ray absorption near edge structure (XANES) to determine valence state, and extended X-ray absorption fine structure (EXAFS) to determine compound structure.

MATERIALS AND METHODS

Compounds

The following uranium compounds U-metal (α-phase), UO_2, U_3O_8, γ-UO_3, uranyl acetate, uranyl nitrate, uranyl sulfate (Atomergic Chemicals, Farmingdale NY), aqueous and solid forms of 1:1 U:citric acid and 1:1:2 U:Fe:citric acid mixed-metal complexes, and a precipitate obtained by photodegradation of the U-citrate complex were used.

Equimolar 1:1 U(VI):citric acid complex was prepared by combining equal volumes of uranyl nitrate and citric acid (pH 6.0) solutions, in a beaker with continous stirring. The pH

Synchrotron Radiation Techniques in Industrial, Chemical, and Materials Science
Edited by D'Amico *et al.*, Plenum Press, New York, 1996

159

was adjusted to 6.0 with sodium hydroxide and the ionic strength was maintained at 0.1 M by adding KCl. The 1:1:2 U(VI):Fe(III):citric acid complex was prepared in a similar fashion using uranyl nitrate, ferric nitrate, and citric acid. The final concentration of each complex was 6.5 mM. Iron and U stock solutions were standardized by visible spectrophotometry, and citric acid by high pressure liquid chromatography. Solid forms of the U- and U-Fe-citrate complexes were prepared by crystallization from the aqueous state. Uranium precipitate from photodegradation of U-citrate complex was collected after exposure of the aqueous complex to a light intensity of 0.04 mEinstein m^{-2} s^{-1} in an incubator at 26±1°C[5].

Uranium contaminated sludge and sediment samples were obtained from Y-12 Plant, Oak Ridge, TN. To determine the extent of uranium transformations in the waste due to anaerobic microbial activity, samples were inoculated with *Clostridium* sp. (ATCC 53464) and incubated for several days[6]. Speciation of uranium in the waste material before and after microbial treatment was determined by XPS and XANES.

Uranium   Characterization

XPS. A VG Scientific ESCA MK II spectrometer at SUNY-Stony Brook was used to determine peak positions at the U 4f$_{7/2}$ and U 4f$_{5/2}$ binding energy of U-metal, UO_2, U_3O_8, γ-UO_3, uranyl acetate, uranyl nitrate, and uranyl sulfate[7]. The shift in peak position with oxidation state due to changes in uranium binding energy was determined after curve fitting.

XANES. XANES, usually within 40 eV of the absorption edge, provides data on the local structure and oxidation state of uranium. U-metal, UO_2, γ-UO_3, and uranyl acetate were analyzed at the National Synchrotron Light Source (NSLS) on beam line X-19A. The sample was mounted on Mylar foil and the XANES spectrum was collected at the M_V absorption edge (3.545 keV) using an electron yield detector and a double-crystal Si(111) monochromator[7]. The spectrum was normalized after background subtraction, the curve was fitted using a least-squares program, and the absorption peak maximum of the sample determined. XANES information at the L_{III} edge was collected for hexavalent uranium γ-UO_3, uranyl acetate, and 1:1 U:citric acid and 1:1:2 U:Fe:citric acid complexes.

EXAFS. EXAFS measurements provide information on the local three-dimensional environment surrounding a central atom and are obtained using a multi-step analysis procedure to identify nearest neighbor atoms, determine atomic distances, coordination number of central atom, and coordination geometry. This information can be obtained regardless of the physical state of the sample (e.g. solid, aqueous). Solid γ-UO_3, uranyl nitrate, uranyl acetate, ferric acetylacetonate, and equimolar U:citric acid and U:Fe:citric acid complexes were mixed with boron nitride (10% w/v), placed in heat sealed polypropylene bags (0.2 mil) and analyzed on beamline X-23A2 at the NSLS using a fluorescence detector at the uranium L_{III} edge (17.165 keV) and at the iron K edge (7.1112 keV). Aqueous U-citrate complexes were transferred to polypropylene bags and analyzed. Data was analyzed using the UW EXAFS analysis software MacXAFS 3.6.

RESULTS   AND   DISCUSSION

XPS. Table 1 shows the U4f$_{7/2}$ and U4f$_{5/2}$ binding energies for U-metal, UO_2, U_3O_8, γ-UO_3, uranyl acetate, uranyl nitrate, and uranyl sulfate. Uranium metal exhibited the lowest binding energy at 376.6 eV for the U4f$_{7/2}$ spin state. With increase in oxidation state to +4 for uranium dioxide, the binding energy increased to 380.4 eV. The +6 oxidation state however, exhibited a range of energies from 381.9 eV for uranium trioxide to 382.2 eV for uranyl

Table 1. Peak binding energies for uranium 4f electrons in model compounds.

Compound	Oxidation state	Binding energy (eV) U $4f_{7/2}$	U $4f_{5/2}$
U metal	0	376.6	387.1
UO_2	+4	380.4	391.3
U_3O_8	+4	381.5	392.5
	+6	382.9	394.1
γ-UO_3	+6	381.9	391.9
Uranyl acetate	+6	381.5	392.4
Uranyl nitrate	+6	382.0	392.1
Uranyl sulfate	+6	382.2	395.5

sulfate. In the mixed-valent uranium oxide, U_3O_8, two peaks at 381.5 and 382.9 eV were fitted to the main peak envelope; these were due to the presence of +4 and +6 oxidation states, respectively. There was a larger shift in the U $4f_{7/2}$ peak for +6 uranium compared to the pure hexavalent compounds. There also was a large shift in binding energy for the U $4f_{5/2}$ spin state of uranyl sulfate. These differences may be a result of the coordination geometry of the oxygens which increase the effective charge on the uranium.

The ability of this technique to differentiate oxidation states of uranium is of particular importance in the bioremediation of wastes. We previously showed using XPS that uranium was reduced from the soluble U^{6+} to insoluble U^{4+} and U^{3+} as a result of anaerobic bacterial activity[8]. Microbial reduction of uranium from soluble U^{6+} to insoluble U^{4+} increases its stability in the environment by forming insoluble uranium (IV) species.

XANES. Figure 1 shows the absorption spectra for a series of uranium compounds collected at the M_V and L_{III} edges. Uranium metal at the M_V absorption edge showed a maximum at 3549.6 eV. An increase in oxidation state to U^{4+}, uranium dioxide, resulted in a shift to higher energy of 0.8 eV (3550.4 eV). Analysis of U^{6+} species, uranium trioxide and uranyl acetate, showed an identical shift in absorption to 3551.1 eV. The presence of a shoulder on the high energy side of the hexavalent uranium compounds was attributed to multiscattering processes involving the uranyl ion[1] and it was more pronounced in uranium trioxide than uranyl acetate. In the 1:1 mixture of +4 and +6 uranium the shift was intermediate between the two indicating the XANES spectrum in this region is sensitive to the oxidation state.

The normalized absorption peak maxima for uranyl acetate, uranium trioxide, 1:1 U:citric acid and 1:1:2 U:Fe:citric acid complexes obtained at the uranium L_{III} edge are broad with a low white line ratio compared to the uranium M_V absorption maxima making it difficult to interpret shifts in peak position resulting from changes in oxidation state. This is due to the increased density of states at the 5f (M_V) compared to the 6d (L_{III}) final states[2]. However, the L_{III} uranium edge is useful for obtaining EXAFS information since it is relatively free from interfering absorption edges. The M_V edge exhibits interference due to presence of the

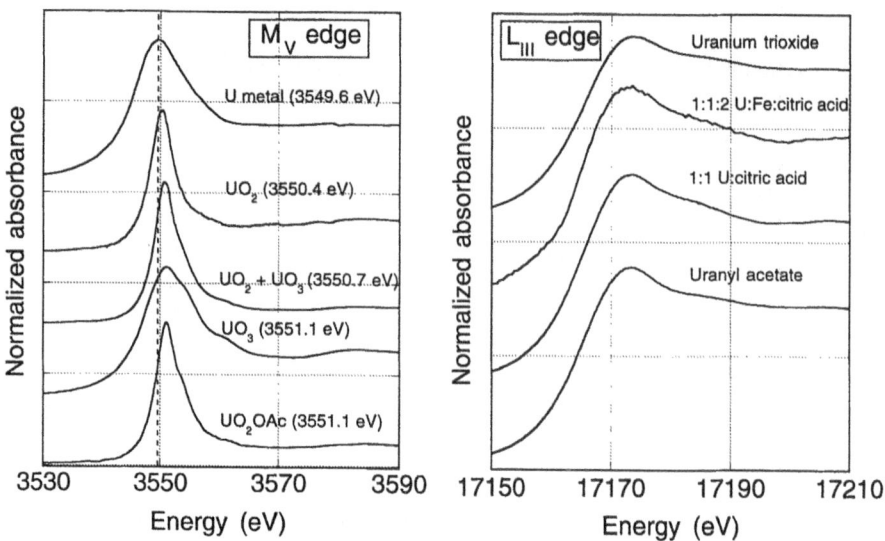

Figure 1. XANES absorption spectra at the M_V absorption edge for compounds of uranium showing a shift in absorption peak maximum with increasing oxidation state, and L_{III} absorption edge for hexavalent uranium compounds. The white line ratio is more pronounced for the M_V compared to the L_{III} absorption edge.

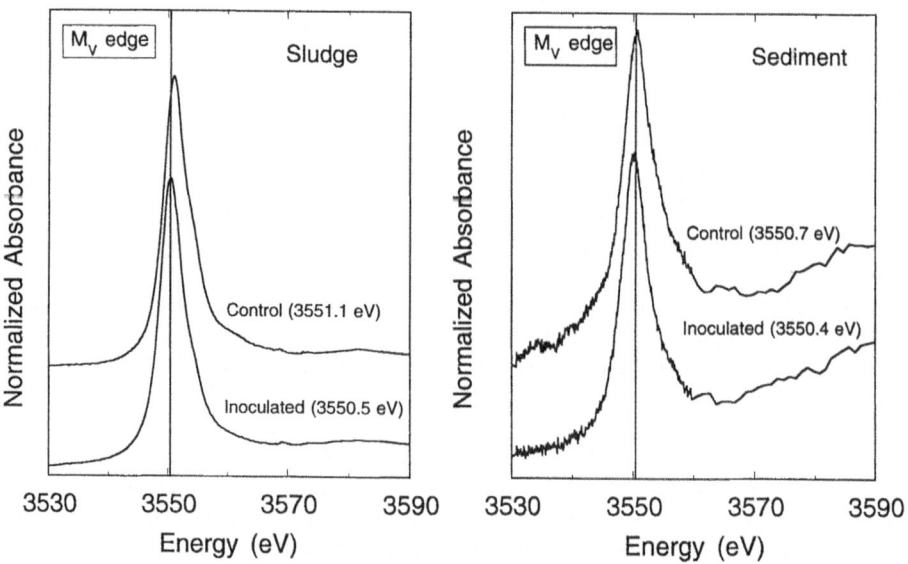

Figure 2. Comparison of XANES absorption spectra at the M_V edge for uninoculated (control) and inoculated samples of sludge and sediment. A decrease in the position of the absorption peak maximum in the bacterially treated samples indicates partial reduction of uranium in the sludge and complete reduction of uranium in the sediment to U^{4+}.

potassium K absorption edge at 3.6078 keV and adjacent U M_{IV} absorption edge at 3.720 keV.

Figure 2 shows XANES analyses of the sludge and sediment at the M_V edge before and after bacterial activity. The sludge contained 3080 ppm uranium and the absorption maximum for the untreated (control) sample was 3551.1 eV, indicating that the uranium was predominantly in the hexavalent form. After bacterial treatment there was a shift in the absorption maximum to 3550.5 eV, which is slightly higher than tetravalent uranium (3550.4 eV), but much less than U(VI); this indicates that there was a significant reduction to the U^{4+} state. The sediment contained 640 ppm uranium and the untreated (control) sample exhibited an absorption maximum at 3550.7 eV, which corresponds to a mixture of U^{4+} and U^{6+}. The high organic content (11.3%) may have contributed to maintaining the sediment in a reduced state. In the presence of bacteria the sample peak was shifted to 3550.4 eV, which was identical to U^{4+}, indicating all of the uranium was reduced to the tetravalent state. The reduction of uranium in sludge and sediment by an anaerobic bacterium formed the basis for a unique treatment by enzymatic reduction and subsequent immobilization of the uranium, as well as overall waste volume reduction due to dissolution of the acid soluble components in the waste by the production of organic acid metabolites[9].

EXAFS. *1:1 U:citric acid complex.* EXAFS spectra at the uranium L_{III} edge were collected for uranyl acetate and for the solid and aqueous forms of U-citrate complex. Citric acid, a naturally occurring chelating agent which is also present in low-level and transuranic wastes, forms multidentate complexes with metals, and is used to extract uranium and toxic metals from wastes[10]. Uranium forms a stable complex with citric acid which is resistant to biodegradation. It was suggested that the binding of uranium to the hydroxy group of citric acid plays a key role in its stability[11].

Figure 3 compares the Fourier transforms for both the solid and aqueous form of 1:1 U:citric acid complex at pH 6.0 over the k range of 3-14 Å⁻¹. The first peak (phase shifted) at 1.8 Å corresponds to the nearest neighbor oxygen ligands of the uranyl ion. Beyond the first shell, information on the aqueous complex is difficult to obtain due to thermal broadening. However, the more ordered solid state shows additional fine structure. The peak at 2.25 Å corresponds to the second shell equatorial oxygen atoms surrounding the uranium. A large amplitude peak also is observed at 5.2 Å. Fourier filtering of the isolated shell displays a large amplitude function with high sinusoidal frequency indicative of the high Z element U (Figure 4). Comparing the backtransformed peak of U-citrate with uranyl acetate dihydrate, a binuclear compound with a $(UO_2)_2(\mu,\eta^2\text{-acetato})_2$ core and a U-U distance of approximately 5.5 Å, shows a similar spectrum (slightly phase shifted), thereby confirming the peak is due to U-U interaction .

Figure 5 shows the k^2 weighted raw EXAFS data after normalization and background subtraction indicating that the U:citric acid complex has a similar structure in both the aqueous and solid state. Although a binuclear structure has been proposed for the complex[12] involving a $(UO_2)_2(\mu\text{-OH})_2$ core, this configuration was based on titration methods and suggests a U-U distance of about 4.1 Å in the complex. EXAFS analysis results shows that the U-U distance in the complex is approximately 5.2 Å. From this data, we propose a new structure consisting of a $(UO_2)_2(\mu,\eta^2\text{-citrato})_2$ core (Figure 6A). The structure is stabilized by the formation of 6 and 7 member rings involving citric acid with uranium.

1:1:2 U:Fe:citric acid complex. The influence of ferric iron, a ubiquitous component in soils and wastes, on the U-citrate complex was investigated to see whether a mixed-metal citrate complex was formed. The ability of citric acid to form complexes with more than one metal has been established[13]. A mixed metal U-Fe complex was confirmed by calorimetric and spectrophotometric methods but its exact structure was not elucidated[14]. Figure 7 shows the Fourier transform spectra of 1:1:2 U:Fe:citric acid at the U L_{III} absorption edge. Absence of a large magnitude peak at 5.2 Å compared to the 1:1 U:citric acid complex indicates ferric iron affects the structure of the U-citrate complex. Fourier-filtering over the k range 3-15 Å⁻¹ at the

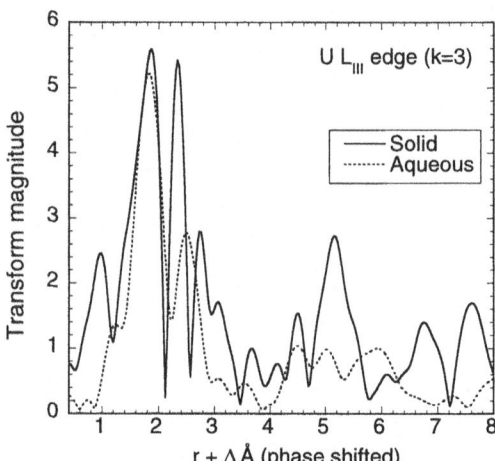

Figure 3. Comparison of phase-shifted Fourier transforms of background-subtracted and k^3-weighted EXAFS spectra of aqueous and solid forms of 1:1 U:citric acid complex at pH 6.0. The large magnitude peak at 5.2 Å for the solid state reveals U-U interaction.

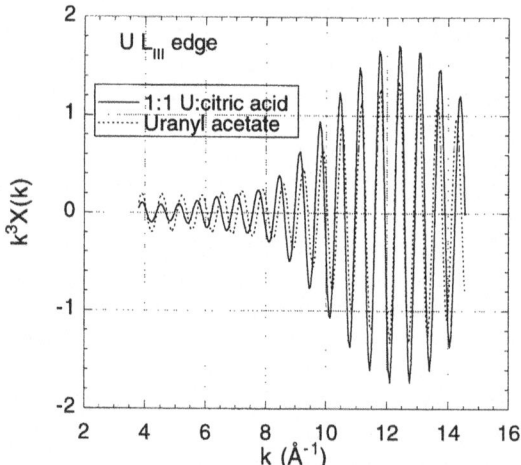

Figure 4. Comparison of Fourier-filtered EXAFS spectra of isolated shells of 1:1 U:citric acid 5.2 Å and uranyl acetate at 5.5 Å plotted as $k^3\chi(k)$ vs. $k(\text{Å}^{-1})$. The high frequency function, as well as the large amplitude peak at high k, indicate the presence of uranium.

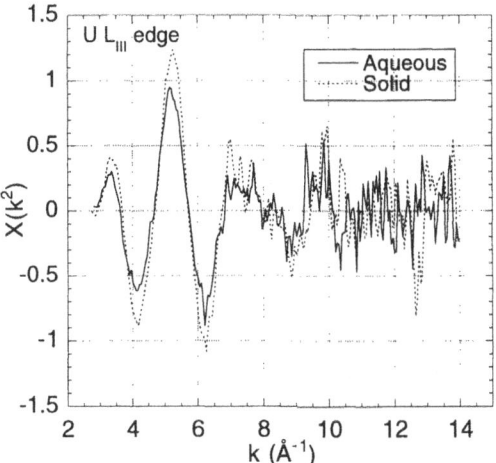

Figure 5. Normalized and background-subtracted EXAFS spectra shown as $\chi(k^2)$ vs. k (Å^{-1}) at the L_{III} edge for aqueous and solid forms of 1:1 U:citric acid complex. The similarity in the beat pattern indicates that there is no difference in structure between the two states.

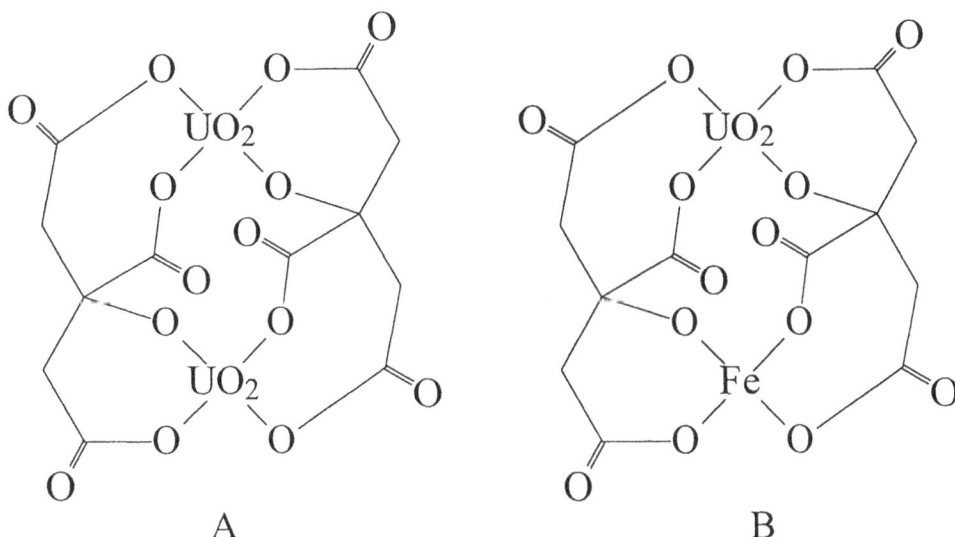

Figure 6. Proposed structures for 1:1 U:citric acid (A) and 1:1:2 U:Fe:citric acid mixed metal (B) complexes based on EXAFS analysis of U-U and Fe-U interactions at 5.2 and 4.8 Å, respectively.

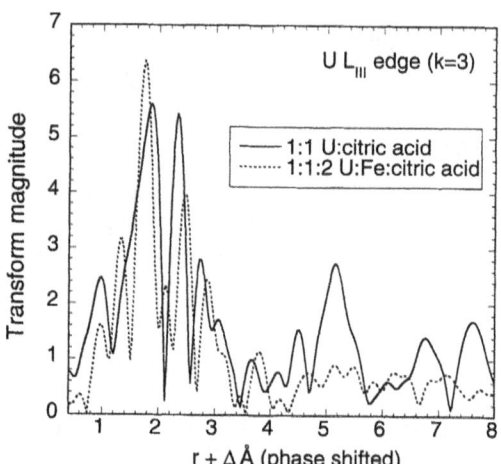

Figure 7. Comparison of phase-shifted Fourier transforms of background-subtracted and k^3-weighted EXAFS spectra of solid 1:1 U:citric acid and 1:1:2 U:Fe:citric acid complexes at pH 6.0. The disappearance of large magnitude peak at 5.2 Å in the presence of ferric iron indicates the formation of a mixed-metal complex with U-Fe interaction.

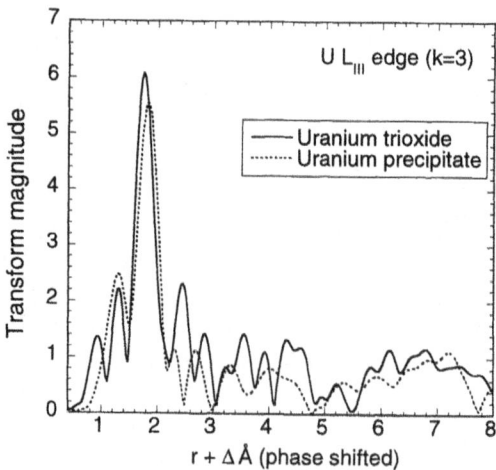

Figure 8. Comparison of phase-shifted Fourier transforms of background-subtracted and k^3-weighted EXAFS spectra of solid uranium precipitate from photodegradation of 1:1 U:citric acid, and uranium trioxide. The lack of structure in the second and third shells for the precipitate suggests that it is slightly amorphous.

Fe edge of the isolated shell at 4.8 Å shows a high frequency function similar to that observed for the U-U interaction (data not shown). These results confirm the presence of a mixed metal binuclear complex with U-Fe distance at 4.8 Å. The difference of 0.4 Å between U-U and U-Fe is approximately the difference in distance for U-O (2.5 Å) and Fe-O (2.0 Å) bonding and indicates that U is replaced by Fe in the complex. Figure 6B shows the proposed structure.

Photochemical Oxidation of Uranyl Citrate Complex. Figure 8 shows the EXAFS spectra of the precipitate obtained from the photochemical degradation of 1:1 U:citric acid complex and uranium trioxide at the L_{III} edge. The structure of the precipitate is similar to uranium trioxide but more amorphous, as revealed by the position of the second and third shells representing the equatorial oxygens surrounding uranium. X-ray diffraction, XPS, and XANES analyses of the precipitate confirmed that it was predominantly uranium trioxide dihydrate with a small amount of the mineral schoepite[5]. The formation of this insoluble compound suggests that photodegradation is an effective treatment for removing uranium from ssolution complexed with citric acid. The uranium oxide is then collected and disposed of, or recycled.

SUMMARY

Several uranium compounds U-metal (α-phase), UO_2, U_3O_8, γ-UO_3, uranyl acetate, uranyl nitrate, uranyl sulfate, aqueous and solid forms of 1:1 U:citric acid and 1:1:2 U:Fe:citric acid mixed-metal complexes, and a precipitate obtained by photodegradation of the U-citrate complex were characterized by X-ray spectroscopy using XPS, XANES, and EXAFS. XPS and XANES were used to determine U oxidation states. Spectral shifts were obtained at the U $4f_{7/2}$ and U $4f_{5/2}$ binding energies using XPS, and at the uranium M_V absorption edge using XANES. The magnitude of the energy shift with oxidation state, and the ability to detect mixed-valent forms make these ideal techniques for determining uranium speciation in wastes subjected to bacterial action.

The structure of 1:1 U:citric acid complex in both the aqueous and solid state was determined by EXAFS analysis of hexavalent uranium at the L_{III} absorption edge and suggests the presence of a binuclear complex with a $(UO_2)_2(\mu,\eta^2$-citrato$)_2$ core with a U-U distance of 5.2 Å.

The influence of Fe on the structure of U-citrate complex was determined by EXAFS and the presence of a binuclear mixed-metal citrate complex with a U-Fe distance of 4.8 Å was confirmed.

The precipitate resulting from photodegradation of U-citrate complex was identified as an amorphous form of uranium trioxide by XPS and EXAFS.

ACKNOWLEDGEMENTS

We thank Dr. Lars Furenlid (NSLS) for helpful discussions, Sanjay Kagwade (SUNY-SB) for XPS analysis, and Dr. F.J. Wobber, Program Manager, for continued support. This research was performed under the auspices of the Environmental Sciences Division's Subsurface Science Program, Office of Health and Environmental Research, U.S. Department of Energy, under Contract No. DE-AC02-76CH00016.

REFERENCES

1. G. Kalkowski, G. Kaindl, W.D. Brewer, and W. Krone, Near edge X-ray absorption fine structure in uranium compounds, Phys. Rev. B 35:2667 (1987).

2. J.M. Lawrence, M.L. den Boer, and R.D. Parks, X-ray absorption spectra in rare-earth and uranium intermetallics: localized versus itinerant final states, Phys. Rev. B 29:568 (1984).

3. J. Petiau, G. Calas, D. Petit-Marie, A. Bianconi, M. Benfatto, and A. Marcelli, Localization of 5f states in various uranium and thorium oxides and glasses, J. Phys. 47:C8-949 (1986).

4. W.H. Hocking, A.M. Duclos, and L.H. Johnson, Study of fission product segregation in used CANDU fuel by X-ray photoelectron spectroscopy (XPS) II, J. Nucl. Materials 209:1 (1994).

5. C.J. Dodge and A.J. Francis, Photodegradation of uranium-citrate complex with uranium recovery, Environ. Sci. Technol. 28:1300 (1994).

6. A.J. Francis, C.J. Dodge, J.B. Gillow, and J.E. Cline, Microbial transformations of uranium in wastes, Radiochim. Acta 52/53:311 (1991).

7. C.J. Dodge, A.J. Francis, F. Lu, G.P. Halada, S.V. Kagwade, and C.R. Clayton, Speciation of uranium after microbial action by XANES and XPS, in: "Applications of Synchrotron Radiation Techniques to Materials Science," D. Perry, ed., Materials Research Society, Pittsburgh (1993).

8. A.J. Francis, C.J. Dodge, F. Lu, G. Halada, and C.R. Clayton, XPS and XANES studies of uranium reduction by Clostridium sp., Environ. Sci. Technol. 28:636 (1994).

9. A.J. Francis, C.J. Dodge, and J.B. Gillow, Microbial stabilization and mass reduction of wastes containing radionuclides and toxic metals, US Patent No. 5,047,152, September, 1991.

10. A.J. Francis and C.J. Dodge, Waste site reclamation with recovery of radionuclides and metals, US Patent No. 5,292,456, March, 1994.

11. A.J. Francis, C.J. Dodge, and J.B. Gillow, Biodegradation of metal citrate complexes and implications for toxic-metal mobility Nature 356:140 (1992).

12. K.S. Rajan and A.E. Martell, Equilibrium studies of uranyl complexes. II. Interaction of uranyl ion with citric acid, Inorg. Chem. 4:462 (1965).

13. A. Adin, P. Klotz, and L. Newman, Mixed-metal complexes between Indium(III) and Uranium(VI) with malic, citric, and tartaric acids, Inorg. Chem. 9:2499 (1970).

14. E. Manzurola, A. Apelblat, G. Markovits, and O. Levy, Mixed-metal hydroxycarboxylic acid complexes: Formation constants of complexes of U^{VI} with Fe^{III}, Al^{III}, In^{III}, and Cu^{II}, J. Chem. Soc., Faraday Trans. 1 85:373 (1989).

APPLICATION OF SYNCHROTRON RADIATION TECHNIQUES TO CHEMICAL ISSUES IN THE ENVIRONMENTAL SCIENCES

Patrick G. Allen,[1,2] Jerome J. Bucher,[2] Melissa A. Denecke,[3]
Norman M. Edelstein,[2] Nik Kaltsoyannis,[2,*] Heino Nitsche,[3]
Tobias Reich,[3] and David K. Shuh[2]

[1]G.T. Seaborg Institute for Transactinium Science
Lawrence Livermore National Laboratory
Livermore, CA 94551

[2]Chemical Sciences Division
Lawrence Berkeley National Laboratory
Berkeley, CA 94720

[3]Forschungszentrum Rossendorf e. V.
Institute of Radiochemistry
P.O. Box 510119, D-01314
Dresden, Germany

*Present Address: Dept. of Chemistry
University College London
London, England WC1H OAJ

INTRODUCTION

The problem of hazardous waste contamination in the environment has become an issue of central importance in society. As the number of remediation efforts taking place worldwide grows and the public awareness of these problems increases, it is clear that the problems faced are enormous in terms of effort and cost. With existing chemical technologies being applied to well over one thousand contaminated sites within the U.S. alone, cost estimates for the cleanup efforts range well into the tens of billions of dollars. Furthermore, because many of the problems are still being assessed and identified, there is no accurate time estimate for the completion of these projects, with currently proposed restoration programs ranging well into the mid-21st century.

The types of contamination that have been recognized to date extend over the entire periodic table and include: Cd, Pb, Cr, Se, As, Zn, Cu, Ba, Ni, and Hg; radioactive Tc, Sr, and Cs as products of nuclear fission; radioactive Co resulting from activation; radionuclides U, Np, Pu, and Am; inorganic anions (NO_3^-, F^-, and CN^-); and a wide array of organics.[1-3] Depending on the physical form and chemical type of contamination, remediation strategies generally involve one or more of following technological disciplines: 1) characterization of the types of contaminants and their concentrations; 2) extraction, separation, conversion, or immobilization; and 3) containment and storage (especially for radionuclides). Since many of the existing technologies, particularly those related to radionuclides, are still in their infancy, it is important to recognize that research and development can play a significant role in these long-term efforts. That is, with the tremendous expenditures of resources already planned, it is beneficial to conduct fundamental and applied research in each of the areas listed above to make the cleanup efforts more efficient, thereby reducing costs.

The purpose of this paper is to provide the potential synchrotron radiation (SR) environmental science user with an introductory level description of x-ray absorption spectroscopy and a brief overview of representative environmental investigations that have been reported to date. The narrative addresses the need, the utility, and fundamental principles of x-ray absorption

Synchrotron Radiation Techniques in Industrial, Chemical, and Materials Science
Edited by D'Amico *et al.*, Plenum Press, New York, 1996

169

spectroscopy techniques. The basic considerations for experimental work with samples of an environmental nature are also presented. This is followed by several short descriptions of the results from selected studies which are organized by the element of the environmental contaminant involved. The authors apologize in advance for the omission of the descriptions of a great body of superlative synchrotron radiation experimental work. This paper is not meant to be a comprehensive review, just an informative one. Many of these reviews emphasize the application of x-ray absorption spectroscopy techniques to materials containing radionuclides since this is the area of our expertise. There are also several examples taken from our research that were presented at the 1994 ACS Washington D.C. Meeting. Lastly, the future role of synchrotron radiation techniques in the environmental sciences is discussed before concluding.

CHARACTERIZATION

At each stage of remediation, a number of standard analytical techniques are commonly employed to determine the presence and concentrations of contaminants. Some examples of sensitive analytical methods are: inductively-coupled plasma atomic emission spectroscopy (ICP-AES), x-ray fluorescence (XRF), α, β, and γ spectroscopies (for radionuclides). However, in order to evaluate and improve upon existing remediation technologies, a more detailed knowledge of the speciation is required. Such information is generally obtained through the use of traditional spectroscopies such as; multi-nuclear nuclear magnetic resonance (NMR), UV-visible absorption/fluorescence, electron paramagnetic resonance (EPR), electron microscopies such as scanning electron microscopy (SEM) and transmission electron microscopy (TEM), photoacoustic spectroscopy (PAS), photoemission, and infrared and Raman vibrational spectroscopies. In the pre-treatment stage, determining the chemical form (i.e., phase, local environment, oxidation state) of the contaminant is important because the most effective remediation treatment is highly dependent on the exact form of the waste. During treatment, identifying the speciation is also required to determine the efficacy of the process being employed. Once a contaminant has been isolated for long-term storage, its speciation directly determines whether it can become mobile or reactive. In many instances, some of these methods may fail to differentiate between the same atom residing in different environments or oxidation states. In other cases, some techniques may be incapable of detecting a species in a particular form or oxidation state.

In addition to the traditional spectroscopic techniques mentioned above, methods based on the use of synchrotron radiation have proven quite valuable in addressing questions regarding speciation. In particular, x-ray absorption spectroscopy (XAS) techniques have been used extensively. XAS is an element specific structural technique which gives information on the average local structure around almost any atom. Since XAS originates from core level atomic-like transitions, a given atom will always exhibit characteristic x-ray absorption regardless of its speciation. In this regard, XAS may be viewed as a complementary tool to other molecular spectroscopies. The information derived from XAS studies is similar to crystallographic experiments (i.e., bond lengths and coordination numbers) with the important exception that XAS does not require long-range order (crystalline samples) and can be performed on amorphous solids or even on species in solution. This is a key point for the study of environmental samples where the element of interest often exists either as a surface adsorbed complex, in aqueous solution, or as a colloidal suspension. The short description that follows is intended only to familiarize the introductory reader with the aspects of the basic technique.

Numerous detailed reviews of the XAS technique have been published elsewhere and the reader is referred to these for a more complete description of the experimental methods and procedures for data analysis.[4-6]

XAS BACKGROUND

Theory

The fundamental components of an x-ray absorption spectrum are centered around a sharp increase known as the absorption edge, which arises from a core-level electronic excitation. The detailed features that appear within ~20 eV above and below the main absorption edge are referred to as x-ray absorption near edge structure (XANES), and the features which lie ~20-1200 eV past the edge are referred to as extended x-ray absorption fine structure (EXAFS). The absorption edge is specific for a given element due to the unique core-level binding energies for each atom. The XANES region typically contains structure associated with transitions from core atomic states to bound molecular orbitals (valence or Rydberg) as well as to unbound states (continuum resonances). Because many of the transitions are between states where electronic selection rules may apply, XANES features can yield information about the electronic structure (density of states, oxidation state) and local geometry. At energies above the edge, absorption results in a transition to an unbound state of a photoelectron. The fine structure observed in the EXAFS region arises from a modulation in the absorption cross-section caused by interference between the outgoing and the backscattered photoelectron waves. The mathematical form describing the EXAFS which arises from N identical backscattering atoms at a distance R from the absorbing atom can be written as:

$$\chi(k) = \frac{N}{kR^2}|f(\pi, k)|e^{-\sigma^2 k^2} \cdot \sin[2kR + \phi_{as} + \pi]$$

where $|f(\pi, k)|$ is the backscattering amplitude, σ is the Debye-Waller damping factor, ϕ_{as} is the phase shift associated with the absorbing and backscattering atoms, and k is the photoelectron energy expressed as a momentum wave-vector in units of Å^{-1}. Curve-fitting is performed on raw $\chi(k)$ data or Fourier-filtered components of the raw data using $|f(\pi, k)|$ and ϕ_{as} functions that are derived either from model compounds of known structure or from theoretical calculations. Because the amplitude and phase functions are a function of atomic number, the types of near-neighbors may be determined to within $Z \pm 2$-6. However, more often the identification of near-neighbor atom type can be made exactly with the help of other techniques or prior chemical knowledge. In general, curve-fitting analyses are able to determine the coordination numbers (N) and bond lengths (R) to within $N \pm 15\%$ and $R \pm 0.02$ Å. In the case of two shells of the same backscatterer, Z, that are located at distances R_1 and R_2 from the central absorbing atom, the resolution is limited by $\Delta R \sim \pi/2k_{max}$. There has been great progress made in the theoretical modeling codes used to fit both EXAFS and XANES spectra. Perhaps the most widely known is FEFF6 developed by Rehr et. al.[7] Additionally, the computer codes used to reduce the raw absorption data continues to improve and an excellent example is the EXAFSPAK program suite developed by G. George.[8]

Data Acquisition

XAS spectra are typically measured either in transmission or as excitation spectra using fluorescence detection as shown schematically in Fig. 1. In transmission mode, a sample is placed between two gas-filled ionization chambers designated as I_0 and I_1. X-rays first pass through I_0, then the sample and subsequently through I_1. The natural logarithm of (I_0/I_1) as a function of x-ray energy gives the characteristic absorption spectrum. Alternatively, the excitation spectrum is collected by measuring the x-ray fluorescence, which is proportional to the x-ray absorption coefficient. This is typically done by placing a detector, I_f, at a right angle to the axis of the incident beam. In this case, in principle, a plot of I_f/I_0 versus x-ray photon energy yields the same absorption spectrum as the one taken in transmission mode (the experimental geometry allows for simultaneous measurement). XAS spectra are energy calibrated by measuring the absorption spectrum of a well-known reference compound placed between I_1 and an additional ion chamber, I_2. For dilute samples (i.e., concentration < 1 wt.% or where the absorbance from the entire

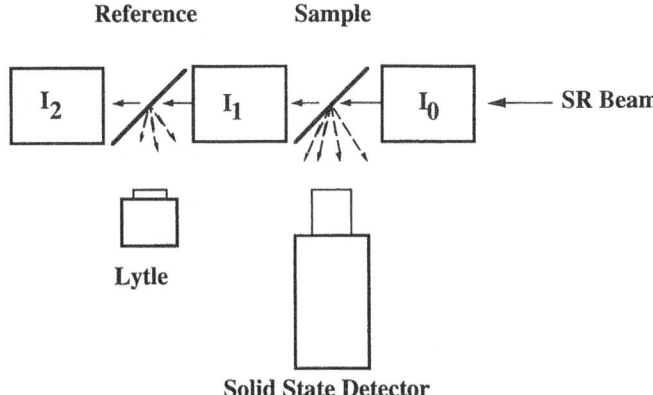

Figure 1. Schematic of an XAS experiment utilizing two fluorescence detectors.

sample is significantly greater than the absorbance from the element of interest) fluorescence detection is the detection method of choice due to its enhanced sensitivity over the transmission mode. In many environmental systems, fluorescence detection is the only suitable means of detection because of the extremely dilute concentrations of contaminants that are typically dealt with ($\sim$10-10^3 ppm).

The simplest means for detection of x-ray fluorescence from an element (atomic number Z) consists of a Soller-type slit and filter assembly in combination with an ionization chamber.[9] The combination Soller slit/filter unit (containing the element Z-1) preferentially absorbs scattered background radiation while transmitting the fluorescence line from element Z. This works well for systems that contain relatively few absorbers with well-separated fluorescence lines. However, in complex heterogeneous systems, there are frequently a multitude of elements at concentrations equal or greater to that of the contaminant which have fluorescence lines falling close to each other. In these cases, an energy dispersive detector may be better suited for resolving and isolating the desired fluorescence emission. Determination of the optimal detector system for a particular application depends on many factors and a complete analysis is presented in Ref. 10 by Warburton.

To date, the most commonly employed energy dispersive detector has been the thirteen element solid-state Ge array.[11] There are variations of these solid state arrays available depending on the energies and the counting rates being detected. However, features that are common to all include: pulse height processing, pulse shaping, and electronic discrimination. Single channel analyzers are used to define the appropriate energy windows for fluorescence signal collection. At low count rates, these detector systems can achieve resolutions of $\sim$150 eV. At count rates approaching $\sim$75 kHz, however, the resolution is degraded in favor of higher throughput. In general, these characteristics combined with signal averaging and the use of a high flux wiggler insertion device allow for collection of useful EXAFS data at concentrations as low as about 50 μM. Recently, Bucher et al.[12] have tested a novel design of a solid-state Ge detector and have demonstrated a practical detection limit approaching 5 μM. Development of new multi-element Ge and HgI systems possessing 25-100 detection channels, faster counting electronics, and third generation high brightness sources with increased flux should lower detection limits to further aid the study of environmental systems.

Sample Handling Requirements

The focus of this discussion is on environmental XAS techniques, thus many of the systems under investigation often contain toxic or hazardous species. Additional experimental safety requirements must be considered to ensure that experiments are conducted safely with minimal risk of exposure to anyone in the synchrotron radiation user facility or to the SR facility itself. Regardless of the exact nature of the environmental contaminant, the primary concerns involve proper documentation of all potential hazards, effective sample containment during the experiment, review of experimental procedures, and some level of monitoring for detecting accidental releases of the hazardous material.

Proper sample containment may be accomplished through the use of a multi-layer container system which is kept sealed at all times. The operative design principle is the presence of redundant, engineered multiple barriers so that in the event that one layer of protection should fail the sample isolation and integrity will be maintained. The challenge is to design and engineer a sample cell out of materials that will simultaneously allow x-rays to enter and exit without appreciable attenuation, while at the same time maintaining an adequate hermetic seal. Many common designs used to date have borrowed from cryogenic and vacuum techniques where the problems associated with materials compatibility (mating plastic or glass windows to metal frames) have already been solved.

The issue of sample containment and monitoring is particularly relevant to working with radionuclides. Monitoring is necessary before, during, and after an experiment. The degree of monitoring effort depends on the nature of the sample and the experimental conditions under which the study is performed. Thus, specialized experiments utilizing temperature as a variable or experiments involving *in-situ* chemical modification of the sample can become more time consuming from a safety engineering perspective. Monitoring of radionuclides is complicated further when considering the different isotopes and their modes of decay. For example, ^{239}Pu is primarily an alpha emitter which makes it less of a radioactive hazard if exposure is only external since the alpha particles are stopped by the outer dead layer of skin. However, ^{239}Pu absorbed internally is highly toxic as the alpha particles are now in contact with living tissue and can cause cancer. In contrast, ^{137}Cs is a strong gamma emitter and thus poses a significant external hazard resulting from the large radiation field. The type of monitoring (α, β, or γ) needs to be tailored to the specific radionuclide. Therefore, the overall safety measures required for each experiment must be reviewed and examined on a case-by-case basis for each experimental system.

ENVIRONMENTAL APPLICATIONS OF XAS

Selenium

Hayes and coworkers have reported a study of selenium adsorption on α-FeOOH (goethite) using XAS at the Se K-edge.[13] In studying the adsorption of selenite and selenate at the α-FeOOH-water interface, they found that selenate adsorbs as a weakly bonded outer-sphere complex while in contrast, selenite adsorbs as a strongly bonded inner-sphere complex. Pickering and coworkers investigated selenium on contaminated soils from the Kesterson Reservoir in California.[14] These results demonstrated that elemental selenium is present in the contaminated soils and in simulated soil experiments. These findings are important in determining transport properties because the mobility (solubility and binding constants) is largely dependent on the chemical form (i.e., reduced selenium is less mobile than the oxidized forms).

Cobalt

O'Day, Chisholm-Brause, and coworkers have reported several XAS investigations of Co chemisorption on oxide surfaces.[15-19] The primary goal of this work was to characterize the dominant form of surface complex under specific coverage conditions and to gain some insight into the mechanisms of partitioning at the solid/water interface. They studied Co(II) sorbed on γ-Al_2O_3, rutile (TiO_2), α-SiO_2, and kaolinite ($Al_2Si_2O_5(OH)_4$). The results established the presence of multi-nuclear as well as mononuclear sorption complexes, whose relative amounts depended on surface coverage. Furthermore, the bond lengths and coordination numbers of the surface complexes were found to be different depending on the type of surface. For example, on kaolinite at low coverages, Co(II) binds strongly as an mononuclear inner-sphere complex. At higher coverages (still below one monolayer of Co), oxy- or hydroxy-bridged polynuclear species are formed.

Hayes and coworkers have also investigated Co(II) adsorption on oxide surfaces with an emphasis on the effect of mixed cation/anion systems (i.e., Co/Se).[20] The utility of acquiring accurate structural data on surface complexes lies in being able to predict transport behavior using macroscopic surface complexation models (SCM). In these examples, the transport properties are dramatically different for mononuclear versus multi-nuclear complexes, and as a result the information on the nuclearity of the surface species is critical to developing accurate models. This type of approach is essential for evaluating and devising innovative waste remediation techniques.

Chromium

Bidoglio and others have recently studied the mechanism of Cr adsorption on iron oxide minerals (goethite).[21] Using XANES measurements at Cr K-edge, they determined that adsorbed Cr(VI) ions are partially reduced to Cr(III) on the α-FeOOH surface under oxygenated conditions. The structure of Cr(III) ion sorbed on goethite has been described by Charlet and Manceau.[22] In this work, it is shown that a γ-CrOOH precipitate is formed at the surface under conditions well below saturation. A similar result was found by Fendorf et al.[23] for Cr(III) adsorbed on silica in that a γ-CrOOH surface precipitate was formed below monolayer coverage and at conditions where solution precipitation does not occur.

Lead

Chisholm-Brause, Hayes et al. have investigated the structure of Pb(II) surface complexes at the γ-Al_2O_3 and α-FeOOH (goethite)-water interfaces.[24-25] The Pb L_{III}-edge XAS results on α-FeOOH indicated that the sorption mechanism is strongly influenced by the concentrations of the sorbate. At low coverages, monomeric surface complexes are formed, while the formation of extended polymeric species is indicated at higher coverages. Similarly, Pb(II) bonds directly to γ-Al_2O_3 as a mononuclear, inner-sphere complex, with the formation of small polynuclear species occurring only as the coverage is increased. In either case, the data did not reveal any evidence that indicated the formation of a solid precipitate. These results demonstrate the utility of XAS for differentiating between mechanisms of adsorption, (i.e., inner-sphere versus outer-sphere complex formation).

Inorganic Elements From Fossil Fuels and Sediments

The presence of toxic inorganic elements in fossil fuels is a problem since the combustion process provides a mechanism for the release of these elements into the environment, either directly into the air, or by the leaching of combustion waste into the groundwater. Knowledge of speciation is crucial since the toxicity of many of these elements depends on their oxidation state. Silk and coworkers[26] have used XAS to investigate the chemical form of arsenic, nickel, and vanadium in oil fly-ash samples. Their results showed that the principal oxidation states that are present for each element are V(IV), Ni(II), and As(V).

Huggins and coworkers[27] have studied the speciation of chromium and arsenic in coal and coal ash. The results show that the chromium exists in the trivalent oxidation state in all samples. The Cr K-edge XANES of the coal samples demonstrate the presence of an oxyhydroxide, i.e., CrOOH, while the speciation in the ash is dominated by a form indicative of chromium in a silicate mineral. Arsenic K-edge XAFS show that As exists substitutionally for the element sulfur in pyrite (FeS_2). Arsenate (AsO_4^{3-}), possibly derived from the oxidation of arsenic contaminated pyrite, is also found in some coals as well as ash.

Vairavamurthy et al.[28] have studied the effects of hydrogen sulfide and thiols on the speciation of chromium, nickel, and copper in anoxic marine sediments. Since such sediments may serve as a sink for toxic metals, marine sediments can act as a source and carrier for the redistribution of toxic species. XAS studies at the Cr, Ni, and Cu K-edges showed that various redox processes were occurring. For example, Cr(VI) is transformed to Cr(III) by a commonly occurring thiol, while Ni(II) and Cu(II) remain inert to the same treatment. Hydrogen sulfide apparently does not reduce Cr(VI), Ni(II), or Cu(II), but instead causes some structural rearrangements to occur primarily in Ni(II) and Cu(II).

RADIONUCLIDES

Strontium

Radioactive ^{90}Sr produced from nuclear fission poses another health hazard in the environment. Pingitore and co-investigators have studied the mode of incorporation of Sr^{2+} into calcite using XAS.[29] They have found that Sr^{2+} enters the calcite lattice substitutionally adopting a structural environment with six-fold coordination similar to that of the Ca^{2+} lattice sites. The results suggest that incorporation of Sr^{2+} occurs through formation of a dilute solid solution rather than through adsorption or occlusion. In addition, since it is substitutional for Ca, it is not likely to be significantly more mobile than Ca. Preferential leaching of Sr from calcite is not anticipated and any encapsulated Sr should be relatively immobile which has important implications for ^{90}Sr in similar environments. Kohn et al.[30] have reported a similar study relevant to Sr partitioning in dry and hydrous silicate glasses. They find that the Sr-O bond distances are shorter than for crystalline silicates and appear to shorten with decreasing polymerization. Hydration of the dry glasses seems to increase the Sr-O bond lengths. These minor changes in the first shell coordination suggest that the Sr partition coefficients are not strongly dependent on melt composition.

Uranium In Contaminated Soils

XAS has recently been used by Conradson et al.[31] to study the adsorption of uranium on contaminated soils at the Feed Materials Production Center in Fernald, Ohio. This site served as a uranium processing plant operating as part of the DOE weapons program. Over a 30 year period, varying amounts of uranium were distributed around the site leaving concentrations in excess of 1000 ppm at some areas. Along with laser-induced luminescence spectroscopy, XAS measurements at the U L_{III}-edge showed that ~80-90 % of the uranium from all of the areas on the site was present in the uranyl form (uranium in a (VI) oxidation state containing the linear UO_2^{2+} moiety). Based on these and other results, numerous treatments were considered and tested for their efficiency of uranium removal. Most of the methods that were tested utilized a chelating agent such as carbonate, citrate, or Tiron as a method for extracting the uranyl species. Variations of these processes involved the addition of a reducing agent like sodium dithionite to promote dissolution. Based on characterizations of the uranium that remained on the soil subsequent to the various treatments, several different effects were observed in the soils. Treatment with water or dithionite had essentially no effect on removal and served as a control. All other treatments served to decrease the total amount of uranium in addition to increasing the relative amount of U(IV)-containing species (such as UO_2) that remained on the soil.

Bertsch et al.[32] have also used XAS to determine the speciation of uranium in contaminated soils and sediments from the Fernald Environmental Management Project and the Savannah River Site. In this work, the x-ray beam was adjusted from 50-300 μm spot sizes, which enabled the experimenters to spatially resolve differences in the distribution of uranium oxidation states. Based on the edge position and spectral signatures associated with the U L_{III}-edge, they found that the uranium in all of the clay fractions was predominantly in the U(VI) form. The lack of spatial variation was attributed to the individual clay particles being much smaller than the beam size. In contrast, spatial variability was seen on the sand samples taken from both sites. They observed two general types of domains: 1) regions that contained primarily U with unusually small amounts of other elements such as Fe and Mn; and 2) regions that were depleted in uranium yet relatively enriched with Fe and Mn. In addition, they also observed variations in the oxidation state among the sand and silt fractions taken from the Fernald site. These amounted to regions of predominantly U(VI) or U(IV), as well as mixtures of these two oxidation states. The presence of U(IV) in the Fernald sand and silt samples was attributed to U(IV) being in the source term as an airborne particulate rather than as soil-reduced U(VI).

Uranium Sorption

Another example of the application of XAS to determine speciation is given in a series of investigations of uranium adsorbed on the mineral montmorillonite. Chisholm-Brause et al.[33] studied the sorbed complex structure as a function of uranium surface coverage. They found that over a 20-fold range of surface coverages, there were systematic changes in the coordination numbers and bond lengths for the oxygen atoms in the equatorial planes of the sorbed uranyl species. This behavior was attributed to the existence of at least three structurally distinct sites on the clay that were interacting with uranyl over the range of coverages studied. Dent et al.[34] also found that the uranyl ion retains its structure after adsorption onto montmorillonite. However, in their study they observed changes in the surface complex structure as a function of varying pH. At pH <3 a monomeric aquo-complex consisting of the uranyl ion with 5-6 equatorial oxygens is observed, while at pH of 4.3 the data indicate the presence of a multi-nuclear species. These results taken together have significant implications for developing accurate transport models and illustrate the valuable microscopic information that is readily available from XAS.

Uranium Bioremediation

Dodge and coworkers[35] recently investigated uranium reduction by *Clostridium* sp. using XANES and x-ray photoelectron spectroscopy (XPS). Comparisons of XPS spectra for U(VI) in the control sample and the U in the sample inoculated with *Clostridium* sp. showed that U(VI) was reduced to U(IV). Analysis of the uranium XANES at the M_V-edge for U metal, UO_2, and the inoculated sample also confirmed the reduction of uranium. Their results further showed that the reduction only occurred in the presence of live or growing cells, suggesting the reduction occurs through enzymatic action.

Actinides In Silicate Glasses

Farges et al. have studied the local structure of trace to minor levels of uranium in various silicate glass/melt systems using XAS.[36] Depending on the glass compositions and the oxidizing or reducing conditions, the syntheses generally yielded uranium with oxidation states of (IV), (V), or (VI). They found that the U(VI) sites were composed of uranyl groups with two axial oxygen atoms at 1.77-1.85 Å and four to five equatorial oxygens at 2.21-2.25 Å. The U(V) was found to occur in distorted polyhedra with six oxygen atoms at 2.19-2.24 Å. The U(IV) sites, although similar to the U(V) sites, have less distortion about the octahedra and have six oxygens at 2.26-2.29 Å. In the syntheses which included minor amounts of F or Cl, no evidence for uranium-halogen coordination was found. In most cases, there was no indication of uranium second nearest neighbors that would suggest any amount of local clustering. In addition, more detailed XAS analyses of the bond lengths and strengths afforded some plausible explanations of the compatibility and solubility of uranium in the melt.

Eller, Larson, and coworkers[37-38] performed XAS investigations of the coordination environment of uranium, thorium and selected lanthanides (Ce, Pr, Gd, and Lu) in borosilicate glass, a medium which has been chosen for the storage of high-level radionuclide waste. The study of lanthanides is relevant because they serve as structural models for the trivalent actinides. Based on the U L_{III}-edge XANES position as well as optical measurements, uranium was found to be present in the glass in the hexavalent form, although there was no evidence for the uranyl structure (2 short axial oxygens) as is found for most U(VI) oxide species. Such a finding is not without precedent in that cubic UO_3 also exists in a non-uranyl, non-localized form. Under conditions of higher uranium loadings in silicate glasses, other groups[39-41] have found that U(VI) formed uranyl-type multi-nuclear clusters. Np(V) apparently behaves similarly to U(VI) and also forms actinyl structures in these glasses. The structure of the tetravalent ions of U, Np, and Pu resemble the fluorite structure of UO_2 with coordination numbers ranging from 6-8. In the case of the lanthanides, each was found to enter the glass primarily in the trivalent oxidation state (Ce may exist as 4+), with well-ordered first shells of oxygen neighbors.

Barret et al.[42] used fluorescence XAS at grazing incidence to investigate changes at the surface of borosilicate glasses containing 3 wt.% uranium. Comparison of the spectra taken at grazing incidence with those of the bulk glass allows detection of structural differences at the surface. Performing the XAS measurements *in-situ* during aqueous leaching at 100°C, they found a local increase in uranium concentration at the surface during the first 30 minutes. The surface structure was composed of polynuclear clusters composed of uranyl-like sub-units. The clusters were partially removed after 30 minutes.

Neptunium Sorption On Goethite

Brown and co-investigators reported a study of Np(V) adsorption at the α-FeOOH (goethite)/water interface.[43] Based on a comparison with Np and U L_{III}-edge XAS data taken for several model systems, the Np(V)-goethite sample was determined to sorb as the neptunyl moiety (NpO_2^+). Curve-fits to the EXAFS data gave two oxygens at 1.85 Å and ~5 equatorial oxygens at 2.51 Å indicative of a distorted pentagonal-bypyramid. While the data indicated no formation of multi-nuclear species, a second shell feature in the data was tentatively assigned to a Np-Fe interaction which would suggest an inner-sphere coordination complex.

Actinide Near-edge Structure

Several groups have performed detailed investigations of U, Np, and Pu XANES (L_{III}; $M_{III,IV,V}$; $N_{IV,V}$; and $O_{IV,V}$ edges) for a variety of solid compounds with different oxidation states.[44-47] One of the key issues under consideration is the assignment of oxidation state based on the L_{III} absorption edge position. In nonmetallic systems (particularly oxides and halides), differentiating between the oxidation states (III), (IV), and (VI) for Np and U is fairly well-established by chemical shifts plus spectral signatures. Another issue studied in some detail is the characteristic shoulder present on the high-energy side of L_{III}-edges of uranyl- or neptunyl-containing compounds. Because it is consistently observed in the L_{III}-edges of these compounds, this spectral feature has been used as a signature for the uranyl or neptunyl groups. It has also been postulated that this feature arises from a multiple-scattering resonance associated the linear UO_2^{2+} or NpO_2^+ groups.[44] Since the uranyl- or neptunyl-containing systems also possess the U(VI) or Np(V, VI, VII) oxidation states, the shoulder has also been used as a indicator for oxidation state. Along these lines, it has been argued that the shoulder results from an electronic effect related to multiple $5f$ final state configurations rather than a multiple scattering resonance.[47]

Plutonium In Concentrated Nitric Acid

Information on the speciation of Pu(IV) in nitric acid is required in order to improve the efficiency of current aqueous separations techniques (extractions and ion-exchange) of the actinides. Veirs et al.[48] have recently reported the structural determination of the aqueous nitrate complexes of Pu(IV). Solutions of the Pu(IV) ion were studied in nitric acid concentrations of 3, 8, and 13 M. Systematic changes in the EXAFS demonstrated the trends of increasing nitrate ligation and decreasing water ligation as a function of increasing nitric acid concentration. The coordination numbers of the nitrogens and non-coordinating oxygens were found to be consistent with previous UV-visible absorption studies in which the principal species were found to be the di-, tetra-, and hexanitrato complexes in these solutions. The Pu(IV) complexes in nitric acid have structures which are similar to analogous solid state compounds of thorium and neptunium nitrate compounds. The results indicated that the nitrate ligands are planar and bidentate. Furthermore, the nitrate structure was found to be distorted with respect to that of the free nitrate in that the nitrogen-oxygen bond length of the non-coordinating oxygen is significantly shorter than the nitrogen-oxygen bond lengths of the coordinated oxygens.

Plutonium Metallurgy

Another important area of plutonium research involves aspects of the stability of its metallic phases. Plutonium has a complex phase diagram possessing six phase transformations and several different crystal structures between room temperature and its melting point. Phase stabilization at low-temperature is an important requirement for fabrication and long-term storage. At ambient temperature, the fcc δ-phase is stabilized by adding low concentrations of gallium or aluminum. The mechanism of this stabilization is still unknown. Cox et al.[49] have investigated the local structure of the gallium and plutonium environments in 1 wt.% Ga δ-stabilized plutonium. XAFS measurements were performed at the Ga K-edge and Pu L_{III}-edges at 80 K. The results of this study showed that the Ga substitutes isomorphically in the fcc lattice. The structure around the Ga atoms is well-ordered with a first shell of 12 Pu neighbors at 3.14 Å. In contrast, the Pu environment is extremely disordered with an average first shell distance of 3.29 Å, which is close to the value expected for the pure δ-phase metal. A model for the fcc lattice stabilization is proposed in terms of the local GaPu$_{12}$ clustering.

RESEARCH FROM THE LBNL ACTINIDE CHEMISTRY GROUP

Selenium Bioremediation

The concentration and immobilization of hazardous materials by microbial agents offers great potential for use in environmental remediation technologies.[50,51] In the case of selenium possible future applications of bioremediation technologies range from oil refinery waste streams containing selenite to agricultural drainage which has contaminated sites such as the Kesterson Reservoir in California. The aerobic soil bacterium, *Bacillus subtilis*, has been found to detoxify selenite, Se(IV), by reductive metabolization although the exact mechanism is not fully understood.[52] As in other cases selenium oxyanions appear to be reduced to a red elemental form, although there is no direct *in situ* spectroscopic evidence to support this contention.[53,54]

Vegetative cells of *B. subtilis* and of an unidentified bacillus isolated from the Kesterson Reservoir were exposed to aqueous growth media containing either selenite or selenate. The bacteria metabolized the selenite growth solution, however, there was no uptake of selenium in the selenate solution. In addition to the isolated bacterium, selenium K-edge XANES spectra were obtained for elemental selenium (red and gray allotropes), Na$_2$SeO$_3$, and Na$_2$SeO$_4$ powders, all of which contain selenium in well-defined oxidation states. The edge positions in selenite and selenate are shifted to higher energy by 4.2 and 9.0 eV respectively, relative to elemental selenium (at 12658 eV). Figures 2-3 (Se K-edges and chemical shifts, respectively) show the XANES of red and gray selenium together with those of the bacteria, and clearly demonstrate that the selenium is present in elemental form in both *B. subtilis* and the Kesterson bacillus. There is also a discernible difference between the red and gray allotropes of selenium in the near edge region (12665-12675 eV). The results suggest that both *B. subtilis* and the Kesterson bacillus reduce selenite to red selenium, a finding consistent with the assertion that red selenium is the biological end product. These results support the view that organisms like *B. subtilis* offer a promising means for the removal of selenite from contaminated aqueous environments such as oil refinery and agricultural waste streams.

The origin of the XANES difference between gray and red selenium is not easy to ascertain. Although the theoretical modeling of XANES has become widespread, it is very difficult to achieve without well-established structural parameters. Unfortunately these are not available for selenium. Elemental selenium occurs in a number of structural modifications, not all fully characterized, and often in a mixture of phases. The gray "metallic" form is thermodynamically the most stable and consists of unbranched helical chains with Se-Se distances of 2.37 Å. Red selenium has three

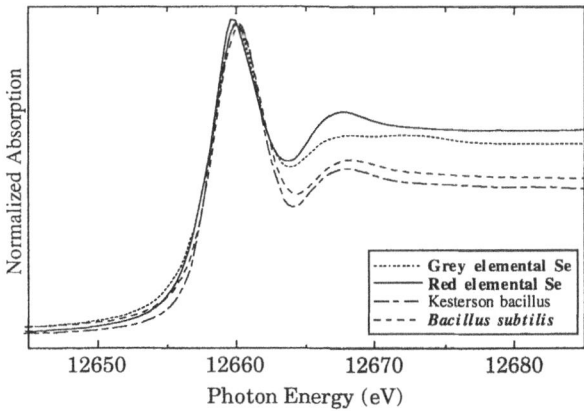

Figure 2. Selenium K-edge XANES spectra of amorphous red and gray selenium and from *B. subtilis* and the Kesterson bacillus after growth in a selenite medium. The spectra have been normalized to equivalent peak heights. The energy scales have been aligned with respect to the gray selenium absorption edge at 12658 eV.

crystalline modifications, all of which feature Se_8 rings and differ only in packing. The average ring Se-Se separation is 2.34 Å. The more common, amorphous form of red selenium consists of a mixture of distorted chains and Se_8 rings.

In view of the observed slight difference in the XANES spectra of gray and amorphous red selenium, EXAFS was employed in an attempt to distinguish the two forms. EXAFS was measured out to k of 12.7 $Å^{-1}$. The Fourier transform revealed a single frequency in both cases, which fit well with essentially the same parameters. Thus, the EXAFS results do not distinguish between the gray and red allotropes of elemental selenium.

Speciation of Tc in Wasteform Cement

^{99}Tc is a by-product of nuclear fission, and its long half-life ($t_{1/2}$ = 2.13x10^5 years) necessitates its careful consideration in the long-term disposal of nuclear waste. The principal Tc-containing component of nuclear wastes is TcO_4^-. The propensity of TcO_4^- to migrate into the geosphere has led to research aimed at the facile conversion of pertechnetate into other, presumably less mobile compounds of ^{99}Tc (i.e., chemical reduction of the mobile TcO_4^- to less soluble Tc^{4+} species). One method for disposal involves microencapsulation of the liquid waste in a cement matrix with a reducing agents such as blast furnace slag (BFS). A preliminary XANES investigation of a process designed to reduce TcO_4^- to TcO_2 in wasteform cement has been described in Ref. 55. The line shapes of the XANES spectra and the chemical shifts of the Tc K-edge in TcO_4^-, TcO_2, and Tc metal were used to establish the oxidation state of Tc.

The Tc cements were prepared and packaged at LBNL. The Tc reference materials were prepared at both LBNL and LANL. The untreated cement was prepared by combining 50 wt.% Ordinary Portland Cement (OPC) and 50 wt.% of a simulated aqueous waste solution containing 1200 ppm ^{99}TcO$_4^-$. The composition of the simulated wastestream has been reported elsewhere.[56] The cement containing the reducing agent was composed of 10 wt.% simulated waste solution, 80 wt.% OPC, and 10 wt.% BFS. Slag is a source of reduced iron (ferrous ion) and complex reduced sulfides, which is believed to reduce TcO_4^- to TcO_2. The cement samples were cured for four days at 60°C. The TcO_2 reference material was prepared by the thermal decomposition of ammonium pertechnetate. The samples were contained in polyethylene tubes or polystyrene cuvettes, followed by the application of several heat sealed polyethylene film barriers.

XANES spectra were collected from cements prepared at LBNL, with and without slag, using a simulated nuclear waste stream spiked with pertechnetate ion. The XANES results of the Tc-containing cements and the Tc reference materials (Tc metal, TcO_2(IV), and a TcO_4(VII) solution) are presented in Figure 4. The range of chemical shifts for the Tc K-edge has been characterized and spans ~15 eV between the metallic and the Tc^{7+} state. The chemical shift from TcO_4^-(VII) to tetravalent TcO_2(IV) is -7 eV, therefore resolving the difference in oxidation state solely on the basis of edge position is relatively straightforward. Additionally, there is a characteristic pre-edge feature present in the TcO_4^- spectra that clearly fingerprints the tetrahedral TcO_4^- moiety. The spectra of Tc metal and TcO_2 have no such pre-edge features and exhibit edge positions

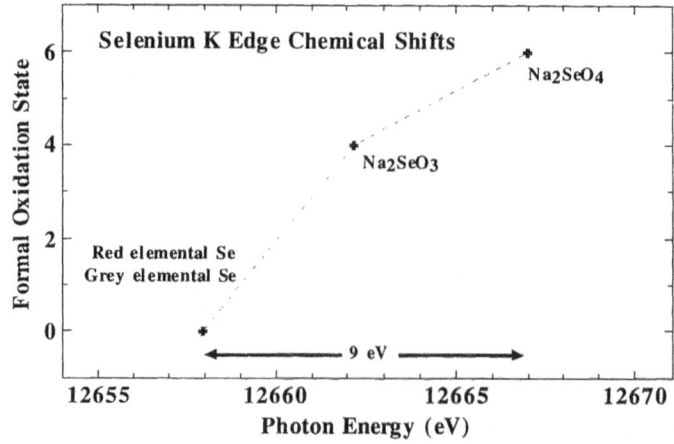

Figure 3. Chemical shifts at the Se K-edge.

significantly shifted to lower binding energies from that of TcO_4^-. Comparison of the XANES spectra of Tc in untreated cement (not shown-identical to that of the TcO_4^- spectra) to spectra of Tc in slag treated cement, shows that the primary oxidation state of Tc in the cements is different. Several locations within the treated cement were characterized and the results were the same throughout. Thus, the addition of slag to the cement formulation, under these curing conditions, was essential to significantly reduce the TcO_4^- to TcO_2 (or some other Tc^{4+} compound) in thecement samples used in this investigation. The analysis of the EXAFS from the Tc metal and TcO_2 also provided some of the first detailed structural information about these materials. The full details of the results of the on-going Tc cement investigations will be covered in a future publication.

To complement the Tc cement studies, a XANES investigation on a wider range of model Tc compounds with formal oxidation states 0, +2, +4, +5 and +7 and an EXAFS characterization of the compounds $Tc_2(CO)_{10}$ and TcO_2 were performed.[57]   Tc K-edge positions relative to NH_4TcO_4 are shown in Figure 5. There is a strong correlation between formal oxidation state and K absorption edge position. An increase in formal oxidation state implies an increase in effective charge, and it has been shown that x-ray absorption edge energies are governed mainly by the effective charge on the absorbing atom or ion. In general, chemical shifts are toward the high energy side of the elemental edge and increase progressively with the formal oxidation state of the absorbing atom. The ~15 eV K-edge shift between Tc metal and NH_4TcO_4 provides a substantial range over which the relationship of oxidation state to absorption edge energy may be examined. Of the eight systems studied, six display an almost linear relationship, with $Tc_2(CO)_{10}$ and $Tc(ArN)_3I$ showing more significant differences. $[TcCl_2(PR_2R')_2]_2$ and $TcCl_2py_4$ contain Tc as Tc(II), and their edge positions are 13.2 and 11.2 eV respectively lower than NH_4TcO_4. That $[TcCl_2(PR_2R')_2]_2$ has a 2.0 eV greater shift than $TcCl_2py_4$ may be a result of the stronger s-donor and weaker p-acceptor nature of PMe_2Ph versus C_5H_5N.

Both $Tc(ArN)_3I$ and NH_4TcO_4 contain Tc in a formal oxidation state of +7, but $Tc(ArN)_3I$ has a Tc K-edge energy which is 5.6 eV lower than NH_4TcO_4. This is probably a reflection of the relative electron withdrawing properties of the ligands in each case, with the oxygens inducing a greater effective charge on the Tc than the ArN/I combination. The magnitude of the difference is somewhat surprising, as $Tc(ArN)_3I$ has an edge position which is further from NH_4TcO_4 than the Tc(V) $[Tc(ArN)_3]_2Hg$. In light of the preceding discussion, however, it is impossible to determine if this discrepancy is a genuine indication of differences in the effective charge at the metal atom. The most significant deviation from the formal oxidation state/K edge energy relationship is that of $Tc_2(CO)_{10}$, which has an edge position closer to NH_4TcO_4 than either $TcCl_2py_4$ or $[TcCl_2(PR_2R')_2]_2$. This is almost certainly the result of the electron withdrawing ability of the carbonyl unit, and indicates that p backbonding from metal to ligand dominates ligand to metal s donation.

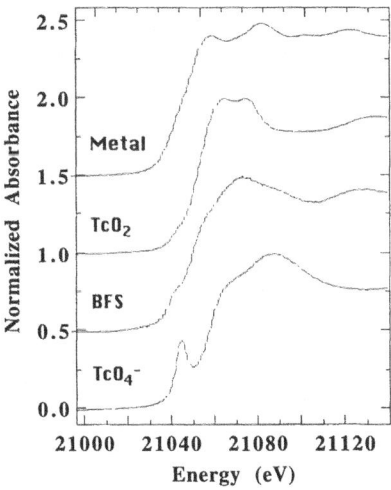

Figure 4.   Normalized Tc K-edge XANES spectra of Tc model compounds (Tc metal, Tc dioxide, and Tc pertechnetate) and a spectrum from a single TcO_4^-/cement mixture cured at 60°C for four days with blast furnace slag (BFS). There was no reduction in mixtures consisting solely of Ordinary Portland Cement and TcO_4 (spectra identical to TcO_4^-).

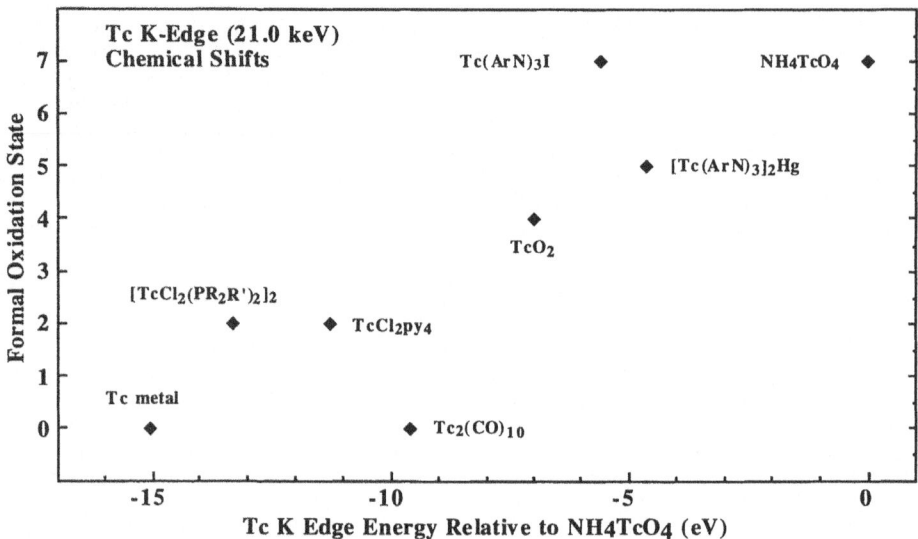

Figure 5. Chemical shifts at the Tc K-edge versus formal oxidation state for Tc metal, $[TcCl_2(PMe_2Ph)_2]_2$, $TcCl_2(py)_4$ (py=C_5H_5N), $Tc_2(CO)_{10}$, TcO_2, $Tc(ArN)_3I$ (ArN=2,6-$Pr^i_2C_6H_3N$), $[Tc(ArN)_3]_2Hg$, and NH_4TcO_4. All near edge spectra were collected with NH_4TcO_4 as a reference, and the chemical shifts are therefore reported relative to NH_4TcO_4.

Uranyl Ion Complexed With Tartaric, Citric, and Malic Acids

Actinide speciation information is essential for assessing and developing long-term strategies addressing problems such as migration in nuclear waste repositories or improvements in the processing of nuclear waste and materials. Relative to the latter, one method for removing uranium contamination from soils involves extraction using a chelating agent such as Tiron or citrate. Martell et al.[58,59] and Markovitz et al.[60] have published a series of articles detailing the complexation of the uranyl ion with tartaric, malic, and citric acids as a function of pH. From the analysis of potentiometric titration results, they showed that in the pH range 2-4, the uranyl ion forms a 2:2 dimeric species, $(UO_2)_2(L)_2$, where L= tartrate, malate, or citrate ligands. In considering the possible 2:2 structures shown in Fig. 6, both of which involve the α-hydroxyl group of the ligand(s) for bridging of the uranyl ions, Martell et al. stated a preference for the structure in the right panel Fig. 6.[59] However, it is not possible from the titration data to directly distinguish between the structures shown in Fig. 6 as well as other plausible 2:2 structures.

A 0.08 M stock solution of the UO_2^{2+} ion was used to prepare 1:1 mixtures of ligand to UO_2^{2+} ion. The final pH readings of the 0.04 M uranyl tartrate, malate, and citrate solutions were

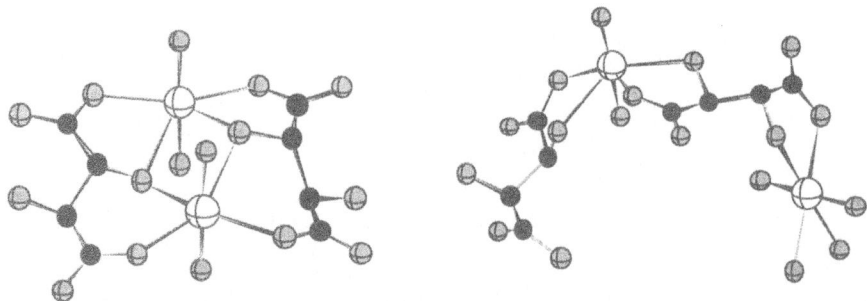

Figure 6. Possible structural models for the uranyl tartrate dimer having the empirical formula $(UO_2)_2(L)_2$ where L=tartrate; (left) uranyl groups are bridged by α-hydroxyl groups, (right) bridging occurs through a single α-hydroxyl group.

2.2, 2.0, and 3.8, respectively. The raw k^3-weighted EXAFS data for the uranyl tartrate, malate, and citrate solutions show strong similarities in phase as well as amplitude. The FTs as shown in Fig. 7 (uncorrected for phase shift) illustrate that the EXAFS is dominated by scattering from the O atoms of the linear UO_2^{2+} group (peak at 1.30 Å) and scattering from the O atoms lying in the equatorial plane of the UO_2^{2+} ion (peak at 1.90 Å). In addition, the FTs show the presence of another interaction at ca 3.75 Å which is reproduced in each solution. The portion of the FT above 4 Å can be used to gauge the noise level in the EXAFS signal. Considering the fact that the FT above 4 Å is featureless and smooth, the 3.75 Å peak is the result of a real interaction, rather than an experimental artifact.

Modeling and curve-fitting analyses were performed on the raw EXAFS data to determine the bond lengths and coordination numbers, and to examine the origin of the peak at 3.75 Å. The data range used in the fitting procedure was 3-14 $Å^{-1}$ for the tartrate and malate solutions, and 3-12 $Å^{-1}$ for the citrate solution. Each sample shows ~N=2 oxygen atoms at 1.78 Å indicative of axial oxygen atoms on the uranyl group. A relatively weak multiple scattering (MS) interaction at 3.56 Å, originating from the 4-legged path along the O–U–O vector (twice the U–O_{ax} distance), was also included in the fits. Rather than being varied independently, this path was linked directly to floating bond lengths (R) and coordination number (N) values of the axial oxygen shell. The equatorial plane around the uranyl group contains ~N=2 oxygen atoms at ca 2.40 Å in each of the samples.

In an investigation of the nature of the 3.75 Å peak, this contribution was isolated by taking the difference between the uranyl tartrate data and a fit which included only the near-neighbor contributions mentioned above. The residual obtained by this procedure contains only those contributions unaccounted for in the fit-specifically the interaction appearing at 3.75 Å. The residual EXAFS signal was Fourier-filtered over the range of 3.2–4.2 Å to remove noise in the fit. Even at this level of analysis, the increasing EXAFS amplitude as a function of k is a signature of backscattering from an atom of relatively high Z. The corresponding single U shell fit confirms this assignment. The only other possible source of this peak would be scattering along paths involving the low Z ligands C or O centers. Low Z atoms are not normally detected in room temperature solutions at R>3 Å, unless a MS enhancement of the amplitude is operative. Examples for uranium systems have been observed when ligands like carbonate or nitrate adopt a symmetric bidentate geometry where the distal O atom is collinear with the absorbing atom and the C or N atoms. Thus, the most plausible alternative explanation for the peak at 3.75 Å is bidentate ligation of the carboxylate groups (i.e., a U–C_1–C_2 path). However, the phase and amplitude derived from a sum of 3-legged (U→C_1→C_2→U) and 4-legged (U→C_1→C_2→C_1→U) MS paths offer a poor match to the data. In addition, the resulting U–C_2 bond length of 4.24 Å would give the C_1–C_2 bond length at 1.31 Å which is inconsistent with the C–C bond length of 1.53 Å determined for $UO_2(CH_3COO)_2 \cdot 2H_2O$.

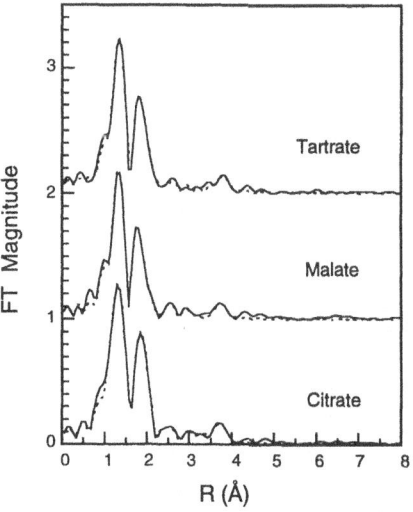

Figure 7. Fourier transforms (uncorrected for phase shift) of U L_{III} EXAFS for 1:1 mixtures of uranyl polycarboxylates. The solid line is the experimental data, and the dashed line corresponds to the best theoretical fit of the data. The U–U peak at 3.75 Å is evidence for the dimeric structure in the left panel of Fig. 6.

Apart from the structure shown in left panel of Fig. 6, the only other structure that could possess a U–U interaction at 3.95 Å would be a hydrolysis product, $(UO_2)_2(OH)_2^{2+}$. However, equilibrium calculations using the computer code HYDRAQL[61] and considering the numerous equilibria involved show that the concentrations of $(UO_2)_2(OH)_2^{2+}$ and other hydrolysis products are several orders of magnitude smaller than the concentration of $(UO_2)_2(L)_2$. Thus, combining the results with those of Martell et al.,[58,59] it is now possible to assign the structure in the left panel of Fig. 6 to the 2:2 dimer present in these solutions. The structure in the right panel of Fig. 6 is ruled out on the basis that a U–U interaction at 3.9 Å, if configurationally allowed, would not be detected due to dynamic disorder in the solution.

SYNCHROTRON RADIATION ENVIRONMENTAL WORK IN THE FUTURE

The utility of SR techniques in a wide range of environmental science applications, some of which have been briefly alluded to herein, clearly supports the contention that there will be a substantial amount of growth and influx of new users in this increasingly important field of SR. It is an understatement to mention that there are a significant number of existing environmental problems of national importance and the economic costs of remediating these problems is enormous. The scientific issues involved are challenging and the severity of many situations actually provides direct impetus for research motivated by potential future cost savings. It is clear to federal science funding agencies that there are significant benefits from performing both fundamental and applied research to support national environmental priorities. Similar to many other scientific fields at this time, the competition for research support is also very competitive. However, the future funding outlook is rather positive, a situation which differs markedly from that of many other disciplines. The utility of SR in environmental research, combined with the expanding research funding, has already led to increased research activities and promises to keep this type of work at the forefront of SR-based science in the near future.

A unique aspect of environmental SR work is that it is interdisciplinary and brings together scientists with widely different backgrounds to focus on a similar set of scientific issues at the end of a beamline. Many of the SR techniques and technologies have matured sufficiently to permit new environmental users to achieve a large degree of success without experiencing a lengthy start-up period. The impact that x-ray SR techniques can have on environmental research programs has been recognized by environmental research groups. These groups have rapidly utilized existing SR facilities to become one of the newest blocks of SR users. Rather than employing SR techniques as the sole means of experimental investigation, SR methods have been used to largely complement laboratory-based scientific methods.

There will be further growth areas within the envelope of SR environmental science. Other SR techniques such as x-ray scattering and standing waves,[62] will undoubtedly become important methodologies for examining specific environmental topics. The most important development may be the common use of experimental microscopy techniques to gain simultaneous spatial and x-ray absorption information. Emerging bioremediation programs will also utilize SR techniques to characterize the fundamental biological/chemical principles of bioremediation and to further develop bioremediation technologies.

The availability of beamtime at many of the best beamlines is quite limited, however the construction of third generation light sources along with the installation/improvement of beamlines at existing synchrotrons promises to make SR techniques more accessible for environmental studies. The inherent flux and brightness characteristics of the third generation sources will permit whole new classes of environmental experiments, until now not possible. Furthermore, the third generation sources are driving improvements in many areas (e.g. fluorescence detector developments) that will also enhance beamtime use at other facilities as well.

The x-ray synchrotron radiation light sources, cognizant of the needs and scientific requirements of the growing but diverse environmental science user community, have proposed the construction of new experimental beamline facilities dedicated to environmental science investigations. This is clearly demonstrated at the APS (Advanced Photon Source, located at Argonne National Laboratory) where there are a number of collaborative access team (CAT) beamlines organized and instrumented to investigate specific classes of scientific issues, a large number which are germane to the environmental sciences. There are also proposals to build new environmental beamline facilities at the ESRF (European Synchrotron Radiation Facility, located in Grenoble, France) and SSRL (Stanford Synchrotron Radiation Laboratory, located at the Stanford Linear Accelerator Center) storage rings to provide increased beamtime and capabilities for environmental studies. Both of these proposed facilities are to be built with the capabilities to safely handle radioactive and other hazardous materials.

Throughout the presentation and discussion herein, the focus has been on research conducted in the x-ray regime alone. Concomitant with the continued growth of environmental programs in the x-ray regime, there will be a corresponding breakthrough of environmental programs into the soft x-ray/vacuum ultraviolet (VUV) part of the spectrum. The development of environmental science programs utilizing VUV SR techniques has been limited to some degree by the ultra-high vacuum restriction of a complex electron spectrometer endstation. Many fundamental environmental processes take place at interfaces and VUV techniques have traditionally been used for the investigation of surface science phenomena. The development of fluorescence detection techniques, photoemission microscopes, and the possibilities of working behind windows (where vacuum restrictions are relaxed or unnecessary) will provide the means for working with "real" samples in the near future. As in the x-ray regime, many of these improvements will be made available by developments at third generation VUV sources such as the Advanced Light Source (ALS, located at Lawrence Berkeley National Laboratory).

CONCLUSIONS

The application of SR x-ray absorption spectroscopy techniques to a wide variety of scientific problems in the environmental sciences has proven to be extremely valuable for determining both oxidation state and local structural arrangements that are either difficult or impossible to obtain by other means. XAS is especially well-suited to environmental science investigations since materials may be examined *in-situ* and hazardous materials can be safely contained during experiments. This is clearly evident from the brief research synopses presented herein. As a result, the environmental science user belongs to one of the fastest growing segments of the SR community. The management of the SR sources has fully recognized the utility of SR techniques to environmental science programs and has made beamtime available to new users in this field. There have also been new, specialized beamline facilities proposed for several x-ray SR sources to provide additional beamtime dedicated to environmental science users. There will likely be a corresponding utilization of VUV techniques and VUV SR sources for environmental research in the near future. The advent of the third generation x-ray light source at the APS and the realization of environmental beamline facilities will provide new state-of the-art capabilities for environmental research. Thus, the future of SR environmental science is bright and will experience continued growth.

ACKNOWLEDGMENTS

This work was supported in part by the Director, Office of Energy Research, Office of Basic Energy Sciences, Chemical Sciences Division of the U.S. Department of Energy under Contract No. DE-AC03-76SF00098. The work was performed in part at SSRL and NSLS which are operated by the U.S. Department of Energy, Office of Basic Energy Sciences, Divisions of Chemical Sciences and Materials Science. The SSRL Biotechnology Program is supported by the N.I.H., Biomedical Resource Technology Program, Division of Research Resources. Part of this work was also supported by the Bioremediation Education Science & Technology (BEST) Consortium at LBNL. NK would like to thank the EPSRC/NATO for a postdoctoral fellowship.

REFERENCES

1. R.G. Riley, J.M. Zachara, and F.J. Wobber, "Chemical Contaminants on DOE Lands and Selection of Contaminant Mixtures for Subsurface Science Research," **1992,** DOE/ER-0547T.
2. J.D. Spencer, editor, "Remediation, "Chapter V in: Joint DOE-EPA Report on Collaborative Environmental Research, **1993**.
3. H. Nitsche, *J. Alloys and Compounds*, to be published.
4. R. Prins and D.E. Koningsberger, "X-ray Absorption: Principles, Applications, Techniques for EXAFS, SEXAFS, and XANES", Wiley, New York, 1988.
5. P.A. Lee, P.H. Citrin, P. Eisenberger, and B. M. Kincaid, *Rev. Mod. Phys.* **1981**, *53*, 769.
6. J.R. Helliwell, *Reports on Prog. in Phys.* **1984**, *47*, 1403-1497.
7. J. Mustre de Leon, J.J. Rehr, S. Zabinsky, and R.C. Albers, *Phys. Rev. B*, **1991**, *44*, 4146.
8. Graham. N. George, EXAFSPAK computer code program suite, SSRL, (1994).
9. E. Stern and S. Heald, *Rev. Sci. Instr.* **1979**, *59*, 1579.
10. W.K. Warbuton, *Nucl. Instr. and Meth.* **1986**, *A246*, 541.
11. S.P. Cramer, O. Tench, M. Yocum, and G.N. George, *Nucl. Instr. and Meth.* **1988**, *A266*, 586.

12. J.J. Bucher, P.G. Allen, N.M. Edelstein, D.K. Shuh, N.W. Madden, C. Cork, P. Luke, D. Pehl, and D. Malone, to be submitted to Rev. Sci. Instrum.

13. K.F. Hayes, A.L. Roe, G.E. Brown Jr., K.O Hodgson, J.O. Leckie, and G.A. Parks, *Science,* **1987**, *238*, 783.

14. I.J. Pickering, G.E. Brown, Jr., and T. Tokunaga, "X-ray Absorption Spectroscopy of Selenium-contaminated Soils," Stanford Synchrotron Radiation Laboratory Activity Report, 1993, p. 92.

15. C.J. Chisholm-Brause, P.A. O'Day, G.E. Brown, Jr., and G.A. Parks, *Nature,* **1990**, *348*, 528.

16. C.J. Chisholm-Brause, G.E. Brown, Jr., and G. A. Parks, *XAFS VI Sixth International Conference on X-ray Absorption Fine Structure*, 1991, (ed. S.S. Hasnain), New York: Ellis Horwood Ltd, Publishers, pp. 263-265.

17. P.A. O'Day, G.E. Brown, Jr., and G.A. Parks, *XAFS XAFS VI Sixth International Conference on X-ray Absorption Fine Structure*, 1991, (ed. S.S. Hasnain), New York: Ellis Horwood Ltd, Publishers, pp. 260-262.

18. P.A. O'Day, G.A. Parks, and G.E. Brown, Jr., *Clays and Clay Minerals,* **1994**, *42*, 337.

19. P.A. O'Day, G.E. Brown, Jr., and G.A. Parks, *J. Colloid and Inter. Sci.,* **1994**, *165*, 269.

20. K.F. Hayes, L.E. Katz, and J.E. Penner-Hahn, "XAS Study of Metal/Ion Partitioning at Water/Mineral Interfaces," Stanford Synchrotron Radiation Laboratory Activity Report, 1993, p. 96.

21. G. Bidoglio, P.N. Gibson, M. O'Gorman, and K.J. Roberts, *Geochim. Cosmochim. Acta,* **1993**, *57*, 2389.

22. L. Charlet and A.J. Manceau, *Colloid and Inter. Sci.,* **1992**, *148*, 443.

23. S.E. Fendorf, G.M. Lamble, M.G. Stapleton, M.J. Kelley, and D.L. Sparks, *Environ. Sci. Technol.,* **1994**, *28*, 284.

24. C.J. Chisholm-Brause, K.F. Hayes, A.L. Roe, G.E. Brown, Jr., G.A. Parks, J.O. Lekie, *Geochim. Cosmochim. Acta,* **1990**, *54*, 1897.

25. A.L. Roe, K.F. Hayes, C.J. Chisholm-Brause, G.E. Brown, Jr., G.A. Parks, K.O. Hodgson, and J.O. Leckie, *Langmuir,* **1991**, *7*, 367.

26. J.E. Silk, L.D. Hansen, D.J. Eatough, M.W. Hill, N.F. Mangelson, F.W. Lytle, and R.B. Greegor, *Physica B,* **1989**, *158*, 247.

27. F.E. Huggins, J. Zhao, N. Shah, F. Lu, and G.P. Huffman, *Energy and Fuels,* **1993**, *7*, 482.

28. A. Vairavamurthy, W. Zhou, and B. Manowitz, "Speciation of Chromium, Nickel, and Copper in Anoxic Sediments: Effects of Hydrogen Sulfide and Thiols," National Synchrotron Light Source Annual Report, Brookhaven National Laboratory, 1992.

29. N.E. Pingitore, F.W. Lytle, B.M. Davies, M.P. Eastman, P.G. Eller, and E.M. Larson, *Geochim. Cosmochim. Acta,* **1992**, *56*, 1531.

30. S.C. Kohn, J.M. Charnock, C.M.B. Henderson, and G.N. Greaves, *Contributions to Mineralogy and Petrology,* **1990**, *105*, p. 359.

31. P.G. Allen, J.M. Berg, C.J. Chisolm-Brause, S.D. Conradson, R.J. Donohoe, D.E. Morris, J.A. Musgrave, C.D. Tait, "Determining Uranium Speciation in Contaminated Soils by Molecular Spectroscopic Methods: Examples from the Uranium in Soils Integrated Demonstration." (Waste Management '94 Meeting Proceedings, Tucson, AZ.)

32. P.M. Bertsch, D.B. Hunter, S.R. Sutton, S. Bajt, M.L. Rivers, *Environ. Sci. Technol.* **1994**, *28*, 980.

33. C.J. Chisholm-Brause, S.D. Conradson, C.T. Buscher, P.G. Eller, and D.E. Morris, *Geochim. Cosmochim. Acta,* **1994**, *58*, 3625.

34. A.J. Dent, J.D.F. Ramsay, and S.W. Swanton, *J. Colloid and Interface Sci.,* **1992**, *150*, 45.

35. A.J. Francis, C.J. Dodge, F. Lu, G.P. Halada, and C.R. Clayton. *Environ. Sci. Technol.,* **1994**, *28*, 636.

36. F. Farges, C.W. Ponader, G. Calas, and G.E. Brown, Jr., *Geochim. Cosmochim. Acta.,* **1992**, *56*, 4205.

37. P.G. Eller, G.D. Jarvinen, J.D. Purson, R.A. Penneman, R.R. Ryan, F.W. Lytle, and R.B. Greegor, *Radiochim. Acta,* **1985**, *39*, 17.

38. E.M. Larson, F.W. Lytle, P.G. Eller, R.B. Greegor, and M.P. Eastman, *J. Noncryst. Solids,* **1990**, *116*, 57.

39. G.S. Knapp, B.W. Veal, D.J. Lam, A.P. Paulikas, and H.K. Pan, *Mater. Lett.,* **1984**, *2*, 253.

40. J. Petiau, G. Calas, T. Dumas, and A.M. Heron, in: *EXAFS and Near-Edge Structure III,* Springer Proceedings in Physics, Vol. 2, eds. K.O. Hodgson et al., **1984**, Springer-Verlag, New York, p. 291.

41. G.S. Knapp, B.W. Veal, A.P. Paulikas, A.W. Mitchell, D.J. Lam, and T.E. Klippert, in: *EXAFS and Near-Edge Structure III*, Springer Proceedings in Physics, Vol. 2, eds K.O. Hodgson et al., **1984**, Springer-Verlag, New York, p. 305.

42. N.T. Barret, G.M. Antonini, F.R. Thornley, G.N. Greaves, and A. Manara, *Mat. Res. Soc. Symp. Proc.* **1987**, Vol. 84, p. 571; F.R. Thornley, N.T. Barret, G.N. Greaves, and G.M. Antonini, *J. Phys. C: Solid State Phys.*, **1986**, *19*, L563.

43. J.-M. Combes, C.J. Chisholm-Brause, G.E. Brown, Jr., G.A. Parks, S.D. Conradson, P.G. Eller, I. Triay, D.E. Hobart, and A. Meier, *Environ. Sci. Technol.*, **1992**, *26*, 376.

44. J. Petiau, G. Calas, D. Petitmaire, A. Bianconi, M. Benfatto, and A. Marcelli, *Phys. Rev. B*, **1986**, *34*, 7350.

45. G. Kalkowski, G. Kaindl, S. Bertram, G. Schmiester, J. Rebizant, J.C. Spirlet, and O. Vogt, O. *Solid State Comm.*, **1987**, *64*, 193.

46. G. Kalkowski, G. Kaindl, W.D. Brewer, and W. Krone, *Phys. Rev. B*, **1987**, *35*, 2667.

47. S. Bertram, G. Kaindl, J. Jove, M. Pages, J. Gal, *Phys. Rev. Lett.*, **1989**, *63*, 2680.

48. D.K. Veirs, C.A. Smith, J.M. Berg, B.D. Zwick, S.F. Marsh, P.G. Allen, and S.D. Conradson, *J. Alloys and Compounds*, **1994**, *213*, 328.

49. L.E. Cox, R. Martinez, J.H. Nickel, S.D. Conradson, and P.G. Allen, *Phys. Rev. B*, **1995**, *51*, 751.

50. F.A. Tomei, L.L. Barton, C.L. Lemanski, and T.G. Zocco, *Can. J. Microbiol.*, **1992**, *32*, 1328.

51. R.S. Oremland, J.T. Hollobaugh, A.S. Maest, T.S. Presser, L.G. Miller, and C.W. Culbertson, *Appl. Environ. Microbiol.*, **1994**, *55*, 2333.

52. B.B. Buchanan, T. Leighton, J. Liu, B.C. Yee, S. Jovanovich, A. Yee, W.-S. Yang, S. Ekune, and B. Chapman, in *Abstracts of Fifth International Symposium on Selenium in Biology and Medicine;* Vanderbilt University School of Medicine, Nashville, TN, **1992**, p 30.

53. A.O. Summers and S. Silver, *Ann. Rev. Microbiol.*, **1978**, *32*, 637.

54. D.R. Lovley, *Ann. Rev. Microbiol.*, **1993**, *47*, 263.

55. D. K. Shuh, N. Kaltsoyannis, J. J. Bucher, N. M. Edelstein, S. B. Clark, H. Nitsche, T. Reich, E. A. Hudson, I. Almahamid, P. Torretto, W. Lukens, K. Roberts, B. C. Yee, D. E. Carlson, A. Yee, B. B. Buchanan, T. Leighton, W.-S. Yang, and J. C. Bryan, *Mater. Res. Soc. Symp. Proc.*, **1994**, *344*, 323.

56. S. Bajt, S.R. Sutton, M.L. Rivers, and J.V. Smith, *Anal. Chem.*, **1993**, *65*, 1800.

57. I. Almahamid, J. J. Bucher, J. C. Bryan, S. B. Clark, N. M. Edelstein, E. A. Hudson, N. Kaltsoyannis, H. Nitsche, T. Reich, P. Torretto, W. Lukens, K. Roberts, and D. K. Shuh, *Inorg. Chem.*, **1995**, *34*, 193.

58. K.S. Rajan, and A.E. Martell, *J. Inorg. Nucl. Chem.*, **1964**, *26*, 1927.

59. K.S. Rajan and A.E. Martell, *Inorg. Chem.*, **1965**, *4*, 462.

60. G. Markovits, P. Klotz, and L. Newman, *Inorg. Chem.*, **1972**, *11*, 2405.

61. C. Papelis, K.F. Hayes, and J.O. Leckie, *HYDRAQL: A Program for the Computation of Aqueous Batch Systems Including Surface Complexation Modeling of Ion Adsorption at the Oxide/Solution Interface*: Tech. Rep. 306, Dept. of Civil Eng., Stanford Univ, 1988.

62. Y. Qian, N.C. Sturchio, R.P. Chiarello, P.F. Lyman, T.L. Lee, and M.J. Bedzyk, *Science*, **1994**, *265*, 1555.

STUDIES OF MAGNETIC MATERIALS WITH
CIRCULAR POLARIZED SOFT X-RAYS

V. Chakarian,[1†] Y. U. Idzerda,[1] C. T. Chen,[2] G. Meigs,[2] and C.-C. Kao[3]

[1]Naval Research Laboratory, Code 6345, Washington DC 20375
[2]AT&T Bell Laboratory, 600 Mountain Ave., Murray Hill, NJ 07974
[3]National Synchrotron Light Source, Brookhaven National Laboratory,
Upton, NY 11973

INTRODUCTION

The prospect of obtaining *element- and site-specific* magnetic information has prompted a fury of experiments taking advantage of circular polarized (CP) photons. These experiments, whether performed in absorption, photoemission, or reflection, offer the possibility of determining the magnitude and direction (with respect to a reference direction) of the magnetic moment of each element in the material being investigated. In addition, in principle it is possible to further dissect the individual magnetic moments into the separate contributions made by the electron spin and electron orbit; to determine the magnetic structure of a multilayer (including interfacial magnetic roughness); and to elucidate the spin-dependent electronic structure of a material.

Among these experiments, magnetic circular dichroism (MCD) has begun to have wide application to technologically important systems. MCD is an element-specific magnetic spectroscopic tool in which the difference in the absorption of left- and right-circularly polarized photons are measured at the absorption edges of the constituent elements. This difference in absorption cross-sections, appropriately normalized, is a measure of the average magnetic moment at a given atomic site.[1,2] To date, measurements have been made in the hard X-ray,[3] soft X-ray,[4] and the VUV[5] spectral regions. In the case of thin films and multilayers, soft X-ray magnetic circular dichroism (SX-MCD) has received particular attention. This is because, those states which are primarily responsible for magnetism, the $3d$ states of transition metals[6-12] and the $4f$ states of rare-earth elements,[13-15] are accessible at soft X-ray energies via strong dipole transitions from $2p$ and $3d$ core levels, respectively. It has also been shown that SX-MCD can be used as a powerful tool to measure element-specific magnetic hysteresis of heteromagnetic multilayers[16] and vector magnetometry in two- and three-dimensions.[17,18]

In this paper, a brief description and results of recent SX-MCD measurements conducted at the U4B beamline located at the National Synchrotron Light Source[19,20] are presented to demonstrate the recent progress in SX-MCD and its applications in the

---

† Mailing Address: NSLS Bldg. 725A/U4B, Brookhaven National Lab., Upton, NY 11973

Synchrotron Radiation Techniques in Industrial, Chemical, and Materials Science
Edited by D'Amico *et al.*, Plenum Press, New York, 1996

research of magnetic ultra-thin films and multilayers. While extremely interesting, photo-emission based experiments, which offer the possibility of extracting the orbital moment and elucidating spin-resolved electronic structure of magnetic materials, will not be discussed in detail as the treatment of photoemission data is substantially different from that of the photoabsorption data and is outside the scope of this paper.

MEASUREMENT SCHEMES

The X-Ray absorption (XAS) and hence MCD spectra can be collected via a variety of methods: by measuring the transmitted or the reflected flux, or it can be determined indirectly by measuring the decay products of the created core-hole state, e.g., total/partial electron yield, sample photocurrent, and fluorescence yield. Fig. 1 shows a schematic view of the ways in which circular polarized photons can be used to probe the magnetic properties of a material under investigation.

Circular polarized soft X-rays are obtained by collecting synchrotron radiation above and below the orbital plane of the bending magnet source where the emitted radiation is highly circular polarized while sustaining only about a factor of 3 reduction in flux.[20] (Only the photons emitted in the orbital plane are linear polarized.) A schematic of the Dragon beamline is shown in Fig. 2. The photon energy is swept through the L_3 and L_2 edges of the transition metal and the soft X-ray absorption is measured as a function of the photon polarization and remnant magnetization direction. The helicity of soft X-rays can be modulated, by appropriately chopping between the two photon beams collected by the vertical focusing mirrors (VFM),[20] and the magnetization direction can be switched with an *in situ* electromagnet. Both switching the photon polarization and reversing the magnetization direction will separately generate MCD spectra.

For MCD measurements, the energy-dependent absorption cross-section for left- and right-CP light are separately determined. This is most accurately performed by measuring the transmitted photon flux through the sample. Unfortunately, this requires the use of a semi-transparent sample which, for soft X-ray energies, confines the thickness of the sample to a few hundred angstroms (requiring the use of a supporting semi-transparent substrate). To overcome this thickness restriction and measure opaque samples, the absorption cross section can be measured indirectly by measuring the generated core-hole

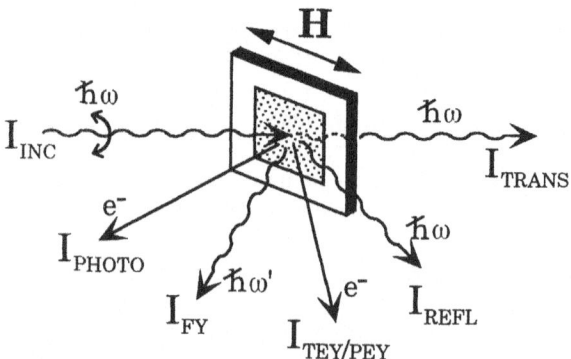

Fig. 1. Experimental geometry showing various ways in which circular polarized photons can be used to extract magnetic information. The incident radiation, I_{INC} or I_0, can be used to normalize the various measured signals.

NSLS U4B DOUBLE HEADED DRAGON MONOCHROMETER

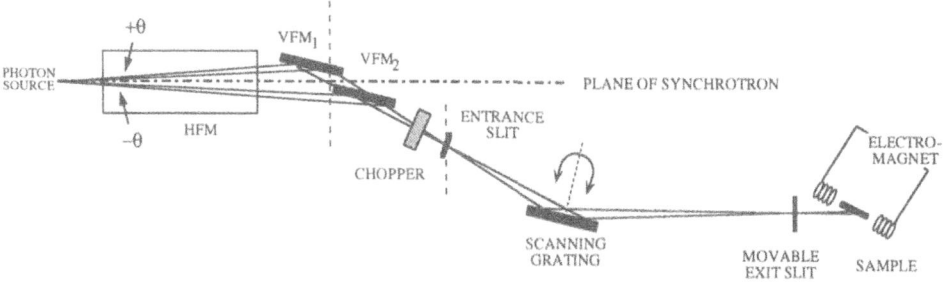

Fig. 2. The layout of double-headed Dragon beamline. The two vertical mirrors, VFM$_1$ and VFM$_2$, can be independently manipulated allowing the collection of circular polarized soft X-rays of both helicity simultaneously.

population. When a circular polarized photon beam is directed onto the sample, its absorption results in the generation of a core hole and the subsequent emission of the of various decay byproducts. Measuring the decay byproducts can be used to indirectly determine the photoabsorption cross-section.

In the case of the resonant scattering scheme, where the energy dependent intensity of the specularly reflected photon is measured as its energy is swept through an excitation edge, the energy-dependent reflectivity is directly related to the photoabsorption cross section by straightforward relations. In this section, a brief description and discussion of various ways in which CP photons can be used to obtain magnetic information on thin films and multilayers is presented. The measurement techniques to be discussed are photoabsorption based where the energy dependent intensity of most or all the photons or electrons is measured.

Except for the transmission yield method, all these x-ray absorption methods are indirect and suffer from a number of experimental difficulties, such as saturation/self-absorption effects and additionally, in the case of electron yield methods, from stray/remnant magnetic field induced asymmetries in the collected spectra. Extreme care has to be taken when using these methods and often the spectra must be corrected for these effects if quantitative analysis is desired. Nevertheless, each of these methods, briefly described below, offer great advantages, especially for cases in which very accurate quantitative information is not crucial.

Transmission Technique

The most straightforward method for measuring the energy dependent XAS, but the most difficult, is the transmission method. The attenuation of the soft X-ray flux is determined by measuring the incident flux, I_0, with a highly transmitting Au grid (either by the photocurrent drain on the grid or by a channeltron), and measuring the transmitted flux from a second foil/channeltron or photodiode apparatus. The probing depth of the transmission-based measurement is the entire sample thickness. Therefore, the MCD spectra in the transmission mode will give the magnetic information representative of the bulk material.

One of the major difficulties with the transmission technique is that at soft X-ray energies usually employed in transition metal studies the photons have a very small penetration depth. For this reason, the magnetic material and the underlying substrate must be sufficiently transparent to allow sufficient transmission of the incident flux. This can be accomplished by either preparing free-standing magnetic films of ~100 Å thick, or by depositing a thin magnetic film onto a suitable substrate. Free-standing films of this thickness are difficult to manufacture. Therefore, the use of a supportive substrate, one which is mechanically strong but is relatively transparent with featureless absorption spectrum in the energy regions of interest, is required. One good candidate for such a substrate is a commercially available ~1μm thick semi-transparent parylene, $(C_8H_8)_n$. Its use is illustrated in the 'Applications' section. These samples can either be grown *in situ* or grown *ex situ* and suitably capped to prevent oxidation.

One experimental difficulty that arises in the transmission technique is that the film thickness variations and pinholes can result in uncertainties in the determined absorption cross sections. A straightforward modeling of the film roughness, by treating the film to be composed of two sections of different thickness, shows that the resulting uncertainty in the absorption cross section is at worst a function of $(\Delta x)^2$, where Δx is the mean thickness variation of the film. For $\Delta x/x \sim 0.1$ or less, the uncertainty is of the order of 0.1% at the L_3 white line, where the errors are most significant, and hence can be ignored in most cases. The uncertainty due to any pinholes that may be present either in the substrate or the film, on the other hand, can cause as much as 1% error in the absorption cross section at the L_3 white line for a 1% pinhole density. However, the actual error, as reflected in the MCD spectra, is much smaller than 1% as MCD involves the *difference* between the absorption cross sections for parallel and anti-parallel configurations. Therefore, for pinhole densities of 1% or less, the as determined MCD spectra are reliable.

Fluorescence Yield Technique

Instead of directly measuring the amount of X-ray absorption, one can indirectly measure the absorption cross section by measuring the core-hole creation rate via the core-hole decay products. One way to measure the core-hole creation rate is to measure the fluorescence yield of the sample associated with the radiative decay of the core hole. Typically, a solid state detector and an energy discriminator circuit is used to separate the fluorescence photon signal from the signals generated by photons from other absorption levels of the same element or other elements. This is, therefore, a partial fluorescence yield measurement. The measurements can be done on opaque samples grown *in situ* or *ex situ* and in the presence of magnetic fields of up to ~2 Tesla. However, the fluorescence yield XAS spectra suffer from significant saturation/self-absorption effects for films of even few tens of angströms thick,[21] especially when either the incident photons or the fluorescence photons have to traverse the sample near grazing angles. This is because the attenuation length of the fluorescence photon is only slightly smaller than that of the incident radiation, the fluorescence technique is sensitive to nearly the entire irradiated volume. Because the dichroism effects are so large, the absorption coefficients at the absorption while line for the two light helicities can differ by 30% or more. As the circular polarized radiation penetrates into the sample, the remaining flux is reduced more quickly for one helicity than the other. For a sample of significant thickness, this results in different photon fluxes though the sample thickness and the total fluorescence yield is no longer proportional to the actual atomic absorption cross section. In the limiting case of an infinitely thick sample, all the incident radiation is absorbed, generating core holes. These core holes can decay via radiative decay and the fluorescence yield is proportional only to the radiation intensity and independent of the light helicity.

Electron Yield Techniques

These techniques encompass the measurement of the energy dependent photoemitted electron intensity (the kinetic energy distribution of the electrons is not determined). The measurements can be made in partial or total electron yield schemes and can be performed via total electron yield or sample photocurrent methods. However, electron yield methods (whether in photocurrent[12,22] or in secondary electron collection modes[23]) suffer from a detection efficiency imbalance in the presence of external/stray magnetic fields.[21] This imbalance can be partially or fully eliminated by using relatively strong electrostatic extraction schemes and paying particular attention to the geometrical arrangement of the sample, photon beam and the applied magnetic field.

Electron yield methods, similar to the fluorescence yield scheme, also suffer from saturation effects.[21] One serious complication caused by this saturation effect is that the measured MCD effect depends on the incidence angle of the photon beam.[24] Furthermore, they may suffer from spin-dependent transmission asymmetries through magnetic layers.[25,26] Therefore, while they are very valuable tools for measuring XAS, the results obtained via electron yield methods should be treated with caution as the above mentioned experimental artifacts are difficult to overcome or corrected.

Resonant Scattering Technique

A very recently developed method utilizes the reflected incident flux of circular polarized soft X-rays to extract magnetic information.[27] The dichroic effects are measured as a function of incidence angle using a θ-2θ spectrometer and a gas proportional avalanche counter. In addition, the helicity dependent intensity of the specular beam at a fixed incidence angle can be measured as the photon energy is swept through the absorption edges of the magnetic material. The energy and angle dependent reflectivity curves are complimentary and display asymmetries as large as 80%![27] Furthermore, variable incidence angle results in a variable probing depth technique, offering the possibility of gaining depth-dependent magnetic information. The reflectivity technique can be used on *in situ* or *ex situ* samples and in an applied field. This is a very promising method for the extraction of magnetic information from multilayer systems. One should note, however, that the dichroic effects exhibited in the reflection are not only due to absorption cross section (i.e., the imaginary part of the dielectric tensor), but also the real part of the dielectric tensor, making the interpretation of the data less straightforward. Simulations using simple electromagnetic theory for layered media utilizing complex dielectric tensors with no free parameters has proven to work very well in the case of films of reasonable quality.[27]

APPLICATIONS

Detecting the presence of ferromagnetism

To demonstrate the utility of element-specific magnetic information, the MCD measurement for a single monolayer of Cr deposited on Fe(001) is presented. Cr is predicted to be ferromagnetic with a large spin imbalance (up to 3.6 μ_B), but anti-aligned to the bulk Fe moment. The samples are generated *in situ* on heat-cleaned (580°C) 1 cm^2 GaAs(001) n$^+$

substrates which were previously lightly sputtered with Ne+ at 500 eV to remove surface carbon. The Fe was evaporated at a vacuum of $2x10^{-10}$ Torr with the substrate held at 175°C followed by Cr evaporation with the substrate at room temperature. The total carbon contamination measured after each deposition was less than 0.01 ML as determined by C_{1s} X-ray photoelectron spectroscopy (XPS). The photon energy is swept through the L_3 and L_2 edges of the transition metal and the soft x-ray absorption is measured as a function of the light polarization and remnant magnetization direction (which can be switched with an *in situ* electromagnet).

Immediately after a 150Å Fe film is deposited, the CP absorption spectra, shown in the upper panel of Figure 3, were measured by total electron yield for the photons incident parallel (solid line) and anti-parallel (dashed line) to the remnant magnetization direction. The dots, which represent the MCD, are the difference between the parallel and anti-parallel spectra. These spectra show a negative MCD signal at the L_3 edge step and a positive signal at the L_2 edge step.

To investigate the magnetic behavior of the first Cr layer, sub-monolayer coverages down to 0.25 monolayer (ML) of Cr/Fe were recorded. The lower panel of Fig. 3 shows Cr XAS and MCD spectra of the lowest coverage, 0.25 ML Cr film. At 0.25 ML coverage, nearly all the deposited Cr should occupy the first layer with very little in second layer sites. Since MCD is element specific, it is immediately evident from the reversal of the Cr MCD intensity at both the L_3 and L_2 white lines (lower panel) in comparison to the Fe MCD spectra (upper panel), that submonolayer coverages of Cr have moments aligned with each other and anti-aligned with the first Fe layer (as depicted in the inset). The Cr data is

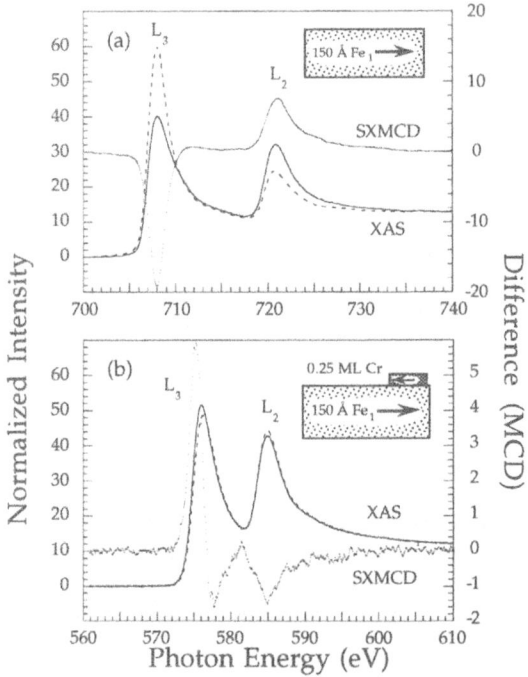

Fig. 3. XAS and MCD spectra at the L_3 and L_2 absorption white lines for (a) 150Å Fe film; (b) 0.25 ML Cr deposited atop the Fe film. The solid (dashed) lines are spectra taken with the spin direction of the incident photons parallel (anti-parallel) to that of the majority electrons of the Fe film. The dots are the MCD spectra obtained from the difference between the parallel and anti-parallel XAS spectra.

distinct from the Fe data in that it shows a strong differentiation-like lineshape at the L_3 white line which is due to an energy shift of the peak between the two XAS spectra.

Separation of Orbital and Spin Moments

The ability to decompose the total magnetic moment of a multi-component, hetero-magnetic system into the orbital and spin moment contributions of each element would be an enormously valuable advance in understanding the technologically relevant aspects of multilayer and alloy magnetic materials. X-ray MCD in conjunction with the recently developed sum-rules offers this possibility, provided that the energy- and photon-polarization-dependent absorption cross sections can be determined via one of the methods described in the previous section. According to these sum-rules the orbital[1] and spin[2] moments of each element in a given system can be determined from the XAS and MCD spectra as follows:

$$m_l = -\frac{4\int_L \delta\mu \, d\omega}{3\int_L (\mu_+ + \mu_-) \, d\omega} \cdot h_{3d} \tag{1}$$

$$m_s = -\frac{2\int_{L_3} \delta\mu \, d\omega - 4\int_{L_2} \delta\mu \, d\omega}{\int_L (\mu_+ + \mu_-) \, d\omega} \cdot h_{3d} \cdot \left(1 + \frac{7\langle T_z \rangle}{2\langle S_z \rangle}\right) \tag{2}$$

Where, m_l and m_s are the orbital and spin magnetic moments in units of μ_B/atom, respectively. $\mu_\pm$ are the absorptivity for left- and right-CP light; h_{3d} is the $3d$ hole count of the specific transition metal; $\delta\mu = \mu_+ - \mu_-$; L, L_2, and L_3 indicate the energy integration ranges. In (1) and (2), the absorption coefficient for linear polarized light, $\mu_0(\omega)$, was replaced with $(\mu_+ + \mu_-)/2$. $\langle T_z \rangle$ is the magnetic dipole term while $\langle S_z \rangle = m_s/2$ in Hartree atomic units.

To demonstrate how the total magnetic moment of a multi-component, heteromag-netic system can be decomposed into the orbital and spin moment contributions of each element, transmission MCD spectra from thin film alloys have been measured. Soft X-ray XAS spectra were obtained by measuring the photon flux transmitted though magnetic thin films using a photodiode mounted behind the sample. The films were grown under ultra-high vacuum conditions by e-beam evaporation from pure metal sources onto ~1μm thick semi-transparent parylene, $(C_8H_8)_n$, substrates.[28,29] The film thicknesses and relative concentrations of constituent elements were determined independently by comparing the measured absorption to the calculated atomic photoabsorption cross-sections and by *ex situ* X-ray fluorescence methods. The photon energy resolution and the degree of circular polarization were set at $\delta E \sim 0.5$ eV and $p_c = 75\%$, respectively. The photon incident angle was fixed at 45°. Magnetic fields were applied along the intersection of film surface and photon incidence planes. MCD data were measured by alternating the applied magnetic field while keeping the polarization of the photon source fixed. The magnetic "quality" and the lack of oxidation in the films were verified via *in situ* element-specific magnetic hysteresis (described further below in this section) and XAS measurements, respectively.

The data reduction process is illustrated in Fig. 4 in which the results for a thin film $Fe_{48}Co_{52}$ alloy are shown. The top panel shows the incident-photon-flux normalized transmission spectra taken with two opposite magnetization directions, $I_\pm$, for the $L_{2,3}$ white lines of both Fe and Co. The corresponding substrate spectrum, I_s, was also measured and is shown in the figure. The absorptivity of the film can easily be derived from these spectra by making use of the absorption equation:

$$\mu_\pm \propto -\log\left(\frac{I_\pm/I_0}{I_s/I'_0}\right) \tag{3}$$

where $I_\pm/I_0$ is the normalized transmission spectrum for the film and the parylene film and I_s/I'_0 is the normalized transmission spectrum for the uncovered parylene film. The resulting spectra are shown in the middle panel of Fig. 4 along with a two-step background function (discussed further below in this section). For these spectra, no additional background corrections were made, and only the representation of the measured data according to Eq. (3) was performed. Any slight errors in a background removal process result in a greatly increased error after the energy integration process.

After taking into account the photon incidence angle and the degree of circular polarization, i.e., multiplying $\delta\mu$ by $1/\cos(45°)/p_c$, the normalized MCD spectra are deduced and shown in the bottom panel of Fig. 4. Also shown in the bottom panel is the integral of the MCD signal. In order to facilitate the evaluation of sum-rules (1) and (2), certain points along the integral curve have been labeled. Since the $\langle T_z \rangle$ term in bulk Fe and Co is very small,[30] neglecting the $\langle T_z \rangle$ term in the spin sum-rule, (1) and (2) can be rewritten in terms of these labeled quantities p and q as follows:

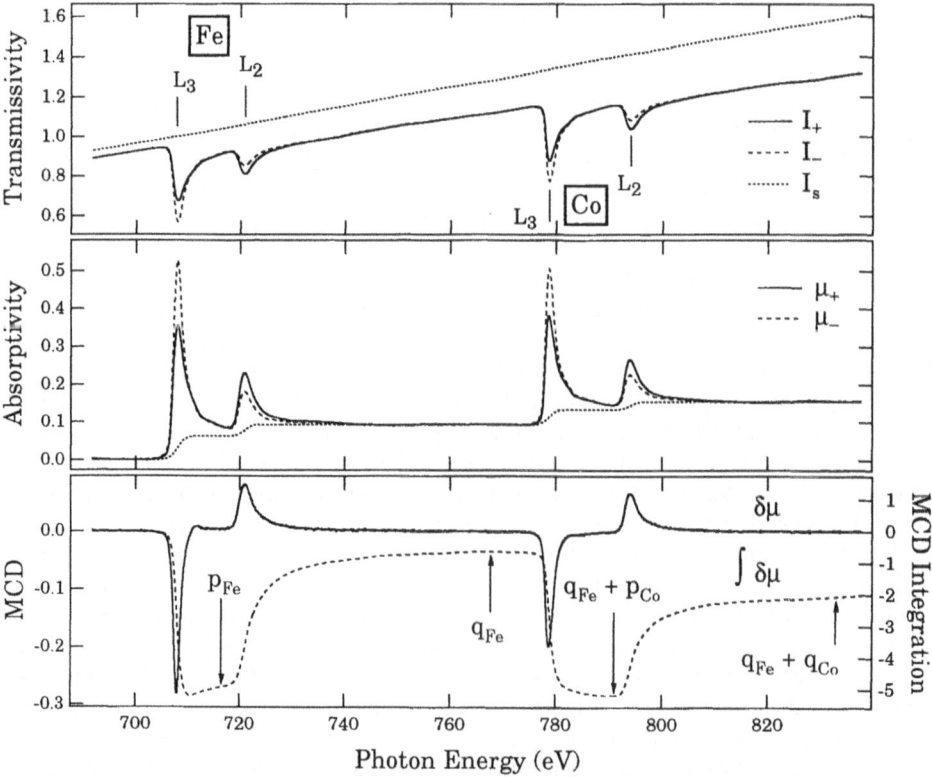

Fig. 4. An illustration of data reduction scheme in the transmission MCD method. The data shown is from a thin film ($\sim$100 Å) $Fe_{48}Co_{52}$ alloy grown *in situ*. See text for the description of the spectra and the symbols.

$$m_l = - \frac{4q}{3 \int_L (\mu_+ + \mu_-) \, d\omega} \cdot h_{3d} \tag{4}$$

$$m_s = - \frac{(6p - 4q)}{\int_L (\mu_+ + \mu_-) \, d\omega} \cdot h_{3d} \tag{5}$$

$$\frac{m_l}{m_s} = \frac{2\left(\frac{q}{p}\right)}{9 - 6\left(\frac{q}{p}\right)} \tag{6}$$

Note that the determination of the individual orbital and spin moments using MCD data in conjunction of the sum-rules requires the knowledge of the total absorption cross section integrated over the L edges of the corresponding element as well as the density of $3d$ hole count for that element. The former quantity can be approximated within an acceptable level of accuracy by adopting a simple no-free-parameter two-step background function for removing the L_3 and L_2 edge jumps. The positions of the steps are set to the peak positions of the L_3 and L_2 white-lines and the heights of the steps are set according to their quantum degeneracy, i.e. $L_3:L_2$ step height ratios are set to 2:1. A typical two-step background function is shown in the middle panel of Fig. 4, where the rigid step jumps were convoluted to a Voigt function to simulate the intrinsic line-width and experimental resolution.

The values of h_{3d} for each element have to be obtained from theoretical calculations and are a significant source of uncertainty. While the procedure as described here has been shown to yield values correct to within 5% for pure Fe and Co thin films,[28,29] the generalization of this method has yet to be shown to work for other transition metals and alloys. Nevertheless, SX-MCD has significant utility since testing the relative ratios of orbital to spin moments requires no additional assumptions to be made and, as shown in (6), only quantities experimentally determined from the MCD spectra are needed. Hence, the SX-MCD measurements, in conjunction with the "combined" sum-rule (6) can provide a stringent test to various theories of magnetism.

Magnetic Extended X-Ray Absorption Fine Structure (MEXAFS)

An inherent assumption in the SX-MCD methods, in which the integrals of the MCD signal are used to determine the orbital and spin moments, is that no MCD signal is present for energies significantly above (>40 eV) the L_2 white-lines. In the hard X-ray energy regime, i.e., at the K-edges of transition-metals and the L-edges of the rare-earths, the presence of spin-polarized EXAFS (SP-EXAFS) has been known for some time.[31,32] SP-EXAFS, more appropriately called magnetic EXAFS (MEXAFS), is speculated to be related to the magnetic short range order and caused by the dependence of the scattering of the photoemitted electron on the spin moment of the neighboring magnetic atoms. The feasibility of MEXAFS was first demonstrated on $3d$ transition metal K-edge and $4f$ rare-earth L-edge measurements,[31] but the lack of a first principles theory allowing for the extraction of tangible quantitative material parameters and the added difficulty of the MEXAFS measurements due to the small size of the effect has hampered the generation of a deeper understanding of the underlying mechanism. In later studies,[31-37] it was almost universally found that, for these edges, the size of the MEXAFS oscillations is linearly proportional to the spin moment of the scattering atom.[33-37] Finally, in the case of pure $3d$

metals Fe, Co, and Ni, it was found that the K-edge MEXAFS oscillations are in phase with the EXAFS oscillations, supporting further the notion that this effect is due primarily to the scattering from magnetic nearest neighbors.[34,36,37]

Until now, due to the lack of a suitable CP photon source as well as the lack of an established data reduction procedure, this type of study has not been extended to the lower binding energy L-edges of the transition metals. The standard EXAFS analysis is complicated by the small energy separation of the L_3 and L_2 edges in transition metals, causing the (M)EXAFS oscillations for these edges to overlap. Furthermore, the MEXAFS oscillations for the well separated L_3 and L_2 edges of rare-earth elements show identical structure, but of opposite sign, prompting prior investigators to suggest that, due to this cancellation, measurements at the overlapping L-edges of the $3d$ elements would yield negligible MEXAFS signal and hence be impractical.[33,34] Nevertheless, due to the strong dipole transition to the magnetically interesting $3d$ states, the MEXAFS effect may be much larger at the L-edges of transition metals than their K-edges, where the earlier studies were conducted. (Similar arguments apply to the M-edges vs. the L-edges of $4f$ rare-earths.)

In order to test the above hypotheses and to demonstrate the feasibility of the MEXAFS measurements for the L-edge of $3d$ transition metals, transmission MCD measurements extended to the EXAFS region have been performed for Fe, Co, and Ni thin films. The films were evaporated from e-beam pure metal sources in UHV onto ~1μm thick semi-transparent parylene, $(C_8H_8)_n$, substrates[28,29] and were verified to be free of oxidation via XAS measurements. The *in situ*, element-specific magnetic hysteresis measurements have shown a square magnetization behavior with remnant magnetizations of 95-100%. The film thicknesses were determined by *ex situ* X-ray fluorescence method and were about 250-300 Å. EXAFS measurements from an Fe thin film similar to the one used in the present work (capped with a V overlayer to prevent oxidation) have shown the film to be in a bcc phase.[38]

Fig. 5 shows the results of the MEXAFS measurements from the Fe, Co, and Ni thin films. The data were collected in a manner similar to that described in the previous section. The top panel shows the XAS spectrum from the film (compare to the middle panel of Fig. 4). The EXAFS region has been enlarged as indicated. Note that the intensity variations of these oscillations are about ± 6% of the edge-jump. Considering that the measurements were made at room temperature and that the films are polycrystalline, this indicates a relatively large coherence length of the crystallites in the film. The second panel shows the corresponding MCD signal (compare bottom panel of Fig. 4). The high energy region of the second panel shows an enlarged view of the MEXAFS oscillations. The data show large MEXAFS oscillations of about ± 5% of the EXAFS intensity in the first 150 eV above the L-edge and much weaker oscillations in the next 150 eV. The oscillations nearly vanish beyond ~300 eV above the L-edge.

The results of similar measurements at the L-edges of Co and Ni thin films grown similarly are shown in the remaining panels of Fig. 5. The data collection times for each film were comparable and were about 30 hours. The XAS spectra display, once again, EXAFS oscillations of about ± 6% of the edge-jump indicating reasonable crystalline quality of the films. There is a substantial difference in the detailed shape of the EXAFS oscillations between Fe, Co, and Ni due to the differences in the local atomic structure of these films and to the changes in the interference between the oscillations associated with the L_3 and L_2 edges (the energy separation between the two edges increases in going from Fe to Ni). Note that although the EXAFS oscillations of the Fe, Co, and Ni films are of comparable intensity, the relative amplitude of the MEXAFS oscillations, compared to the EXAFS oscillations, changes dramatically from about ± 5% for Fe, ± 2.5% for Co, to <± 0.5% for Ni of the respective EXAFS oscillations. Noting that the spin moments of bcc Fe, hcp Co, and fcc Ni are 2.1, 1.5, and 0.6 μ_B, respectively, these measurements

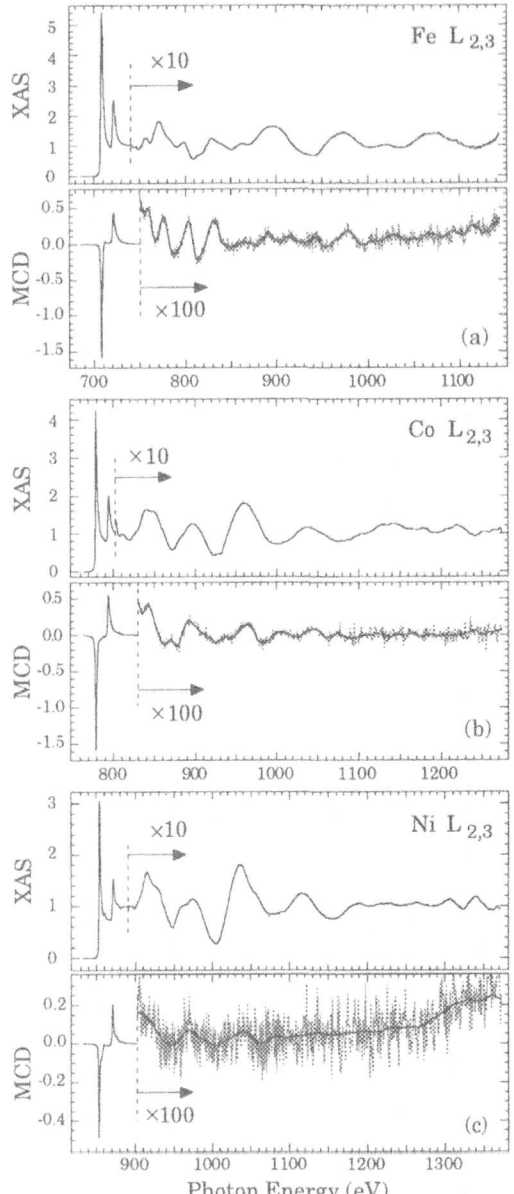

Fig. 5. The $L_{2,3}$ XAS (EXAFS) and MCD (MEXAFS) for Fe (top two panels), Co (middle two panels), and Ni (bottom two panels) thin films. The extended fine structure region has been expanded as indicated.

indicate that the size of the effect does *not* scale linearly with the spin moment of the scattering atom.

It is important to note that these oscillations are up to 50 times larger than those observed at the K-edge of Fe[31,32,35,37] and Co,[34] demonstrating the enhanced sensitivity of the L-edge absorption to the spin-dependent scattering potential of the neighboring atoms. Furthermore, contrary to the K-edge measurements, an inspection of the respective

EXAFS and MEXAFS spectra immediately indicates that the MEXAFS spectra show higher frequency oscillations, and are not in-phase with the EXAFS spectra. This clearly indicates that the scattering paths that dominate MEXAFS oscillations are not the same as those which give rise to the EXAFS oscillations. Indeed, the shorter wavelength, higher frequency MEXAFS signal can not be generated by nearest neighboring single scattering paths, but is dominated by longer scattering path lengths, most likely from multiple scattering events. Multiple scattering is consistent with both the higher frequency of the MEXAFS, and the nonlinear dependence of the MEXAFS oscillations with the magnetic moment of the scattering center.

Element-Specific Magnetic Hysteresis

Traditionally, magnetization reversal in thin ferromagnetic films has been studied via a variety of methods which includes vibrating sample magnetometry (VSM), magnetization loopers, ferromagnetic resonance, magneto-optic Kerr effect (MOKE), magneto-resistance, superconducting quantum interference device susceptometry, and Lorentz electron microscopy.[39] These conventional methods are often used to determine the magnitude and/or the orientation of only a single component of the magnetic moment vector (typically parallel to the applied magnetic field). In the case of MOKE, by using longitudinal and transverse Kerr effect one can obtain the two orthogonal in-plane magnetization components.[40] This approach, however, does not allow a direct comparison of the two components. Recently, Daboo, et al. have demonstrated that using only longitudinal MOKE one can obtain the two in-plane components of the magnetic moment vector,[41] although the measurements are complicated by the presence of the intrinsic small transverse MOKE signal. Nevertheless, in the case of heteromagnetic systems, none of the aforementioned methods discriminate among the various constituent elements of the magnetic material studied. Therefore, most previous studies involving magnetization reversal in mono- and heteromagnetic systems provide valuable but limited and incomplete information regarding the magnetic moment reversal process.

Magnetization reversal processes can also be studied by using SX-MCD. Typically SX-MCD is measured by reversing the polarization direction of the incident photon beam. An alternate, and equivalent, method for measuring SX-MCD is to reverse the applied field direction while keeping the polarization of the incident photon beam fixed.[4] It is therefore possible to obtain element-specific magnetic hysteresis curves by monitoring the peak height of a given elemental absorption white-line as a function of the applied field.[16]

To demonstrate the usefulness of such a measurement, the element-specific hysteresis curves for a Fe (102 Å) / Cu (30 Å) / Co (51 Å) trilayer sample have been measured. The photon energy resolution was set at 0.4 eV and the degree of circular polarization was set at 77%. The angle of the incident photon beam was fixed at 45° with respect to surface normal and the absorption spectra were recorded by monitoring the soft X-ray fluorescence yield with a high sensitivity seven-element germanium detector. The use of the fluorescence yield method in this case allows the measurement of buried layers in opaque samples.

The top panel of Fig. 6 shows the results of such measurements taken with the photon energies tuned at the Fe (solid curve) and Co (dashed curve) L_3 white lines. The bottom panel of Fig. 6, on the other hand, shows the comparison between the hysteresis curve obtained by a VSM (solid curve) and a least-squares-best-fit linear combination of the Fe and Co hysteresis curves shown in the upper panel. (The intensities shown in the upper panel reflect the scaling one obtains from this least-squares-best-fit.) It is clear that the seemingly complicated VSM curve is the result of the different hysteretic behavior of the Fe and Co layers. Furthermore, an inspection of the Co hysteresis curve shows that a portion of the Co layer is ferromagnetically coupled to the Fe layer, possibly via pin holes through the Cu spacer layer. Such subtle hysteresis features, which have important ramifications for understanding other magnetic phenomena involving magnetic heterostructures, particularly

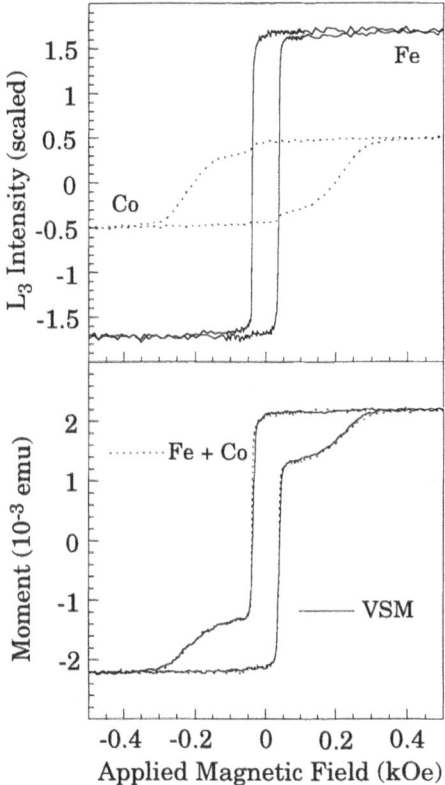

Fig. 6. The top panel shows the Fe and Co L$_3$ white line intensities as a function of the applied magnetic field for the trilayer sample. The bottom panel shows the comparison between the conventional hysteresis curve obtained by a vibrating sample magnetometer (solid line; labeled VSM) and the least-squares-best-fit linear combination of the Fe and Co curves (dashed line; labeled Fe+Co. The magnetic moment shown here is normalized for a 1x1 cm^2 thin film.

giant magnetoresistance effect, are imperceptible in the VSM or other conventional magnetization measurements and demonstrate the usefulness and power of element-specific methods.

2D and 3D Element-Specific Magnetic Hysteresis

The element-specific magnetic hysteresis technique described in the previous section lends itself easily to the determination of *all* three components of the magnetization vector, i.e., 3D element-specific vector magnetometry (ESVM). A brief description of the method is given below.

The measured hysteresis curves are proportional to $k_i \cdot M$, the projection of the magnetization vector, M, along the photon momentum direction k_i. The experimental setup is shown schematically in the inset of Fig. 7. (Only k_1 and k_2 are shown for clarity; k_3 and k_4 are along the mirror plane projection of k_1 and k_2, respectively.) Briefly, four hysteresis curves are collected by azimuthally rotating the analysis direction relative to the magnetization plane, so that the projection of the analysis direction in the plane of the sample is changed successively by 90° increments. This can be accomplished by a concerted rotation of the magnet/sample assembly with respect to the incident photon beam direction. All three components of the magnetization vector can then be obtained by forming sums and differences of the measured hysteresis curves as follows:

$$\mathbf{k}_1 \cdot \mathbf{M} - \mathbf{k}_3 \cdot \mathbf{M} = 2p \sin\theta \, M_x \tag{7}$$

$$\mathbf{k}_2 \cdot \mathbf{M} - \mathbf{k}_4 \cdot \mathbf{M} = -2p \sin\theta \, M_y \tag{8}$$

$$\mathbf{k}_1 \cdot \mathbf{M} + \mathbf{k}_3 \cdot \mathbf{M} = \mathbf{k}_2 \cdot \mathbf{M} + \mathbf{k}_4 \cdot \mathbf{M} = -2p \cos\theta \, M_z \tag{9}$$

where θ is the angle between the photon beam direction and the surface normal, and p is the degree of circular polarization of the incident photon beam ($-1 \leq p \leq 1$). Once M_x, M_y and M_z are determined, a 3D rendering of the moment reversal can then be constructed. Note that for thin films that display only in-plane magnetizations, such as the system described here, $M_z = 0$ so that only two measurements, e.g., along $\mathbf{k}_1$ and $\mathbf{k}_2$, are required for obtaining M_x and M_y.

As a demonstration of the technique, the magnetization reversal of *each* element of a complicated trilayer system, $Fe_{25}Co_{75}$ (106 Å)/Mn (8.7 Å)/$Fe_{25}Co_{75}$ (106 Å), are determined separately. The sample was grown by molecular beam epitaxy (MBE) on a ZnSe(001) buffer layer grown on GaAs(001) substrate in an MBE chamber and capped with a 30 Å Al. The growth morphology was monitored via *in situ* reflection high-energy electron diffraction. Element-specific magnetic hysteresis curves were collected via partial fluorescence yield at the L_3 edges of the relevant elements.[16] The field intensity was calibrated by comparing the ESVM hysteresis loops to the VSM loops obtained previously. The system presented here exhibits in-plane magnetization for all applied fields, i.e., $\mathbf{M} = M_x \hat{\mathbf{x}} + M_y \hat{\mathbf{y}}$. The lack of an M_z component was verified directly by measuring a null hysteresis signal for the photon beam incident normal to the film plane ($\mathbf{k}_\perp \cdot \mathbf{M} = p \, M_z$). In order to determine M_x and M_y, hysteresis curves were collected for incident photon beam directions parallel to the x-z and y-z planes of the sample, i.e., along $\mathbf{k}_1$ and $\mathbf{k}_2$, and at 45° to these planes. The former two orientations yield M_x and M_y (since $M_z = 0$), while the latter yields $\left(M_x + M_y\right)/\sqrt{2}$ and serves as an internal normalization of the M_x and M_y hysteresis curves.

The M_x and M_y components of average Fe moment as a function of applied magnetic field are shown in Fig. 7. The vector nature of the magnetization reversal process is clearly depicted by the interplay of the M_x and M_y components' intensities. Using these data, 2D parametric representations of the moment reversal can be constructed.[17,18] The results indicate that the magnetization reversal process is a combination of coherent rotation toward the magnetically easy <100> axes and domain formation and reversal along these in-plane <100> axes.

Similar measurements of M_x and M_y for Mn, shown in Fig. 8, indicate that Mn has a net moment. A comparison of the switching fields for Mn moment with that of Fe and Co indicates that Mn moment is strongly coupled to the enclosing FeCo layers. The shape of the Mn hysteresis loops are seemingly more complex but detailed analysis reveals that the Mn moment is rotated from that of Fe and Co by ~23° and this angle is largely preserved throughout the reversal process. To demonstrate this fact, also included in Fig. 8 (solid curves) are the corresponding Fe loops (from Fig. 7) numerically rotated by 23°. Note that these "rotated" Fe loops reproduce all the features of the Mn loops. The small discrepancies are due to the fact that the angle between the Mn and Fe moments are not constant but vary as a function of applied field. This angle between the Mn moment and the Fe (also Co) moment can be better demonstrated by comparing the 2D parametric representations of the respective moments, shown in Fig. 9. Both the presence of a net Mn moment and its rotation from the net Fe moment can be understood if the microscopic magnetic structure of the Mn interlayer is an antiferromagnetic helix structure,[42] consistent with the proposed helical structure for a system with an ordered anti-ferromagnetic interlayer bound by ferromagnetic sheets.[43]

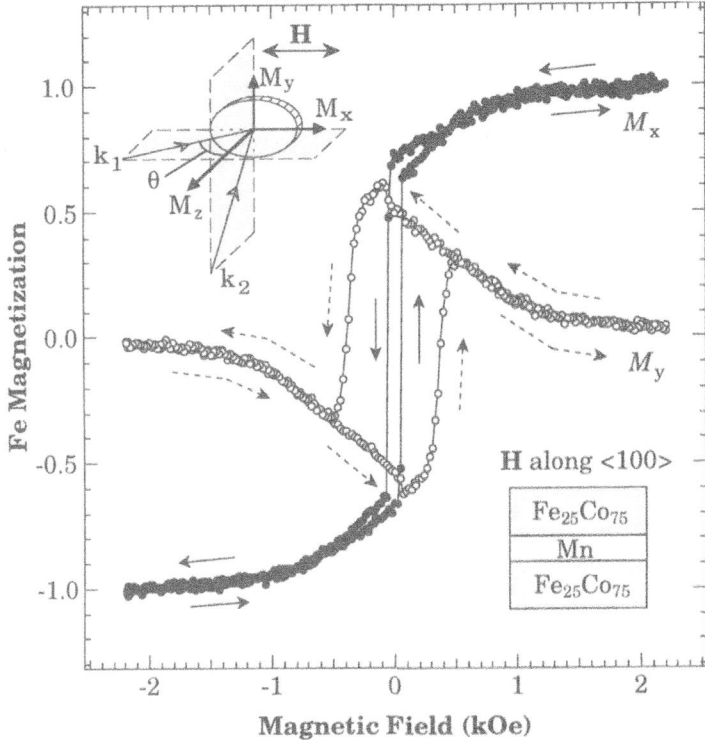

Fig. 7. M_x and M_y hysteresis loops for Fe where the points are the data and the solid curves are provided as a guide to the eye. The arrows indicate increasing and decreasing magnetic field for the M_x (solid arrows) and M_y (dashed arrows) magnetization curves. INSETS: measurement configurations for the determination of M_x and M_y and a schematic view of the trilayer.

Circular Polarized Soft X-Ray Resonant Magnetic Scattering

Circularly polarized soft X-ray resonant magnetic scattering is a technique which combines the structural sensitivity of X-ray scattering and the magnetic sensitivity of soft X-ray magnetic circular dichroism. It offers the possibility of determining layer thickness, interface quality, and the magnetization of a heteromagnetic multilayer in an element- and layer-specific manner.

As a demonstration of the technique, angle (θ–2θ) and energy scans were collected near the L-edges of Fe, Co, and Mn, for a heteromagnetic trilayer: $Fe_{25}Co_{75}$ (106 Å) / Mn (8.7 Å) / $Fe_{25}Co_{75}$ (106 Å) (same as that used for the ESVM experiment). The sample was mounted on the central rotating axis of a high precision, two-axis goniometer and was magnetized as in the previous section along the <100> direction. The intensity of the specularly reflected beam was measured by a gas proportional avalanche counter with an overall efficiency of 10% and an energy bandwidth of ~40%.

Representative energy dependent reflectivity spectra along various incidence angles in the Co L-edge region are shown in Fig. 10. The magnetization direction of the sample was reversed at each energy point. The θ–2θ scans (not shown) display intriguing interference at various photon energies. Comparison with a electromagnetic theory for layered media utilizing complex dielectric tensors with no free parameters shows that the rich and intriguing interference characteristics of the reflectivity curves can be used to determine the multilayer magnetic ordering, film thicknesses, interface quality, as well as the degree of light polarization from the spatial and polarization vector interference effects.[44]

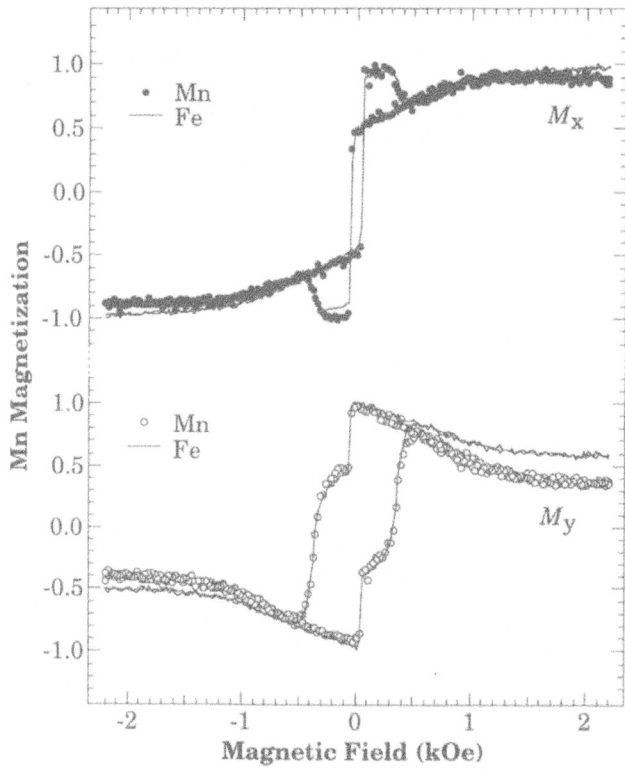

Fig. 8. M_x and M_y hysteresis loops for Mn (points). The solid curves are the corresponding Fe loops generated by the inclusion of a rotation of 23° to the data of Fig. 7.

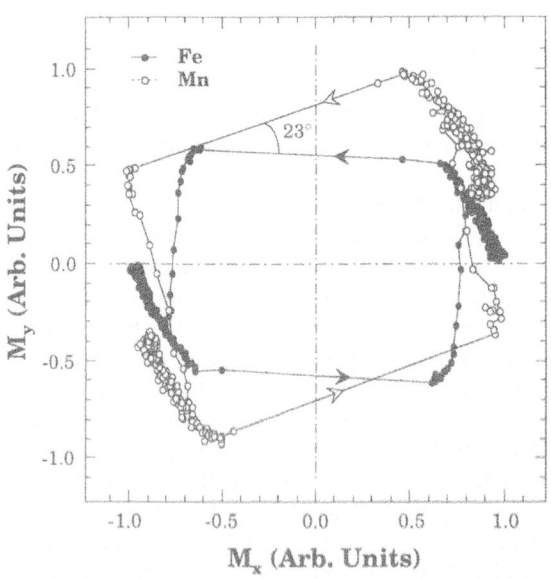

Fig. 9. 2D parametric representations of $\langle M \rangle_{Fe}$ and $\langle M \rangle_{Mn}$ clearly showing the canting of $\langle M \rangle_{Mn}$ with respect to $\langle M \rangle_{Fe}$.

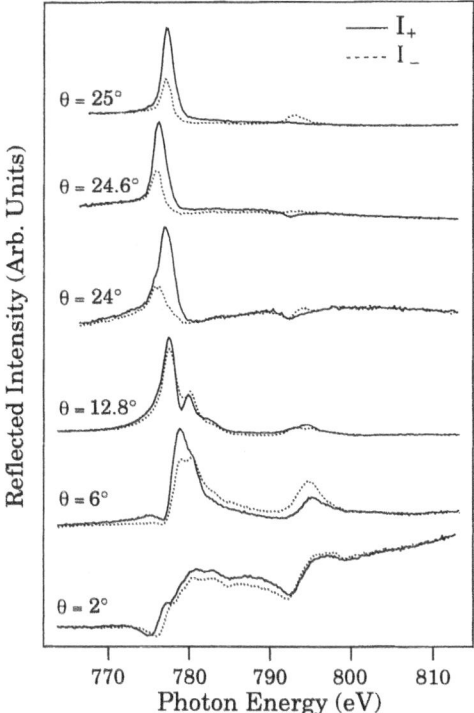

Fig. 10. Energy scans along various incidence angles in the Co L-edge regions. The solid (dashed) lines are spectra taken with the spin direction of the incident photons parallel (anti-parallel) to that of the majority electrons of the trilayer. The spectra were normalized to have equal height at L_3 energy (~780 eV).

SUMMARY

In summary, applicability and limitations of SX-MCD has been described via several applications of this technique. Its usefulness in detecting ferromagnetism in dilute systems has been demonstrated by determining the antiferromagnetic coupling of 0.25 ML Cr to an Fe substrate. A transmission MCD method, which offers the capability of determining element-specific orbital and spin moments for each element of a heteromagnetic or multilayer system, has been illustrated. This method has been extended to show the presence of spin-polarization effect on the L-edge EXAFS of Fe, Co, and Ni thin films. As more practical examples, the use of SX-MCD to measure element-specific hysteresis curves has been illustrated and a method that allows the measurement of 2D and 3D element-specific magnetic hysteresis loops has been described. Using this method, two dimensional magnetic hysteresis loops for each element of a $Fe_{25}Co_{75}/Mn/Fe_{25}Co_{75}$ trilayer have been measured. Finally, a new technique, circular polarized soft X-ray resonant magnetic scattering, that combines the structural sensitivity of X-ray scattering and the magnetic sensitivity of soft X-ray magnetic circular dichroism has been introduced.

ACKNOWLEDGMENTS

One of the authors (VC) is supported by the Office of Naval Research. Work done at National Synchrotron Light Source was supported by DOE, under contract No. DE-AC02-76CH00016.

REFERENCES

1. B. T. Thole, P. Carra, F. Sette, and G. van der Laan, *Phys. Rev. Lett.* 68:1943 (1992).
2. P. Carra, B. T. Thole, M. Altarelli, and X. Wang, *Phys. Rev. Lett.* 70:694 (1993).
3. G. Schütz, W. Wagner, W. Wilhelm, and P. Kienle, *Phys. Rev. Lett.* 58:737 (1989).
4. C. T. Chen, F. Sette, Y. Ma, and S. Modesti, *Phys. Rev. B* 42:7262 (1990).
5. T. Koide, T. Shidara, H. Fukutani, K. Yamaguchi, A. Fujimori, and S. Kimura, *Phys. Rev. B* 44:4697 (1991).
6. L. H. Tjeng, Y. U. Idzerda, P. Rudolf, F. Sette, and C. T. Chen, *J. Magn. Magn. Mater.* 109:288 (1992).
7. Y. Wu, J. Stöhr, B. D. Hermsmeier, M. G. Samant, and D. Weller, *Phys. Rev. Lett.* 69:2307 (1992).
8. J. G. Tobin, G. D. Waddill, and D. P. Pappas, *Phys. Rev. Lett.* 68:3642 (1992).
9. Y. U. Idzerda, L. H. Tjeng, H.-J. Lin, C. J. Gutierrez, G. Meigs, and C. T. Chen, *Phys. Rev. B* 48:4144 (1993).
10. T. Böske, W. Clemens, C. Carbone, and W. Eberhardt, *Phys. Rev. B* 49:4003 (1994).
11. W. L. O'Brien, B. P. Tonner, G. R. Harp, and S. S. P. Parkin, *J. Appl. Phys.* 76:6462 (1994).
12. G. R. Harp, S. S. P. Parkin, W. L. O'Brien, and B. P. Tonner, *Phys. Rev. B* 51:3293 (1995).
13. P. Rudolf, F. Sette, L. H. Tjeng, G. Meigs, and C. T. Chen, *J. Magn. Magn. Mater.* 109:109 (1992).
14. C. Giorgetti, S. Pizzini, E. Dartyge, A. Fontaine, F. Baudelet, C. Brouder, P. Bauer, G. Krill, S. Miraglia, D. Fruchart, and J. P. Kappler, *Phys. Rev. B* 48:12732 (1993).
15. J. P. Schillé, F. Bertran, M. Finazzi, C. Brouder, J. P. Kappler, and G. Krill, *Phys. Rev. B* 50:2985 (1994).
16. C. T. Chen, Y. U. Idzerda, H.-J. Lin, G. Meigs, A. Chaiken, G. A. Prinz, and G. H. Ho, *Phys. Rev. B* 48:642 (1993).
17. V. Chakarian, H.-J. Lin, Y. U. Idzerda, G. Meigs, E. E. Chaban, J.-H. Park, C. J. Gutierrez, G. A. Prinz, and C. T. Chen, *MRS Proceedings* 375:in press (1995).
18. V. Chakarian, Y. U. Idzerda, G. Meigs, E. E. Chaban, and C. T. Chen, *Appl. Phys. Lett.* 66:3368 (1995).
19. C. T. Chen, and F. Sette, *Rev. Sci. Instrum.* 60:1616 (1989).
20. C. T. Chen, *Rev. Sci. Instrum.* 63:1229 (1992).
21. Y. U. Idzerda, C. T. Chen, H.-J. Lin, G. Meigs, G. H. Ho, and C.-C. Kao, *Nucl. Instrum. Methods Phys. Res. A* 347:134 (1994).
22. M. G. Samant, J. Stöhr, S. S. P. Parkin, G. A. Held, B. D. Hermsmeier, and F. Herman, *Phys. Rev. Lett.* 72:1112 (1994).
23. D. Weller, Y. Wu, J. Stöhr, M. G. Samant, B. D. Hermsmeier, and C. Chappert, *Phys. Rev. B* 49:12888 (1994).
24. V. Chakarian, Y. U. Idzerda, G. Meigs, and C. T. Chen, *Phys. Rev. B* to be published (1995).
25. K. Satter, and H. C. Siegmann, *Phys. Rev. Lett.* 29:1565 (1972).
26. J. C. Gröbli, D. Guarisco, S. Frank, and F. Meier, *Phys. Rev. B* 51:2945 (1995).
27. C.-C. Kao, C. T. Chen, E. D. Johnson, J. B. Hastings, H.-J. Lin, G. H. Ho, G. Meigs, J.-M. Brot, S. L. Hulbert, Y. U. Idzerda, and C. Vettier, *Phys. Rev. B* 50:9599 (1994).
28. C. T. Chen, Y. U. Idzerda, C.-C. Kao, L. H. Tjeng, H.-J. Lin, and G. Meigs, *MRS Proceedings* 375:in press (1995).
29. C. T. Chen, Y. U. Idzerda, H.-J. Lin, N. V. Smith, G. Meigs, E. Chaban, G. H. Ho, E. Pellegrin, and F. Sette, *Phys. Rev. Lett.* 75:152 (1995).
30. R. Wu, and A. J. Freeman, *Phys. Rev. Lett.* 73:1994 (1994).
31. G. Schütz, R. Frahm, P. Mautner, R. Wienke, W. Wagner, W. Wilhelm, and P. Kienle, *Phys. Rev. Lett.* 62:2620 (1989).
32. G. Schütz, M. Knülle, and H. Ebert, *Physica Scripta* Vol. T49:302 (1993).
33. G. Schütz, P. Fischer, K. Attenkofer, M. Knülle, D. Ahlers, S. Stähler, C. Detlefs, H. Ebert, and F. M. F. de Groot, *J. Appl. Phys.* 76:6453 (1994).

34. M. Knülle, D. Ahlers, and G. Schütz, *Solid State Comm.* 94:267 (1995).
35. O. Isnard, S. Miraglia, D. Fruchart, C. Giorgetti, S. Pizzini, E. Dartyge, G. Krill, and J. P. Kappler, *Phys. Rev. B* 49:15692 (1994).
36. E. Dartyge, F. Baudelet, C. Brouder, A. Fontaine, C. Giorgetti, J. P. Kappler, G. Krill, M. F. Lopez, and S. Pizzini, *Physica B* 208-209:751 (1995).
37. K. Kobayashi, H. Maruyama, H. Maeda, T. Iwazumi, H. Kawata, and H. Yamazaki, *Physica B* 208-209:779 (1995).
38. K. M. Kemner, private communication.
39. L. M. Falicov, D. T. Pierce, S. D. Bader, R. Gronsky, K. B. Hathaway, H. J. Hopster, D. N. Lambeth, S. S. P. Parkin, G. Prinz, M. Salamon, I. K. Schuller, and R. H. Victora, *J. Mater. Res.* 5:1299 (1990), and references therein.
40. J. M. Florczak, and E. D. Dahlberg, *J. Appl. Phys.* 67:7520 (1990).
41. C. Daboo, J. A. C. Bland, R. J. Hicken, A. J. R. Ives, and M. J. Baird, *Phys. Rev. B* 47:11852 (1993).
42. V. Chakarian, Y. U. Idzerda, H.-J. Lin, C. J. Gutierrez, G. A. Prinz, G. Meigs, and C. T. Chen, *Phys. Rev. Lett.* to be published (1995).
43. J. C. Slonczewski, *J. Magn. Magn. Mater.* in press (1995).
44. V. Chakarian, Y. U. Idzerda, C.-C. Kao, J.-H. Park, G. Meigs, E. D. Johnson, G. A. Prinz, and C. T. Chen, *Phys. Rev. B* to be published (1995).

RESONANT PHOTOEMISSION
IN POLYMERS

J. Kikuma[1,*] , J. D. Denlinger[1,2] , E. Rotenberg[1,3] and B. P. Tonner[2,4]

[1]Lawrence Berkeley Laboratory, Berkeley, CA 94720
[2]Dept. of Physics, Univ. of Wisconsin-Milwaukee, P.O. Box 413
 Milwaukee, WI 53201
[3]Dept. of Physics, Univ. of Oregon, Eugene, OR 97403
[4]Synchrotron Radiation Center, University of Wisconsin-Madison,
 Stoughton WI 53589

INTRODUCTION

Resonant photoemission, the decay processes following the resonant excitation of a core level electron to an unoccupied state, has been studied for various metallic and non-metallic materials[1,2]. Generally, strong enhancement in valence band photoemission due to the large photoabsorption cross section has been observed. Resonant photoemission from small organic molecules has been successfully interpreted as the superposition of "participator" decay and "spectator" decay. It has been shown that organic molecules which have $\pi*$ unoccupied states show strong participator decay as well as spectator decay[3-7], whereas molecules with no $\pi*$ unoccupied states showed only spectator decay[5].

Recently, we have reported the result of resonant photoemission of polystyrene as the first resonant photoemission study on organic polymers[8]. It has been shown that the valence band photoemission is strongly enhanced in the decay following the excitation to the lowest unoccupied $\pi*$ band. The degree of participator contribution can be associated with the spatial distribution of involved orbirtals.

In this work, we show valence-band resonant photoemission results from a polystyrene-like polymer, namely, poly(α-methyl styrene) (PαMS) at the C K-edge, obtained by using third generation synchrotron radiation with a high photon energy resolution and a high photon flux to reveal detailed information about satellites and

* Permanent address : Analytical Research Center, Asahi Chemical Industry Co. Ltd.,
2-1 Samejima, Fuji, Shizuoka, 416 JAPAN

Synchrotron Radiation Techniques in Industrial, Chemical, and Materials Science
Edited by D'Amico *et al.*, Plenum Press, New York, 1996

spectator shifts. The correlation map between occupied and unoccupied states will be also presented.

EXPERIMENTAL METHODS

Experiments were performed at the undulator beamline (BL-7.0) of the Advance Light Source, using a spherical grating monochromator. Photoemission spectra were collected using a 137 mm hemispherical analyzer. In addition to photoemission spectra, x-ray absorption near edge structure (XANES) spectroscopy was performed using total electron yield detection.

The PαMS was commercially obtained and used without further purification. Thin films of PαMS were prepared by spin casting from a toluene solution onto the clean Si substrate[8]. The thickness of the film was estimated to be 200 Å. Prior to the measurement by synchrotron radiation, the quality of the film was checked with a conventional Mg K_α source installed in the same analysis chamber, confirming no signal from the substrate or impurities.

A series of valence band photoemission spectra was obtained in the photon energy range from 282.2 to 302.2 eV with a photon energy step of 0.2 eV to cover the entire XANES region for C K-edge. Because of the high photon flux, each valence spectrum could be obtained in less than 1 minute with very good statistics. The photoelectron take-off angle was set at normal emission, and the overall experimental energy resolution was estimated to be 0.06 eV at an analyzer pass energy of 3 eV. The sample position was shifted by 50 μm for each spectrum acquisition in order to eliminate effects of beam damage. The typical x-ray spot size was about 50 μm in diameter, however in the case of longer data acquisition time, the x-ray beam was defocused to make the spot size of 1.0 mm x 0.2 mm. No significant change in spectral shape was observed during data acquisition. Less than 1 eV of stable charging was observed and corrected by aligning peak F (defined below) to 7.6 eV binding energy relative to vacuum level[9].

The photon energy was calibrated using second order light from the monochromator. The kinetic energy difference of the sharp Fermi edge of a clean Cu single crystal in first and second order was measured and used to determine the absolute photon energy.

All data have been normalized with the incident beam intensity monitored by a transmissive gold grid.

RESULTS AND DISCUSSION

Core excitation spectrum

The XANES spectrum at the C K-edge is shown in Fig. 1. The overall spectral shape is almost identical to that of polystyrene[8]. The spectral shape is also very similar to that of benzene[10-12]. The full width at half maximum (FWHM) of the lowest π* resonance (peak 1) is 0.67 eV, which is better than our previous work for polystyrene (0.9 eV FWHM) because of higher photon energy resolution.

According to the peak assignments for XANES spectrum of benzene[10], absorption peaks 1 and 3 correspond to two π* states, and peaks 4 and 5 correspond to two σ* states, respectively. Peak 2 corresponds to a 3p Rydberg state, but it may also contain the contribution of C-H* transition since this peak is relatively stronger than in the spectrum of benzene[10]. Feature 4' is probably a step-like feature due to excitation to the continuum.

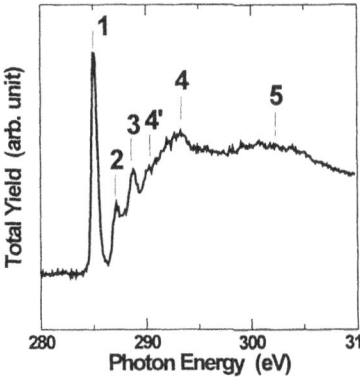

Figure 1. A XANES spectrum at the C K-edge, obtained by the total electron yield detection. Intensity is normalized with incident photon flux.

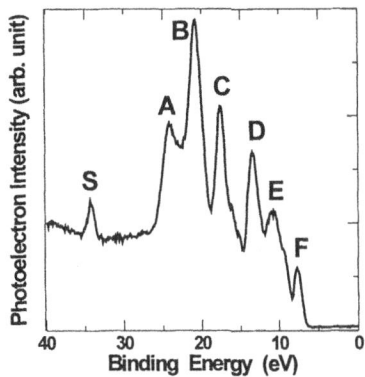

Figure 2. A valence band spectrum obtained at the photon energy below the C K-edge (250 eV). Six features from A to F are resolved. Binding energy scale is calibrated in the way that the position of peak F becomes 7.6 eV. The vacuum level is referred as E=0. Peak "S" stands for C 1s peak excited by the second order light.

We use these assignments in the interpretation of our resonant photoemission data described below.

Valence band spectrum

Non-resonant valence band photoemission spectrum obtained at a photon energy of 250 eV is shown in Fig. 2. Six valence peaks, labeled A to F are clearly resolved. The C 1s peak excited by second order light is also observed (peak labeled S). The spectral shape is in good agreement with XPS data by Beamson and Briggs[13], in which the experimental energy resolution is estimated to be 0.5 eV. The fact that the spectral shape does not change with higher experimental energy resolution indicates that the peak broadening due to experimental conditions is negligible in Fig. 2 and that peak widths in Fig. 2 are intrinsic to the material.

Resonant photoemission : Overview

Fig. 3 shows a complete mapping of the valence band photoemission intensity, displayed as a surface plot, as the photon energy is tuned through the C K-edge.

A strong enhancement of photoemission is observed in addition to "normal" processes such as valence band photoemission (a) and Auger decay process (b). At the photon energy below the ionization threshold, relatively weak non-resonant (normal) valence band photoemission is observed (c,d), which is represented by Fig. 2. As the photon energy approaches to the lowest π^* resonance, a dramatic increase in intensity (nearly 50 times) of the entire valence band photoemission is observed. After passing through this resonance, the intensity once decreases to the non-resonant intensity level, and then rises again due to the (b) shifts to higher binding energy region with increasing photon energy. There are several peaks at the photon energy of 287.2, 289.0, 290.2 and 293 eV corresponding to XANES peaks 2, 3, 4' and 4, respectively.

The feature diverging away from the valence levels (e) is C 1s photoemission excited by the second order light.

Decay spectra and spectator shift

At this point, we discuss the spectra obtained at the photon energies corresponding to the XANES peaks in more detail.

The valence band spectrum obtained at the photon energy of resonance 1 is shown in Fig. 4a. The enhanced valence band intensity consists of (i) normal photoemission, (ii) participator decay contribution, in which the initially excited electron is involved in the core hole decay, and (iii) spectator decay contribution, in which the initially excited electron remains as a "spectator" (not involved in the core hole decay) during the decay process. The participator contribution peaks at the same final state as normal photoemission, while the spectator contribution has the same spectral shape as Auger

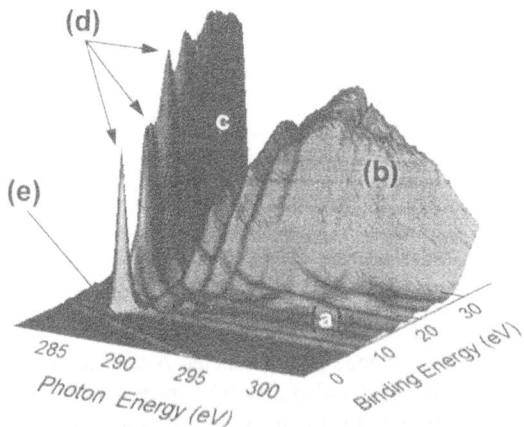

Figure 3. A surface plot of photoemission spectra as a function of binding energy and photon energy. The plot consists of (a) normal photoemission, (b) normal Auger decay, (c) spectator decay and (d) participator decay (see text). C 1s photoemission excited by second order light is also observed in this plot (e).

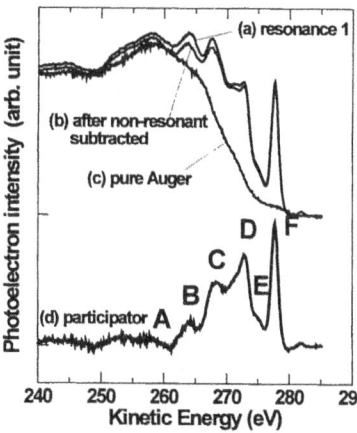

Figure 4. Comparison of (a) valence band spectrum obtained at resonance 1 (b), after subtraction of non-resonant photoemission contribution, and (c) pure Auger spectrum obtained at the photon energy of 418.7 eV. Spectrum (c) has been multiplied appropriately and shifted to the higher kinetic energy by 2.3 eV which corresponds to the spectator shift (see text). Spectrum (d) is the difference of spectrum (b) and (c), which represents the participator contribution of this decay process. Labels on the peaks (A through F) are the same notation as those in Fig. 2.

decay. After subtracting the contribution of normal photoemission (Fig. 4b), we distinguished the contributions from participator decay and spectator decay. A pure Auger spectrum (Fig. 4c) obtained at the photon energy far above the C K-edge (418.7 photon energy) has been used as to represent the spectator decay line shape. The pure Auger spectrum was shifted so that the position of the 'dip' around 250 eV kinetic energy matches to that of resonant spectrum, and then scaled so that the resultant residue becomes minimum. The shift needed to align the position of two spectra reflects the screening effect caused by the spectator electron, and is called 'spectator shift.'[3-7] The spectator shift for this resonance is determined to be 2.3 eV, which is in good agreement with the value for polystyrene[8].

After removing the normal photoemission contribution and the Auger-like spectator part, what remains is the participator spectrum (Fig. 4d). The participator decay channel at resonance 1 exhibits the six valence band features (A - F), but with quite different relative intensities from that of the non-resonant spectrum (Fig. 2). Peak F is particularly enhanced relative to the other valence structures in the resonant spectrum. Peaks C and D also showed strong enhancement. Peaks A and B, on the other hand, are relatively small. The selective enhancement of peak F can be interpreted in terms of the spatial distribution of the molecular orbitals[3,5,6]. For benzene as a model compound, the lowest unoccupied two π^* orbitals (energetically degenerate) are both highly localized on carbon atoms, and lead to the sharp, strong XANES peak corresponding to resonance 1[14]. According to the simple, one electron scheme of the participator decay, both the occupied and unoccupied orbitals of the ground state must have similar spatial extents, in order for there to be comparable overlaps with the core-hole to create a resonant interference. This is the case for peak F of PαMS, which is derived from the benzene highest occupied state, which has a very similar spatial extent to the lowest π^* state according to the SCF calculation[14]. For other valence peaks, the relative intensity cannot be simply explained by the spatial distribution of benzene molecular orbital, implying that the distribution around the carbon backbone chain should be taken into account. Complete theoretical calculation for this or related polymers are desirable to discuss the relative intensity further.

The same analysis procedure was applied to other resonances. The spectator shifts have been determined to be 2.2, 1.5 and 0.6 eV for resonances 2, 3 and 4 respectively. The decrease of the spectator shift for higher unoccupied bands is reasonably explained by the fact that molecular orbitals are more delocalized in the higher levels and the screening effect becomes smaller. There are very small participator decay contribution observed for these resonances, which will be described in the next section.

Correlation plot between occupied and unoccupied states

Here we would like to discuss the physical meaning of the axes of Fig. 3. Similar to XANES spectrum, the photon energy axis represents positions of unoccupied bands into which core electrons are excited. On the other hand, the binding energy axis indicates positions of occupied valence bands. Therefore, this type of surface plot can be interpreted as a correlation map between occupied valence bands and unoccupied bands.

As described above, Fig. 3 consists of normal photoemission, participator decay, spectator decay and normal Auger decay. Among the decay processes shown in Fig. 3, only the participator decay includes direct energy transfer from unoccupied to occupied bands. Thus, extraction of the participator decay contribution from Fig. 3 reveals a "map" indicating the degree of energy transfer between unoccupied and occupied band, which can be considered as a "correlation plot." A surface plot of the participator decay contribution is shown in Fig. 5. Figure 5 was obtained basically in the same manner as in Fig. 4, except that the spectator shift was not taken into account in the subtraction of Auger-like feature.

From Fig. 5, several features including the strongest correlation between the lowest π^* band and the highest occupied band is seen, as described in the previous section. Small correlation peaks are observed in photon energies correspond to resonance 2 and 3, as well. The absence of participator contribution in the decay from unoccupied bands higher than resonance 4 can be attributed to the delocalized nature of σ^* band or Rydberg state. This is consistent with the fact that the spectator shift for these resonance is relatively small.

Satellite feature in Auger spectrum

Figure 6 shows a pure Auger spectrum obtained at a photon energy (418.7 eV) far above the C K-edge. A small extra feature is clearly observed at the higher kinetic energy side of main feature, and can also be observed in the surface plot of Fig. 3. The kinetic energy of this satellite is same as that of the sharpest peak in Fig. 3. It is obvious that this

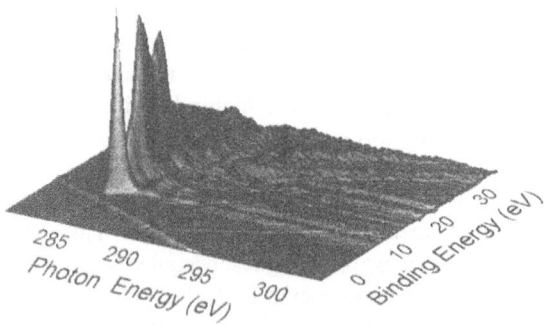

Figure 5. A correlation plot obtained by the subtraction of normal valence photoemission, normal Auger decay, and spectator decay contribution from Fig. 3. The strongest correlation between the lowest unoccupied state and the highest occupied state are clearly seen. Several other features are also seen at the photon energies corresponds to the lowest, second lowest, and third lowest unoccupied state.

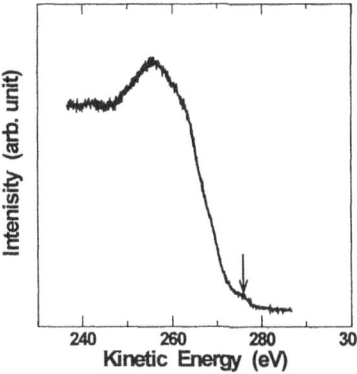

Figure 6. A pure C KLL Auger spectrum obtained at the photon energy of 418.7 eV. A satellite peak is indicated by an arrow.

feature does not stem from normal Auger process because the kinetic energy is too large. The fact that the kinetic energy of this satellite peak coincides with that of participator decay electron from the lowest unoccupied to the highest occupied band energy transfer suggests the energy transfer between these two bands is involved in the satellite process as well.

Similar satellite peaks in C KLL Auger spectra have also been observed in some polymers[13,15] and C_{60} thin films[18]. Ramaker and co-workers interpreted these satellites as a result of secondary electrons produced by either electron excitation or relatively high energy photon excitation[16,17,19]. According to Ramaker et al., secondary electrons, whose kinetic energy matches the excitation energy of the core-to-bound excitation, excite another core electron to unoccupied bound states. Since the core hole created by this process is screened by the electron in unoccupied state, the kinetic energy of ejected Auger electron become larger than normal Auger electron. Ramaker et al. have predicted this satellite should be observed only with electron excitation or high energy photon excitation because this requires high energy secondary electrons. For the present study, however, the satellite peak is observed with the excitation photon energy just above C K-edge, which cannot create the secondary electron with enough kinetic energy.

We tentatively assign this satellite as the result of "shake-up" in initial core excitation, as Brühwiler et al. pointed out in their study of C_{60}[18]. That is, one of the valence band electrons is promoted to the lowest unoccupied band in the event of core excitation, and this electron is involved in the decay process. This explanation, however, does not account for the satellite observed in saturated hydrocarbon systems[15] which do not have strongly localized π^* states.

CONCLUSION

We have performed a detailed measurement and analysis for the resonant photoemission of PαMS at the C K-edge. From the data set which consists of 101 valence band photoemission spectra, the correlation "map" of occupied valence bands and unoccupied bands has been obtained for the first time. The strongest correlation has been observed between the lowest π^* band and the highest occupied band. A satellite in C KLL Auger peak has been interpreted as the result of shake-up in initial core excitation.

Acknowledgment: This work was supported by the National Science Foundation, Division of Materials Research.

References

1. J. W. Allen, Resonant photoemission of Solids with strongly correlated electrons, *in* : Synchrotron Radiation Research, vol.1, R. Z. Bachrach ed., Plenum Press, New York, 1992.
2. W. Wurth and E. Menzel, Near edge x-ray absorption and decay dynamics of adsorbed molecules, *in* : "Application of Synchrotron Radiation," W. Eberthardt ed., Springer-Verlag, Berlin, 1995.
3. D. Menzel, G. Rocker, H.-P. Steinrück, D. Coulman, P. A. Heimann, W. Huber, P. Zebisch and D. R. Lloyd, *J. Chem. Phys.*, **96**, 1724 (1992).
4. T. Porwol, G. Dömötör, H.-J. Freund, R. Dudde, C.-M. Liegener and W. von Niessen, *Physica Scr.*, **T41**, 197 (1992).
5. W. Wurth and D. Menzel, *J. Elec. Spectrosc. Rel. Phenom.*, **62**, 23 (1993).
6. M. Mauerer, P. Zebisch, M. Weinelt and H.-P. Steinrück, *J. Chem. Phys.*, **99**, 3343 (1993).
7. R. Dudde, M. L. M. Rocco, E. E. Koch, S. Bernstorff and W. Eberhardt, *J. Chem. Phys.*, **91**, 20 (1989).
8. J. Kikuma and B. P. Tonner, *J. Chem. Phys.*, submitted
9. N. Ueno, W. Gadeke, E. E. Koch, R. Engelhardt and R. Dudde, *J. Elec. Spectrosc. Rel. Phenom.*, **36**, 143 (1985).
10. J. A. Horsley, J. Stöhr, A. P. Hitchcock, D. C. Newbury, A. L. Johnson and F. Sette, *J. Chem. Phys.*, **83**, 6099 (1985).
11. A. P. Hitchcock and C. E. Brion, *J. Elec. Spectrosc. Rel. Phenom.*, **10**, 317 (1977).
12. S. Aminpirooz, L. Becker, B. Hillert and J. Haase, *Surf. Sci.*, **244**, L152 (1991).
13. G. Beamson and D. Briggs, "High Resolution XPS of Organic Polymers," Wiley , Chichester, 1992.
14. W. L. Jorgensen and L. Salem, " The Organic Chemist's Book of Orbitals," Academic Press, New York, 1974.
15. F. L. Hutson and D. E. Ramaker, *Phys. Rev. B*, **35**, 9799 (1987).
16. F. L. Hutson and D. E. Ramaker, *J. Chem. Phys.*, **87**, 6824 (1987).
17. D. E. Ramaker, N. H. Turner and J. Milliken, *J. Phys. Chem.*, **96**, 7627 (1992).
18. P. A. Brühwiler, A. J. Maxwell, A. Nilsson, N. Mårtensson and O. Gunnarsson, *Phys. Rev. B*, **48**, 18296 (1993).
19. D. E. Ramaker, *Crit. Rev. Solid State Mat. Sci.*, **17**, 211(1991). : D. E. Ramaker, *J. Vac. Sci. Tech. A*, 7, 1614 (1989).

CHARACTERIZATION OF THE INTERACTION OF URANYL IONS WITH HUMIC ACIDS BY X-RAY ABSORPTION SPECTROSCOPY

Tobias Reich,[1] Melissa A. Denecke,[1]
Susanne Pompe,[1] Marianne Bubner,[1] Karl-Heinz Heise,[1]
Maren Schmidt,[1] Vinzenz Brendler,[1] Lutz Baraniak,[1] Heino Nitsche,[1]
Patrick G. Allen,[2] Jerome J. Bucher,[2]
Norman M. Edelstein,[2] David K. Shuh[2]

[1]Forschungszentrum Rossendorf e.V.
Institute of Radiochemistry,
P.O.Box 510119, D-01314 Dresden, Germany
[2]Lawrence Berkeley National Laboratory
Chemical Sciences Division
MS 70A-1150, Berkeley, CA 94720, USA

INTRODUCTION

Humic substances are present throughout the environment in soil and natural water. They are organic macromolecules with a variable structural formula, molecular weight, and a wide variety of functional groups depending on their origin. In natural waters, humic substances represent the main component of the "dissolved organic carbon" (DOC). The DOC may vary considerably from 1 mg/L at sea water surfaces to 50 mg/L at the surface in dark water swamps.[1] There is strong evidence that all actinides form complexes with humic substances in natural waters.[2] Therefore, humic substances can play an important role in the environmental migration of radionuclides by enhancing their transport. Retardation through humic substance interaction may be also possible due to formation of precipitating agglomerates. For remediation and restoration of contaminated environmental sites and risk assessment of future nuclear waste repositories, it is important to improve the predictive capabilities for radionuclide migration through a better understanding of the interaction of radionuclides with humic substances.

A large research program at the Institute of Radiochemistry at the Forschungszentrum Rossendorf is devoted to the study of the influence of humic acids (HA) on the environmental migration of uranium (VI) by a broad range of analytical and spectroscopic methods.[3] The research is related to the environmental problems due to large uranium mining and processing facilities in the Erzgebirge (Ore Mountains) which were operational in the Eastern part of Germany until 1990. Seepage waters from uranium mill tailing piles and

Synchrotron Radiation Techniques in Industrial, Chemical, and Materials Science
Edited by D'Amico *et al.*, Plenum Press, New York, 1996

215

Table 1. Sample composition and calculated speciation for model solutions of uranyl ions with acetic acid (Ac) and malonic acid (Mal).

	UO_2^{2+} (mol/L)	Ligand (mol/L)	pH	I (mol/L)	Calculated speciation
B1	0.05	0.05 Ac	0.5	0.25	UO_2^{2+} 100%
B2	0.05	0.2 Ac	2.9	0.5	UO_2^{2+} 48%, UO_2Ac^+ 35%
B3	0.05	1.1 Ac	3.7	1.3	$UO_2(Ac)_3^-$ 100%
B4	0.05	0.2 Mal	4.0	1.05	$UO_2Mal_2^{2-}$ 100%

from flooded mine shafts, which contain both uranium and humic substances, can act as a transport medium for the migration of uranium and its decay products into our environment.

By an operational definition, HA's are the part of humic substances which is soluble in alkaline and precipitating in acidic medium. The protolytic and complexation behavior of HA's is determined by their functional groups and substituents. The most important functional groups are carboxylic and phenolic OH groups. The amount of carboxylic and phenolic OH groups can vary considerably depending on the natural origin of the HA from 1.5-5.7 meq/g and 2.1-5.7 meq/g, respectively.[4] Several phenomenological parameters, such as stability constant, complexing capacity, and loading capacity, quantitatively describe the complexation behavior of HA's with U(VI). For example, the following stability constants have been reported for the 1:1 and 1:2 complex of uranyl with HA at pH of 4.5 and ionic strength of 0.1: log β_1 5.16 and log β_2 9.31.[5] The maximum amount of U(VI) that can be bound by HA at pH 4 is approximately 10-18% of its proton-exchange capacity (PEC).[6,7]

It is important to describe the complexation behavior of HA not only on a phenomenological but also on a molecular level. Extended X-ray Absorption Fine Structure (EXAFS) analysis is a standard technique which can provide molecular-level information on the nearest-neighbor structure of a chosen absorbing atom.[8] In the present study, EXAFS spectroscopy was applied to measure coordination numbers and distances of U(VI) to its nearest neighbors when it interacts with HA's. The molecular structure of HA's is so complex that only generic structural formulae are discussed in the literature (ca. Schulten et al.[9]). Therefore, our structural analysis of U(VI) interaction with HA's was based on the following strategy: 1) Investigation of solid and aqueous uranyl complexes with simple organic molecules which might represent "building blocks" of HA's. The goal is to identify certain structures of U(VI) complexes, such as chelate rings, monodentate, and bidentate configurations, depending on the presence of carboxylic and phenolic OH groups and their positions relative to each other in a given molecule. 2) Investigation of the U(VI) interaction with synthetic HA model substances. Synthetic HA is characterized by a chemical behavior similar to natural HA but with a considerably simpler overall structure.[10] 3) Comparison of structural parameters obtained from U(VI) model systems with those where U(VI) interacted with HA's isolated from natural sources.

In the present paper, we report EXAFS studies of solid samples of uranyl acetate, salicylate, o-methoxybenzoate, and aqueous solutions of both uranyl acetate and malonate. The structural parameters are compared with those of uranyl humates which were prepared under various conditions from one synthetic and two natural HA's.

SAMPLE PREPARATION

Solid Uranyl Model Compounds

For uranyl acetate (sample A1), we used the commercially available uranyl acetate dihydrate (p.A., Merck) without further purification. Uranyl salicylate (sample A2) and uranyl o-methoxybenzoate (sample A3) were synthesized following a modified procedure used in Ref. 11. The last part of the synthesis was modified to yield anhydrous crystals. If no crystal water is present in the compound, the presence of an additional oxygen coordination in the equatorial uranyl shell can be excluded during the EXAFS analysis. Anhydrous bis(salicylato)dioxouranium(VI) was prepared by dissolving uranyl acetate dihydrate and salicylic acid with a molar ratio of 1:2 in ethanol at 65°C. Anhydrous bis(methoxybenzoato)dioxouranium(VI) was prepared in a similar manner by dissolving uranyl acetate dihydrate and o-methoxybenzoic acid in a molar proportion of 1:2 in ethanol at 60°C. For both cases, acetic acid and ethanol were separated from the mixture by lyophilization. The resulting products were dissolved in ethanol and concentrated again to remove the rest of acetic acid. Finally, the crystals were washed several times in benzene and freeze-dried. Elemental and thermal analysis (EA and TA) confirmed that the synthesis had resulted in a 1:2 uranyl salicylate and 1:2 uranyl mehtoxybenzoate without water of hydration.

Aqueous Uranyl Model Solutions

Table 1 shows the molar ratio, pH, and ionic strength of the aqueous solutions of uranyl acetate (samples B1-B3) prepared. Solution B1 was prepared from a uranyl nitrate stock solution by adding acetic acid. Solutions B2 and B3 were made from uranyl acetate dihydrate and acetic acid. The pH was adjusted by adding NaOH. For each solution the speciation was calculated by using the corresponding complex stability and acid dissociation constants,[12] the speciation modeling software EQ3/6,[13] and the NEA Data Base.[14] From the results of the calculation that are shown in Table 1, one can see that solution B1 contains only the hydrated uranyl species. Solution B2 is a mixture consisting predominantly of the hydrated uranyl ion and the 1:1 complex. The remaining 17% is a mixture of monomeric and dimeric hydroxyl species. Due to a large excess of acetic acid, solution B3 contains only the tris-acetato complex.

The concentration, pH, and ionic strength of a solution containing 100% of the 1:2 uranyl malonate complex (sample B4) is also given in Table 1. The solution was prepared from uranyl nitrate hexahydrate and malonic acid. The pH was adjusted by adding NaOH. The speciation was calculated similar to the uranyl acetate solutions with the corresponding constants for malonic acid and uranyl malonate.[15]

Uranyl Humates

Uranium (VI) was sorbed onto HA's by two different ways: Samples C1-C3 were prepared by shaking solid HA in a solution of uranyl nitrate or uranyl perchlorate for 45-50 hours. Samples D1-D5 are precipitates formed when uranyl solution was added to dissolved HA. We used three different HA's: Fluka HA (untreated and purified), Aldrich HA, and synthetic HA. Details of the purification of Fluka HA and of the synthesis of HA and their characterization by different analytical methods, e.g., elemental analysis, determination of functional groups, IR spectroscopy, and capillary electrophoresis, are given by Pompe et al.[10] Table 2 summarizes the sample preparation conditions. Two sample preparations are described in detail below to illustrate the meaning of the columns in Table 2.

Sample C2 was prepared by shaking synthetic HA in a uranyl nitrate solution at a pH < 1 for 50 hours. The concentrations were 400 mg HA per mL solution and 176 mg U(VI) per gram HA (see columns 3 and 6 in Table 2). After centrifugation to separate the solid and

Table 2. Sample preparation conditions for uranyl humates by sorption on solid HA (samples C) and precipitation out of HA solution (samples D).

	HA source	mg HA/mL soln.	pH	State	mg U/g HA
C1	Fluka	100	1	dry	100
C2	synthetic HA	400	<1	dry	176 (170)*
C3	Fluka (purified)	200	<1	dry	500 (185)*
D1	Aldrich	58	8-10	precip./soln.	92
D2	synthetic HA	173	5.1	precip./soln.	136
D3	synthetic HA	173	5.4	wet paste	137
D4	Fluka (purified)	125	5.3	wet paste	190
D5	Fluka (purified)	7.3	4	dry	540 (440)

* Uranium loading, given in parentheses, was determined by photospectrometric measurement of the uranyl concentration left in solution after separating the uranyl humate.

liquid phases, the uranyl humate was washed several times with triply distilled water. For EXAFS measurements, the sample was dried by lyophilization (column 5 in Table 2). The U(VI) uptake by the HA was 170 mg/g (in parentheses in column 6 of Table 2) which corresponds to 97% of the total U(VI). The uranium uptake was measured only for a few selected samples.

To prepare sample D4, purified HA (Fluka) was first dissolved at pH 12. Then the pH was lowered to 5.7 with $HClO_4$. After adding 0.1 M acidic $UO_2(ClO_4)_2$ solution, the pH was adjusted to 5.3 by adding NaOH. The uranyl humate precipitated immediately upon adding the uranyl solution. The precipitate was separated from the solution by centrifugation and washed several times. This sample was measured by EXAFS as a wet paste.

EXPERIMENTAL

Uranium L_{III} edge X-ray absorption spectra (XAS) were measured at room temperature in transmission mode at the Stanford Synchrotron Radiation Laboratory (SSRL) on wiggler beamline 4-1 using a Si(220) double-crystal monochromator. Several samples were measured at the Hamburger Synchrotronstrahlungslabor (HASYLAB) using a Si(311) double-crystal monochromator at the bend magnet beamline RÖMO II. In order to obtain an energy calibration, the XAS spectrum of a solid uranyl nitrate hexahydrate sample was measured simultaneously. To reduce the higher-harmonic content of the X-ray beam at SSRL, the two Si crystals were detuned by 50%. For the EXAFS data analysis, we used the software package EXAFASPAK.[16] Theoretical scattering amplitudes and phases were calculated with the program FEFF6.[17] The threshold of the uranium L_{III} edge was set at 17185 eV. During the fit, a constant ΔE_0 shift of -13 eV was applied.

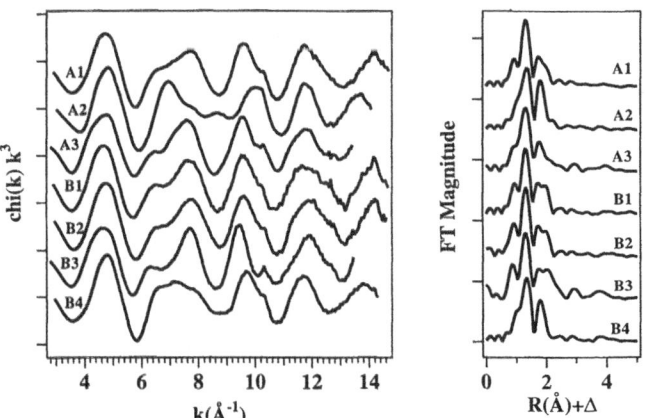

Fig. 1. Left panel: Experimental U L_{III} edge EXAFS of samples A1 - B4. Right panel: Fourier-transformed EXAFS of samples A1 - B4.

In the present work, structural analysis was restricted to the first two uranium coordination shells by fitting the Fourier-filtered EXAFS spectrum. Fourier filtering was performed to isolate the EXAFS contribution of the first two coordination shells from contributions of higher order shells which were observed for several samples. During the fit, the coordination number of the axial oxygen atoms in the uranyl group, UO_2^{2+}, was kept constant at two. The remaining five parameters, e.g., coordination number, distances, and Debye-Waller (DW) factors, were allowed to vary. In a future publication, we will present a more detailed structural analysis of solid and aqueous model compounds where higher coordination shells and multiple scattering effects (see below) are included to fit the experimental EXAFS.

RESULTS AND DISCUSSION

Solid Uranyl Model Compounds

The experimental EXAFS spectra of samples A1-A3 and their corresponding Fourier transforms (FT) are shown in Fig. 1. The structural parameters for the first two uranium coordination shells obtained from the fit are presented in Table 3.

The crystal structure of uranyl acetate dihydrate has been determined by single-crystal X-ray diffraction (XRD).[18] The bond lengths between uranium and the two axial oxygen

Table 3. EXAFS structural parameters for the first two uranium coordination shells in solid uranyl model compounds and uranyl model solutions.

Sample	U-O$_{ax}$			U-O$_{eq}$		
	N	R(Å)	σ^2(Å^2)	N	R(Å)	σ^2(Å^2)
A1 (Uranyl acetate)	2	1.77	0.003	4.8(4)*	2.41	0.007
A2 (Uranyl salicylate)	2	1.78	0.003	3.6(2)	2.32	0.003
A3 (Uranyl methoxybenzoate)	2	1.78	0.002	4.3(5)	2.44	0.010
B1 (UO$_2^{2+}$ 100%)	2	1.77	0.002	4.8(3)	2.42	0.007
B2 (UO$_2^{2+}$ 48%, UO$_2$Ac$^+$ 35%)	2	1.77	0.002	4.9(5)	2.40	0.008
B3 (UO$_2$(Ac)$_3^-$ 100%)	2	1.78	0.002	4.7(5)	2.46	0.006
B4 (UO$_2$Mal$_2^{2-}$ 100%)	2	1.79	0.002	4.6(3)	2.37	0.007

*) Uncertainties, given in parentheses, are statistical errors and do not reflect possible systematic errors.

220

atoms are 1.74 Å and 1.76 Å. The equatorial plane of the uranyl group contains five oxygen atoms: one oxygen at 2.34 Å from a water molecule, two oxygen atoms at 2.37 Å from two different bidentate bridging acetate groups, and two oxygen atoms at 2.45 Å from a chelating acetate group. As shown in Table 3, the uranium-oxygen bond distances agree within ±0.02 Å with the weighted average distances measured by XRD. Due to the limited data range in k space, the different bond lengths of the equatorial oxygen atoms are not resolved by EXAFS. The broader distribution of U-O_{eq} distances is indicated by a larger DW factor of the equatorial shell as compared to the axial shell. A coordination number of 4.8(±15%) for the equatorial shell agrees nicely with the crystal structure. Based on the agreement of structural parameters obtained by EXAFS and XRD, we conclude that theoretical scattering phases and amplitudes calculated by FEFF can be used to fit the EXAFS of organic uranyl complexes of unknown structure.

The experimental EXAFS curve of uranyl salicylate (sample A2) differs from that of uranyl acetate (see Fig. 1). The fit shows that the equatorial coordination of the uranyl group consists of approximately four oxygen atoms at a distance of 2.32 Å. The U-O_{eq} bond distance has a narrow distribution as indicated by the small DW factor of 0.003 (see Table 3). For a discussion of possible uranyl salicylate structures, it is essential to include information obtained by other analytical techniques. TA and EA measurements suggest that the uranyl ion forms an anhydrous bis-salicylato complex. The coordination of uranium with a water molecule, as in the case of uranyl acetate, can be ruled out. The IR spectra of uranyl salicylate did not show evidence of free phenolic OH groups as in the case of salicylic acid. Therefore, it can be concluded that the phenolic OH group of salicylic acid participates in the uranyl complexation. Together with the carboxylic OH group, a six-member chelate ring can be formed in the equatorial plane by each of the two salicylato molecules surrounding the uranium atom.

In o-methoxybenzoic acid the phenolic OH group is replaced by a methyl group and can not participate in the complex formation. By comparing the EXAFS spectra of uranyl salicylate and uranyl methoxybenzoate given in Fig. 1, it can be seen that the two uranyl complexes have different structures. The fit reveals that the complexes A2 and A3 differ in their U-O bond length and DW factor of the equatorial uranium shell. An average U-O_{eq} bond distance of 2.44 Å is close to that of the chelating acetate group of sample A1. Sample A3 in Fig. 1 shows pronounced peaks at 2.8 Å and 3.7 Å in the FT (uncorrected for EXAFS phase shifts). Similar peaks have been reported recently in an EXAFS study of monomeric tris-carbonato and trimeric bis-carbonato complexes with U(VI).[19] It was shown that multiple scattering pathways within the uranyl group and along the linear U-C-O configuration contribute significantly to the EXAFS FT peaks centered at 3.0 Å and 3.6 Å, respectively. If a similar explanation is valid for sample A3, the terminal carbon of the carboxylic group can give rise to the peak at 3.7 Å originating from multiple scattering along the linear U-C-C configuration. The large DW factor for the equatorial oxygen bond distance indicates a broad distribution. This may result from the presence of a bridging configuration of the carboxylic group.

Aqueous Uranyl Model Solutions

According to the calculated speciation described above, uranyl ions in solution B1 are coordinated by water molecules. At a pH of 0.5, ligands such as nitrate, chloride, perchlorate, and acetic acid do not coordinate uranium in a 0.05 M uranyl solution. The best fit of the Fourier-filtered EXAFS is obtained with five oxygen atoms at 2.42 Å, forming the second uranium coordination shell. Similar structural parameters as those listed for sample B1 in Table 3 have been obtained for other uranyl solutions at very low pH. For example, the U-O distances in a 10^{-3} M UO_2Cl_2 solution in 1.2×10^{-2} M HCl are 1.78 Å and 2.41 Å, respectively.[20] In a 0.05 M uranyl nitrate solution at pH 1.8, two U-O coordination shells at distances of 1.77 Å and 2.40 Å were observed.[21]

Due to a higher pH and the larger acetate concentration of sample B2 compared to sample B1, solution B2 should contain the uranyl complex with a 1:1 molar ratio in addition to the hydrated uranyl ion. The measured EXAFS spectrum given in Fig. 1 represents a superposition of all uranyl species present in the solution. As can be seen from Table 3, the nearest-neighbor structure of U(VI) in solutions B1 and B2 is the same within the experimental error. If there is any 1:1 complex present in solution B2, it is either undetectable by EXAFS or exhibits similar structural parameters as $UO_2(H_2O)_5^{2+}$.

Based on thermodynamic data, the tris-acetato complex is the only uranyl species present in solution B3. The EXAFS spectrum of solution B3 differs from those of samples B1 and B2, especially in the 6-10 $Å^{-1}$ region (see Fig. 1). The fit of the Fourier-filtered EXAFS shows an expansion of the equatorial U-O bond by 0.05 Å compared to samples B1 and B2. The value of 2.46 Å is close to the value of 2.45 Å for the chelating carboxylic carbon in solid uranyl acetate. Therefore, it is very likely that the carboxylic groups coordinate as bidentates. In a bidentate configuration, carboxylic carbon and methyl carbon at 2.8 Å and 4.34 Å should be observable by EXAFS, respectively.

The EXAFS spectrum of sample B4 with a 1:2 molar ratio of uranyl to malonic acid is different from those of the other model solutions, as shown in Fig. 1. A two-shell fit of the Fourier-filtered EXAFS gives a slightly longer axial U-O bond of 1.79 Å and a significantly contracted equatorial U-O bond of 2.37 Å compared to solutions B1-B3. The equatorial bond length matches that of a bridging acetic carbon rather than a chelating acetic carbon in solid uranyl acetate dihydrate. This leads to the conclusion that the carboxylic OH groups form a monodentate configuration with the uranyl ion. Note that the FT of sample B4 does not exhibit a pronounced peak at 3.8 Å, as in the case of samples A3 and B3, where chelating carboxylic groups are presumed to dominate the uranium coordination in the equatorial plane. This conclusion also agrees with structures proposed in the literature where the terminal carboxylic groups of two malonic acid molecules form two six-member rings in the equatorial plane of the uranyl group.[22]

Uranyl Humates

Inspection of the EXAFS spectra of uranyl humates given in Fig. 2, one notices that all samples have similar EXAFS except for sample D1 which differs in phase in the 6-11 Å^{-1} range. This situation is more clearly reflected in the structural parameters for the first two uranium coordination shells given in Table 4. Fig. 3 shows the Fourier-filtered EXAFS, the corresponding FT's, and the best fit for the two uranyl humate samples.

The structural parameters of samples C1-C3 and D2-D5 are the same to within the error inherent to EXAFS analysis and can be summarized as follows: The U-O distance to the axial oxygen atoms averages 1.78(2) Å with a DW factor of 0.002 Å^2; and the equatorial shell consists of approximately 5 oxygen atoms at 2.38(2) Å with a DW factor of 0.013(3) Å^2. In sample D1 the bond distance U-O_{eq} of 2.30 Å is much shorter. The longer U-O_{ax} bond of 1.83 Å is explained by the roughly inverse relationship between the electron density along the axial and equatorial U-O bonds. The agreement of the nearest-neighbor structure of U(VI) when it interacts with both Fluka HA and the synthetic HA can be interpreted as a success of the synthesis. Although the synthetic HA differs in its molecular structure and number of functional groups from the natural HA, it shows identical functionality with respect to interaction with uranyl ions. This is a somewhat surprising result if one considers that the uranyl humate samples were prepared from different HA starting materials at different pH and measured as wet pastes or dried powders. How can this be explained and why has one sample different structural parameters?

The main difference between sample D1 and the other uranyl humates is that D1 was prepared at high pH and with a uranium concentration that accounted for the loading capacity

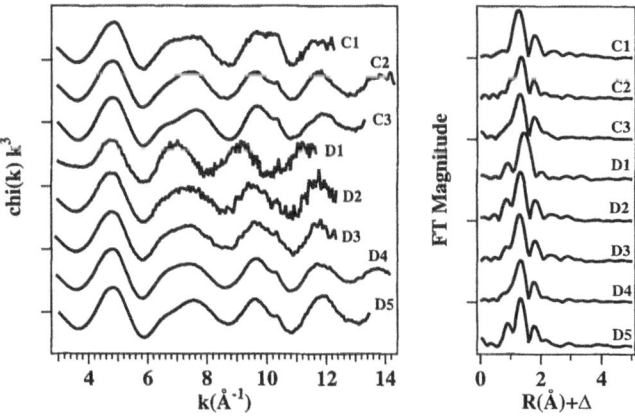

Fig. 2. Left panel: Experimental U L_{III} edge EXAFS of samples C1 - D5. Right panel: Fourier-transformed EXAFS of samples C1 - D5.

Table 4. EXAFS structural parameters for the first two uranium coordination shells in uranyl humates.

Sample name	U-O$_{ax}$			U-O$_{eq}$		
	N	R(Å)	σ^2(Å^2)	N	R(Å)	σ^2(Å^2)
C1	2	1.77	0.002	4.5(6)*	2.37	0.010
C2	2	1.78	0.002	5.2(6)	2.38	0.014
C3	2	1.77	0.002	5.0(5)	2.39	0.010
D1	2	1.83	0.002	4.6(7)	2.30	0.015
D2	2	1.79	0.002	4.5(7)	2.38	0.012
D3	2	1.78	0.002	5.0(6)	2.38	0.012
D4	2	1.78	0.002	4.8(5)	2.38	0.011
D5	2	1.78	0.002	5.2(8)	2.37	0.013

* Uncertainties, given in parentheses, are statistical errors and do not reflect possible systematic errors.

of dissolved HA. The amount of uranyl ions to prepare D1 at pH 8-10 was 15% of the measured PEC at pH of 4. Due to the increased deprotonation of the carboxylic and phenolic OH groups at higher pH, the actual loading capacity of the HA may be even higher than 15%.

For samples C1-C3, which were prepared at pH ≤ 1, fewer functional groups will deprotonate and can complex with uranyl ions. Additionally, the uranyl concentration in solution relative to the HA's PEC at pH 4 exceeded the 15-18% level for samples C1-C3 and D2-D5. For example, the PEC of purified HA (Fluka) was defined by the total number of COOH and phenolic OH groups and was measured to be 8.5 meq/g.[10] The amount of U(VI) during the preparation of sample D5 corresponds to 53% of its PEC. The measured uptake of U(VI) by the HA was 43% (Table 2). High levels of uranium uptake are possible as a result of sorption on non-specific sites in addition to site-specific sorption onto certain functional groups. The EXAFS spectra of such samples represent a superposition of uranium atoms sorbed onto different specific and non-specific sites. For the case of sample D1, the amount of U(VI) was below the saturation level of the specific sorption sites. A similar short U-O$_{eq}$ distance as observed for sample D1 was measured only for solid uranyl salicylate. A comparison with other solid and aqueous uranyl model complexes leads to the conclusion, that the phenolic OH groups in sample D1 participate in the uranyl humate interaction and cause the observed short U-O bond of 2.30 Å.

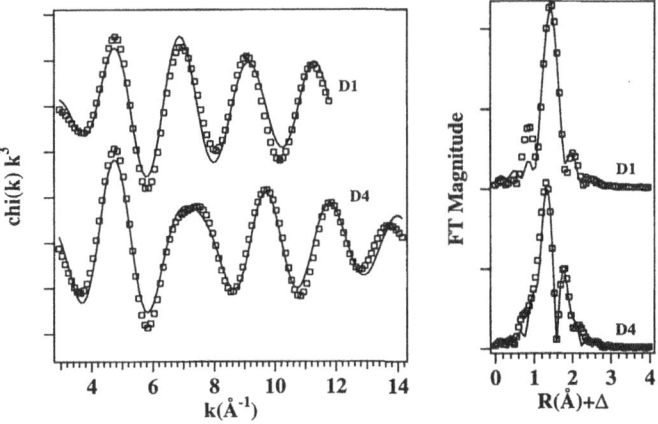

Fig. 3. Left panel: Fourier-filtered (squares) and fitted (solid line) U L$_{III}$ edge EXAFS of samples D1 and D4. Right panel: Fourier-transformed EXAFS of samples D1 - D4. Squares - Fourier-filtered data; solid line - fit.

Clearly, such a conclusion has to be supported by additional measurements of humate samples with uranium concentrations below the loading capacity of the HA to avoid sorption on non-specific sites. This will be a challenging experiment due to the large number of different functional groups of the HA's. One possibility is to block phenolic OH groups of the HA by derivatization with diazomethane.[10] Carboxylic COOH and phenolic OH groups can be derivatized into COOCH$_3$ and OCH$_3$ groups, respectively. In a second reaction with alkali, the carboxylic esters can be specifically hydrolyzed. Therefore, the OCH$_3$ groups will remain unchanged and cannot interact with uranyl ions similar to the o-methoxybenzoic acid described herein.

SUMMARY

Humic acids can play an important role in the transport properties of radionuclides throughout our environment. A molecular-level understanding of the structures formed as a result of the interaction of radionuclides with HA's is essential for the description of the complexation behavior of radionuclides. The complexation of HA's mainly occurs through their carboxylic and phenolic OH groups. In the present work, we obtained EXAFS structural parameters for the nearest-neighbors environment of U(VI) in uranyl humates. The influence of carboxylic and phenolic OH groups on the structure of uranyl complexes was also studied by measuring the EXAFS of selected simple model compounds in solid and liquid phases.

Uranyl acetate, uranyl salicylate, and uranyl o-methoxybenzoate were selected as solid model compounds and studied by EXAFS. From previous XRD analysis it is known that in uranyl acetate dihydrate the carboxylic group coordinates with the uranium in two ways with different U-O distances. For the bridging carboxylic group, the U-O$_{eq}$ bond distance is 2.37 Å. For the chelating carboxylic group, the U-O$_{eq}$ bond is 0.08 Å longer than for the bridging configuration. The agreement between the structural parameters obtained by EXAFS and XRD confirmed the conclusion of earlier EXAFS studies of uranyl compounds that theoretical scattering amplitudes and phases calculated by FEFF can be used successfully in the U L$_{III}$ edge EXAFS analysis.[23]

We are not aware of any XRD studies of crystalline anhydrous bis-(salicylato) dioxouranium(VI) and anhydrous bis-(methoxybenzoato) dioxouranium(VI). These two model compounds were measured with EXAFS to study structural changes of the uranyl complex when the phenolic OH group is replaced by OCH$_3$. In addition, the configuration of a COOH group and a phenolic OH group in ortho position to each other is often considered as an important structural element of HA's. In uranyl salicylate, approximately four short U-O$_{eq}$ bonds were observed at 2.32 Å. By replacing the phenolic OH group of salicylic acid by OCH$_3$, the structure of the resulting uranyl complex changes significantly as indicated by an increase of the U-O$_{eq}$ bond length by 0.12 Å. Based on the EXAFS and IR spectroscopic data, the structure of the uranyl bis-salicylato complex is dominated by the formation of six-member chelate rings involving the carboxylic and phenolic OH groups. The uranyl bis-methoxybenzoato complex is characterized by a bidentate configuration of the carboxylic oxygen atoms.

EXAFS is a valuable tool for structural analysis especially of amorphous systems. Structural parameters of several uranyl species in aqueous solutions with acetic and malonic acids were determined by EXAFS. For EXAFS measurements it is important to prepare samples which contain only one uranyl species. If several species are present simultaneously, the resulting average EXAFS signal may be difficult if not impossible to interpret. Therefore, prior to the preparation of the uranyl solutions, the appropriate conditions, e.g., metal and ligand concentrations and pH, were calculated based on thermodynamic data. For both solid and aqueous uranyl model systems, the bond distance of uranium to the equatorial oxygen atoms is most sensitive to differences in the coordination. A short U-O distance of 2.37 Å is observed for end-on coordination of the carboxylic group of the bis-malonato complex. For chelating bidentate coordination of the carboxylic group in the case of the uranyl tris-acetato complex, the U-O bond length is approximately 0.07 Å longer. A medium U-O bond distance of 2.41 Å is observed for the $UO_2(H_2O)_5^{2+}$ species.

HA's are characterized by a complicated structure and one could expect that the EXAFS signal of different HA's would look similar due to averaging over the large number of functional groups. Out of eight samples studied by EXAFS, seven uranyl humates exhibit similar structural parameters. Under the chosen sample preparation conditions, no difference was observed between dried powders and wet pastes and natural and synthetic HA's. Only

226

one uranyl humate sample (D1) exhibits significant different axial and equatorial U-O bond lengths. Although the interpretation of this result is limited by the number of samples studied, this difference is discussed in terms of uranyl sorption on specific and non-specific sites depending on proton exchange capacity and loading capacity of the HA's, and the pH and uranium concentration during the sample preparation. For the majority of uranyl humates, the structural parameters represent an average over uranyl sorbed on specific and non-specific sites. Based on the comparison of structural parameters observed in solid and aqueous uranyl model compounds, the short U-O_{eq} bond distance in sample D1 is interpreted as site specific sorption involving phenolic OH groups. This preliminary conclusion needs to be supported further by systematic EXAFS experiments on a series of uranyl humates prepared from chemically modified HA's to block certain functional groups and with lower uranium concentrations in solution.

In summary, pronounced changes observed in the equatorial uranyl coordination of organic uranyl model compounds and uranyl humates as a function of the uranyl speciation is encouraging for further EXAFS studies to obtain a better molecular-level understanding of the uranyl humate interaction. For certain complex configurations, e.g., chelating carboxylic groups, the amount of structural information will be enhanced further by including more distant coordination shells and multiple scattering effects in the EXAFS analysis.

ACKNOWLEDGMENT

We thank M. Meyer, R. Nicolai, R. Ruske, and G. Schuster for their valuable help in the sample preparation and characterization. EXAFS measurements were performed in part at SSRL, which is operated by the U.S. Department of Energy, Office of Basic Energy Sciences, Division of Chemical Sciences.

REFERENCES

(1) Leenher, J. A.; Malcolm, R. C.; McKinley, P. W.; Eccoles, L. A. *J. Res. U.S. Geol. Survey* **1974**, *2*, 361-369.

(2) Choppin, G. R.; Allard, B. In *Handbook of Physics and Chemistry of the Actinides Vol. 3*; A. J. Freeman and C. Keller, Eds.; North Holland: 1985.

(3) Nitsche, H. "Institute of Radiochemistry, Annual Report 1994" Forschungszentrum Rossendorf e.V., Dresden, Germany, 1995.

(4) Grauer, R. "Zur Koordinationschemie der Huminstoffe, Report No. 24," Paul Scherrer Institut, 1989.

(5) Munier-Lamy, C. *Organic Geochim.* **1986**, *9*, 285.

(6) Czerwinski, K. R.; Buckau, G.; Scherbaum, F.; Kim, J. I. *Radiochim. Acta* **1994**, *65*, 111-119.

(7) Schmidt, M.; Baraniak, L.; Bernhard, G.; Nitsche, H. In *Institute of Radiochemistry, Annual Report 1994*; H. Nitsche, Ed.; Forschungszentrum Rossendorf e.V.: Dresden, Germany, 1995; pp 58-59.

(8) Koningsberger, D. C.; Prins, R. *X-Ray Absorption: Principles, Applications, Techniques of EXAFS, SEXAFS and XANES*; John Wiley & Sons: New York, 1988.

(9) Schulten, H. R.; Schnitzer, M. *Naturwissenschaften* **1993**, *80*, 29-30.

(10) Pompe, S.; Bubner, M.; Brachmann, A.; Geipel, G.; Reich, T.; Denecke, M. A.; Nicolai, R.; Heise, K. H.; Nitsche, H. *Radiochim. Acta* **1995**, *submitted*.

(11) Kim, B.-I.; Miyake, C.; Imoto, S. *J. Inorg. Nucl. Chem.* **1974**, *36*, 2015-2021.

(12) Rabinowitch, E.; Belford, R. L. *Spectroscopy and Photochemistry of Uranyl Compounds*; Pergammon Press: New York, 1964, pp 125.

(13) Wolery, T. J. "EQ3/6, A Software Package for the Geochemical Modeling of Aqueous Systems," Lawrence Livermore National Laboratory, Livermore, CA, USA, 1992.

(14) *OECD Nuclear Energy Agency. Thermodynamic Data Base. Chemical Thermodynamics of Uranium*; Elsevier: Amsterdam, 1992.

(15) Paramonova, V. I.; Mesewitsch, A. N.; Tsi-Guan, M. A. *Radiokhimia* **1964**, *11*, 682-694.

(16) George, G. N.; Pickering, I. J. "EXAFSPAK: A Suite of Computer Programs for Analysis of X-ray Absorption Spectra," Stanford Synchrotron Radiation Laboratory, Stanford, CA, USA, 1995.

(17) Mustre de Leon, J.; Rehr, J. J.; Zabinsky, S. I.; Albers, R. C. *Phys. Rev. B* **1991**, *44*, 4146-4156.

(18) Howatson, J.; Grev, D. M. *J. Inorg. Nucl. Chem.* **1975**, *37*, 1933-1935.

(19) Allen, P. G.; Bucher, J. J.; Clark, D. L.; Edelstein, N. M.; Ekberg, S. A.; Gohdes, J. W.; Hudson, E. A.; Kaltsoyannis, N.; Lukens, W. W.; Neu, M. P.; Palmer, P. D.; Reich, T.; Shuh, D. K.; Tait, C. D.; Zwick, B. D. *Inorg. Chem.* **1995**, *34*, 4797-4807.

(20) Hudson, E. A.; Terminello, L. J.; Viani, B. E.; Reich, T.; Bucher, J. J.; Shuh, D. K.; Edelstein, N. M. In *Application of Synchrotron Radiation Techniques to Materials Science: MRS Symposium Proceedings, Vol 375*; D. L. Perry, N. D. Shinn, K. L. D'Amico, G. Ice and L. J. Terminello, Eds.; Materials Research Society: Pittsburgh, 1995; pp 235.

(21) Dent, A. J.; Ramsay, J. D. F.; Swanton, S. W. *J. Colloid Interface Sci.* **1992**, *150*, 45-60.

(22) Rajan, K. S.; Martell, A. E. *J. Inorg. Nucl. Chem.* **1967**, *29*, 523-529.

(23) Thompson, H. A.; Brown, G. E.; Parks, G. A. *Physica B* **1995**, *208/209*, 167-168.

EXAFS STUDIES OF LANTHANIDE COORDINATION IN CRYSTALLINE PHOSPHATES AND AMORPHOUS PHYTATES

Lester R. Morss,[1] Mark A. J. Schmidt,[1] Kenneth L. Nash,[1] Patrick G. Allen,[2] Jerome J. Bucher,[2] Norman Edelstein,[2] David K. Shuh,[2] Melissa A. Denecke,[3] Heino Nitsche,[3] and Tobias Reich[3]

[1]Chemistry Division, Argonne National Laboratory, Argonne, IL 60439
[2]Lawrence Berkeley National Laboratory, Berkeley, CA 94720
[3]Institut für Radiochemie, Forschungszentrum Rossendorf, Postfach
    510119, D-01314 Dresden, Federal Republic of Germany

INTRODUCTION

As part of the Efficient Separations and Processing Integrated Program of the U.S. Department of Energy, technologies are being developed that can stabilize radioactive and hazardous contaminants in order to reduce their potential to migrate from buried waste matrices. This report is part of a study to immobilize actinide ions in the near-surface environment by reacting them with organophosphorus complexants that decompose to inert, stable phosphates. In the initial phase of this study, the lanthanide ions Nd^{3+} and Gd^{3+} were used as models for the trivalent actinides Pu^{3+}, Am^{3+}, and Cm^{3+}. Phytic acid, *myo*-inositol hexakis(dihydrogenphosphate), was chosen as the organophosphorus complexant. The goal of this part of the project was to determine the bonding in precipitated lanthanide phytates.

Inositol phosphates are the major phosphorus-containing components of cereal grains, and they participate in transport of calcium in cell metabolism. Phytic acid complexes and precipitates divalent and trivalent metal ions, but the solubility of these salts varies with pH and concentration.[1] Phytic acid is most readily handled in its completely neutralized (basic) form as the dodecasodium salt hydrate. Two conformations of phytate have been described in the literature: one (5a/1e or axial) has five axial phosphates and one equatorial phosphate around the C_6H_6 ring, and the other (5e/1a or equatorial) has five equatorial phosphates and one axial phosphate around the ring.

(5e/1a), equatorial   (5a/1e), axial

Synchrotron Radiation Techniques in Industrial, Chemical, and Materials Science
Edited by D'Amico *et al.*, Plenum Press, New York, 1996

229

The structure of the hydrate $C_6H_6(PO_4Na_2)_6 \cdot 38H_2O$ has been determined[2] to be (5a/1e). In aqueous solution, ^{13}C and ^{31}P NMR studies have shown[3-5] that the (5e/1a) conformation predominates in acid and that the (5a/1e) predominates in base. Another ^{31}P NMR study[6] showed more complex behavior. Since the sodium phytate is a highly basic species, its (5a/1e) conformation in the solid is consistent with the aqueous species at high pH characterized by most authors.

Studies of phytic acid-metal ion equilibria and structure models are limited to Mg^{2+}, Ca^{2+}, and some transition metal ions. Martin and Evans found up to 4.8 Ca^{2+} bound to one phytate, with the maximum ratio at 0.006 M $[Ca^{2+}]$ and 0.001 M phytate[7] and they propose a 6:1 Ca^{2+}/phytate ratio with the (5a/1e) conformation:

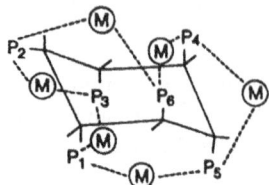

For the 6:1 Cu^{2+}/phytate species these authors[8] propose the (5e/1a) conformation:

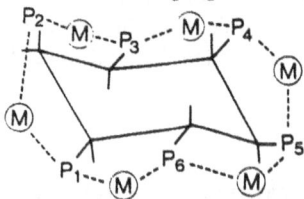

The protonation constants of phytic acid and metal-ligand stability constants for several lanthanide ions with phytic acid in aqueous solution have been determined by potentiometry.[9] The strongest phytic acid-lanthanide complexes are formed in acid solution, as is normally the case.

The infrared spectrum of phytic acid[10] has strong absorption bands at 1060 cm^{-1}, which was assigned to v(P-O-C) and 1400 cm^{-1}, which was assigned to v(P=O). Phytin is the neutral mixed Mg^{2+}/Ca^{2+} salt of phytic acid. Its infrared spectrum was reported[10] to show a metal-oxygen vibration at 460 cm^{-1} and a strong absorption band at 980 cm^{-1} that is characteristic of the P-O-P bond stretch in pyrophosphate ($P_2O_7^{4-}$).

We were not aware of any spectroscopic or crystallographic studies of solid metal phytates other than sodium phytate. Therefore we undertook to prepare solid lanthanide phytates and to characterize their lanthanide coordination by EXAFS. If the bonding in lanthanide phytates is similar to that in lanthanide phosphates, then the phytates should be insoluble and should decompose gradually to crystalline and inert phosphates.

SAMPLE PREPARATION AND CHARACTERIZATION

Crystalline orthophosphates $NdPO_4 \cdot xH_2O$ and $GdPO_4 \cdot xH_2O$ were used as standards. The orthophosphates $NdPO_4 \cdot xH_2O$ and $GdPO_4 \cdot xH_2O$ (Table 1) were precipitated by adding 0.05 M lanthanide nitrate solutions to dilute phosphoric acid solutions, buffered to pH 1 or pH 5 with NH_3(aq). The pH of the mixtures was readjusted to pH 5 and the mixtures were held at 20 °C or 50 °C for three days. The precipitates were separated by centrifugation and were dried in air at 100 °C. This temperature is below the temperature of first water loss (200-234 °C) and well below the reported transition temperature of 500-600 °C[11] to the anhydrous LnPO$_4$ (monoclinic, monazite). Our X-ray powder diffraction confirmed that the NdPO$_4$·xH$_2$O and GdPO$_4$·xH$_2$O had the hexagonal structure[11, 12]. Mooney did not locate the water molecules but she proposed that there was room for no more than alternate filling of three sites per unit cell with Z = 3, i.e. (moles H$_2$O)/(moles PO$_4$) = x = 0.5. Other researchers have reported 0 < x < 2. Thermogravimetric analyses of our preparations showed water content 0.5 < x < 1.

To prepare solid phytates, dilute (ca. 0.01 M) aqueous solutions of Nd^{3+} and Gd^{3+} nitrates were reacted with dilute sodium phytate solution at pH 3.2 in a ratio of 1:1 metal:phytate. Immediate precipitates formed, which were centrifuged, washed with water, recentrifuged, and dried at 95 °C in air (Table 1). Both Nd and Gd phytates were amorphous to X-ray powder diffraction.

Table 1. Samples for EXAFS studies

Description	Preparation	Structure (Literature)	Structure (Found by X-ray Diffraction)
NdPO$_4$·xH$_2$O	Precipitate at pH 5, 20 °C	Hexagonal	Hexagonal
Nd phytate	Precipitate 1:1 ratio at pH 3.2, dried air 95°C		Amorphous by powder X-ray diffraction
GdPO$_4$·xH$_2$O	Precipitate at pH 5, 20 °C	Hexagonal	Hexagonal
Gd phytate	Precipitate 1:1 ratio at pH 3.2, dried air 95°C		Amorphous by powder X-ray diffraction

Infrared spectra were taken of all samples (two are shown in Figure 1). The IR spectrum of NdPO$_4$·xH$_2$O is consistent with that reported[13]: the PO$_4^{3-}$ site symmetry is consistent with C$_2$ symmetry (which is lower than the D$_2$ symmetry required by the space group preferred by Mooney). The IR spectrum of Nd phytate is rather similar, showing a distinct peak at 990 cm^{-1} which is similar to that reported[10] in calcium-magnesium phytate at 980 cm^{-1}.

EXAFS DATA ACQUISITION AND ANALYSIS

Appropriate amounts of phosphates and phytates (Table I) were diluted with boron nitride powder in order to give an edge jump of ~1 across the Nd and Gd L$_{III}$ absorption edges. The samples were mounted in 3-mm thick X-ray cells consisting of a 6-mm x 25-mm cavity in a polyethylene frame with Kapton windows on each side. All spectra were acquired at room temperature in transmission mode at the Stanford Synchrotron Radiation

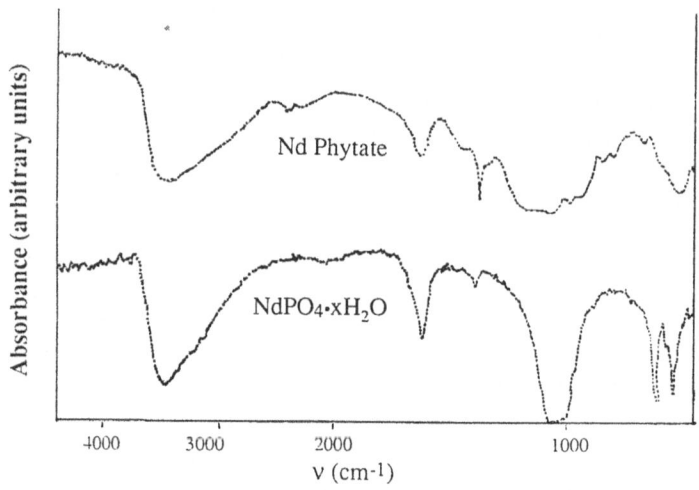

Fig. 1. The infrared spectra of Nd phytate and NdPO$_4$·xH$_2$O.

Laboratory (SSRL) on wiggler beamline 4-1 (unfocused) using a Si(220) double-crystal monochromator. Harmonic rejection was achieved by detuning the monochromator by 50% relative to the maximum incoming flux measured in I_0. The data were calibrated by simultaneously measuring spectra for Nd and Gd metal foils, defining the first inflection point of their L$_{III}$ edges at 6208 and 7243 eV, respectively.

EXAFS data reduction was performed by standard methods using the suite of programs EXAFSPAK developed by G. George of SSRL.[14] Data reduction included pre-edge background subtraction followed by spline fitting and normalization (based on the Victoreen falloff) to extract the EXAFS data above the threshold energies, E_0, defined at 6225 and 7260 eV respectively. Due to the closeness of the L$_{II}$ and L$_{III}$ absorption edges, the EXAFS data range was limited to k = 11 Å^{-1} for Nd and k = 13 Å^{-1} for Gd. Curve-fitting analyses were performed using EXAFSPAK to fit the raw k^3-weighted EXAFS data. The theoretical EXAFS modeling code, FEFF6, of Rehr et al.[15] was used to calculate the backscattering phases and amplitudes of the different neighboring atoms which where included in the fits. To aid in refinement of FEFF calculations, the atomic parameters for NdPO$_4$·xH$_2$O and GdPO$_4$·xH$_2$O were taken from Mooney.[12] EXAFS spectra and Fourier transforms are shown in Figures 2 and 3. Bond lengths and coordination numbers are given in Table 2.

DISCUSSION

For both phosphate samples, the bond distances and coordination numbers found by EXAFS are consistent with crystallographic values. The neodymium-oxygen distances (2.41 and 2.62 Å) in the Nd phytate samples are similar to those in NdPO$_4$·xH$_2$O (2.41 and 2.58 Å). Eight neodymium-oxygen bonds (coordination numbers 4 and 4) were found for NdPO$_4$·xH$_2$O and also for Nd phytate (coordination numbers 6 and 2). The same is true for the Gd samples except that only 7 Gd-O bonds were seen in Gd phytate. If each lanthanide is coordinated to two oxygens of two adjacent phosphates on a phytate ring, the lanthanide must also be coordinated to three or four other oxygens to complete its coordination sphere: to an oxygen on another phosphate in the same ring, to an oxygen on

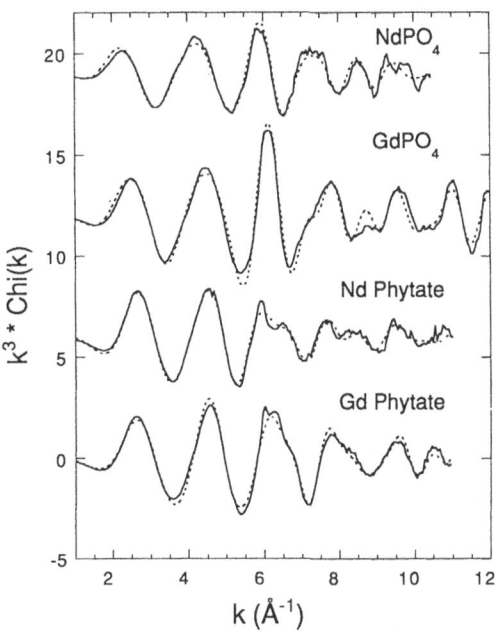

Fig. 2. k^3 weighted EXAFS spectra at Nd or Gd L_{III} edge. Top to bottom: $NdPO_4 \cdot xH_2O$, $GdPO_4 \cdot xH_2O$, Nd phytate, Gd phytate. Solid line, experiment; dashed line, fitted.

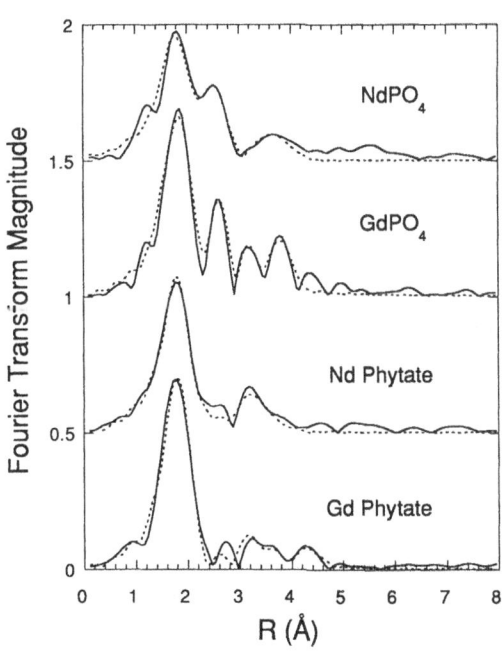

Fig. 3. Fourier transforms of Nd and Gd EXAFS from Fig. 2 (top to bottom, $NdPO_4 \cdot xH_2O$, $GdPO_4 \cdot xH_2O$, Nd phytate, Gd phytate). Solid line, experiment; dashed line, fitted.

233

second phytate ring, and/or to water molecules. It is worth noting that $LnPO_4 \cdot xH_2O$ converts upon heating to monazite (monoclinic structure, for $NdPO_4$ nine Nd-O at distances[16] 2.42-2.76 Å); the phytates do not have the monazite coordination geometry.

The shortest Nd-P distances (3.20 Å) in $NdPO_4 \cdot xH_2O$ were found in the $NdPO_4 \cdot xH_2O$ EXAFS but were not found in the Nd phytate; a parallel situation exists for the Gd samples. This observation is consistent with the expectation that the metals are bridged between two phosphates, as proposed by Martin and Evans (*vide supra*), rather than having bidentate (intraligand) metal-phosphate coordination. Three Gd-Gd interactions at 4.59 Å were seen in the Gd phytate. These near neighbors imply that there are four Gd^{3+} ions preparation ratio and inconsistent with the model of Martin and Evans[7, 8] that metal ions bridge adjacent phosphates on the ring. Perhaps some of the phytate has begun to decompose, although our continuing work has shown that lanthanide phytates require two weeks or more at 85 °C to decompose to crystalline phosphates.

There has been one X-ray absorption spectroscopy study[17] of lanthanide orthophosphates. This was an energy-level study of the 3d levels $3d_{5/2}$ and $3d_{3/2}$ in single-crystal $LnPO_4$ (monazite or zircon structures) at the $M_{4,5}$ edge and did not yield EXAFS

Table 2. EXAFS Results Compared with Crystallographic (Diffraction) Parameters

Bond	EXAFS			Diffraction	
	$d(Å)^a$	CN^b	$\sigma^2(Å^2)^c$	$d(Å)$	CN
$NdPO_4 \cdot xH_2O$, 20 °C					
Nd-O	2.41	4	0.0091	2.32	4
Nd-O	2.58	4	0.0077	2.64	4
Nd-P	3.19	2	0.0080	3.17	2
Nd-P	3.81	4	0.0183	3.65	4
Nd-Nd	4.11	4	0.0175	4.11	6
$GdPO_4 \cdot xH_2O$, 20 °C					
Gd-O	2.39	8	0.0110	2.29	4
Gd-O				2.60	4
Gd-P	3.12	2	0.0070	3.16	2
Gd-P	3.68	4	0.0125	3.60	4
Gd-Gd	4.03	4	0.0123	3.93	6
$GdPO_4 \cdot xH_2O$, 50 °C					
Gd-O	2.39	8	0.0115	2.29	4
Gd-O				2.60	4
Gd-P	3.12	2	0.0059	3.16	2
Gd-P	3.68	4	0.0110	3.60	4
Gd-Gd	4.03	4	0.0102	3.93	6
Nd Phytate					
Nd-O	2.41	6	0.0126		
Nd-O	2.62	2	0.0069		
Nd-P	3.89	2	0.0078		
Gd Phytate					
Gd-O	2.37	7	0.0116		
Gd-P	3.83	1	0.0028		
Gd-Gd	4.59	3	0.0133		

a Bond distances
b Coordination numbers (for M-O) or next-nearest neighbors (for M-P and M-M). Estimated uncertainty ±25% of CN.
c σ = Debye-Waller factor

parameters. Therefore our work appears to be the first EXAFS study of lanthanide phosphates and indicates that lanthanide L_{III}-edge spectra are appropriate for obtaining lanthanide coordination environments in phosphates.

We hope to ascertain whether the lanthanide phytates adopt the (5e/1a) conformation found for phytic acid at low pH by molecular modeling and ^{31}P NMR.

ACKNOWLEDGEMENTS

We thank Dr. John Ferraro for his help in interpretation of the infrared spectra of phosphates and phytates. This work was sponsored by the U.S. Department of Energy, at ANL by the Efficient Separations Program, under Contract W-31-109-ENG-38; and at LBNL by the Office of Basic Energy Sciences, Chemical Sciences Division, under Contract DE-AC03-76SF00098. This work was performed in part at SSRL, which is operated by the U. S. Department of Energy, Office of Basic Energy Sciences, Division of Chemical Sciences.

REFERENCES

1. E. Graf and J. W. Eaton, *J. Nutrition*, 114:1192 (1984).
2. G. E. Blank, J. Pletcher, and M. Sax, *Acta Cryst.*, B31:2584 (1975).
3. L. R. Isbrandt and R. P. Oertel, *J. Am. Chem. Soc.*, 102:3144 (1980).
4. N. Li, O. Wahlberg, I. Puigdomenech, and L.-O. Öhman, *Acta Chem. Scand.*, 43:331 (1993).
5. C. Brigando, J. C. Mossoyan, F. Favier, and D. Benlian, *J. Chem. Soc. Dalton*:575 (1995).
6. J. Elmsley and S. Niazi, *Phosphorus and Sulfur*, 10:401 (1981).
7. C. J. Martin and W. J. Evans, *J. Inorg. Biochem.*, 27:17 (1986).
8. C. J. Martin and W. J. Evans, *J. Inorg. Biochem.*, 28:39 (1986).
9. K. S. Siddiqi, S. A. S. Fathi, M. A. M. Aqra, S. Tabassum, S. A. A. Zaidi, and D. Benlian, *Indian J. Chem., Sect. A*, 32A:421 (1993).
10. K. A. Saburov and K. M. Kamilov, *Khimiya Prirodnykh Soedinenii*, 6:818 (1990).
11. Y. Hitichi, K.-i. Hukuo, and J. Shiokawa, *Bull. Chem. Soc. Japan*, 51:3645 (1978).
12. R. C. L. Mooney, *Acta Cryst.*, 3:337 (1950).
13. A. Hezel and S. D. Ross, *Spectrochim. Acta*, 22:1949 (1966).
14. G. N. George and I. J. Pickering, "EXAFSPAK, A suite of Computer Programs for Analysis of X-ray Absorption Spectra,", Stanford Synchrotron Radiation Laboratory, Stanford, CA, 1995.
15. J. Mustre de Leon, J. J. Rehr, S. I. Zabinsky, and R. C. Albers, *Phys. Rev. B*, 44:4146 (1991).
16. D. F. Mullica, D. A. Grossie, and L. A. Boatner, *J. Solid State Chem.*, 58:71 (1985).
17. D. K. Shuh, L. J. Terminello, L. A. Boatner, and M. M. Abraham, *X-ray absorption spectroscopy of the rare earth orthophosphates*, Mat. Res. Soc. Symp. Proc., vol. 307, Boston, MA, 1993, p 95.

CHEMICAL AND STRUCTURAL ELUCIDATION OF MINOR COMPONENTS IN SIMULATED HANFORD LOW-LEVEL WASTE GLASSES

John G. Darab, Hong Li, Dean W. Matson and Peter A. Smith

Materials Sciences Department
Pacific Northwest Laboratory[1]
Richland, WA 99352

Robert K. MacCrone

Materials Engineering Department
Rensselaer Polytechnic Institute
Troy, NY 12181

INTRODUCTION AND BACKGROUND

This symposium marks to the month the fiftieth anniversary of the beginning of the atomic age with the detonation of the world's first fission weapons in August, 1945. To support this effort, in 1943, the first full-scale nuclear reactors and processing plants needed for the production and isolation of ^{239}Pu were built at the Hanford Engineering Works along the Columbia River in southeastern Washington. Starting in December, 1944, the Hanford Site, as it would later be called, began processing irradiated uranium fuel elements and subsequently isolating ^{239}Pu bearing solutions.[2] Nearly thirty years of specialty nuclear materials production at Hanford as well as a concomitant generation of vast amounts of solid and liquid radioactive waste ensued. With the end of the cold war, the emphasis at the Hanford Site is now directed at remediation of these radioactive waste by-products.

Currently, the Hanford Site wastes are being stored in single-shell and double-shell tanks (SST and DST). SST and DST wastes are planned to be separated into low-level and high-level fractions, pretreated, mixed with glass precursor additives, then vitrified for long term storage. The Hanford low-level waste (LLW) inventory, which represents the major volume of waste at

Synchrotron Radiation Techniques in Industrial, Chemical, and Materials Science
Edited by D'Amico *et al.*, Plenum Press, New York, 1996

about 90%, has been estimated to amount to approximately 8.0×10^4 kg (88 tons) of solids and 6.8×10^8 L (180 million gallons) of aqueous solutions containing predominantly Na^+, K^+, $Al(OH)_4^-$, F^-, NO_2^-, NO_3^-, OH^-, organics, and a plethora of other minor mixed waste ionic species.[3]

Table 1 lists the anticipated composition of one particular Hanford pretreated LLW—that which is based on the double-shell slurry feed (DSSF). Because many of the Hanford Site waste tanks are still not fully characterized and the waste pretreatment details not yet completely developed, the composition listed in Table 1 is only tentative but has been agreed upon to be used for testing purposes. Both radioactive (i.e., containing $^{99}TcO_4^{2-}$) and non-radioactive (i.e., containing the technetium surrogate ReO_4^{2-}) DSSF simulants have been formulated from the composition listed in Table 1. This

Table 1. Anticipated composition of Hanford Site double-shell slurry feed (DSSF) low-level radioactive waste inventory after pretreatment.

Component	Concentration (moles/L)
$Al(OH)_4^-$	1.0
Ca^{2+}	0.0010
$Cr(OH)_4^-$	0.0087
Fe^{3+}	0.00077
K^+	0.50
Mg^{2+}	0.0010
Mn^{2+}	0.00042
MoO_4^-	0.017
Na^+	10.0
Sr^{2+}	0.017
Cs^+	0.017
TcO_4^-	0.017
PO_4^{3-}	0.043
IO_3^{3-}	0.017
CO_3^{2-}	0.27
Cl^-	0.16
F^-	0.25
SO_4^{2-}	0.043
NO_3^-	3.1
NO_2^-	1.7
OH^-	3.8
TOC[a]	1.4
pH	13.5
density	1.42
settled solids (%)	<5

[a] Total Organic Carbon (TOC) from EDTA.

composition does not reflect the tank-to-tank variations in chemical composition which, since blending of tank wastes is not foreseen in the process flow sheet, can significantly affect the vitrification process and the properties of the final glass. For example, variations in certain minor component concentrations in the melter feed can drastically modify the melt viscosity, increase the volatilization of radionuclides and other species during melting, as well as decrease the chemical durability of the final glass.[4,5]

One of the proposed final waste glass compositions, LD6-5412, which is based on a 26.7 wt% oxide-equivalent loading of DSSF LLW (e.g., the Cs^+, Sr^{2+}, SO_4^{2-}, F^-, etc. ions in the LLW are converted to Cs_2O, SrO, SO_3, F, etc. in the final glass), is summarized in Table 2. The nominal oxide-equivalent compositions of the DSSF waste and the glass precursor additives are also included. The glass precursor additives, which generally include SiO_2, Al_2O_3, H_3BO_3 (converted to B_2O_3 during melting), and $CaCO_3$ (converted to CaO during melting), are combined with the pretreated LLW to form a slurry prior to vitrification. Depending on the final melter design, the slurry can be (1) directly fed into the melter, which would typically be operating at temperatures of 1150-1350°C, (2) evaporated to dryness first, then melted, or (3) dried, calcined, then melted.

Laboratory-scale crucible melts involving nitrate-containing liquid feeds, such as DSSF, are often problematic due to processing difficulties associated with feed drying, process off-gassing, and melt foaming.[6] Thus, we have also made glasses from melt feeds based primarily on dry oxides and carbonates. LLW-based glass compositions produced by melting DSSF-based liquid feeds are designated with the nomenclature prefix LD, whereas those produced from dry oxide/carbonate feeds are designated with the prefix L. Using L-series LLW-based compositions, we have been empirically studying the effects minor components have on the properties of the melt and the final glass.[4,5]

The chemical and structural elucidation of minor component (i.e. waste) species such as Cl, Cr, Cs, F, P, S, and Tc/Re in Hanford LLW glass is an important aspect of understanding their solubility, volatility, and effect on glass chemical durability. With such an understanding comes the ability to tailor the glass to accept the desired level of waste loading while maintaining chemical durability and glass formability. Although techniques such as X-ray diffraction and optical and electron microscopies are useful in detecting large-scale separation (i.e., second phase precipitation) once the solubility limits have been exceeded, the determination of the chemical environment around minor component species in the solid solution phase(s) is a more difficult task.

We have undertaken a broad, multi-technique spectroscopic approach to studying the structure of Hanford LLW-based glasses, with the emphasis being placed on associating the structure of a glass to its observed properties. X-ray Absorption Near Edge Structure (XANES), X-ray Absorption Fine Structure (XAFS), Electron Paramagnetic Resonance (EPR), Raman, and Nuclear Magnetic Resonance (NMR) spectroscopies as well as magnetization measurements are currently being used to study the structure and chemistry of minor components in these glasses.

Table 2. Nominal oxide–equivalent composition of the final base glass (LD6-5412) and of the DSSF waste stream and glass precursor additive components that the glass is vitrified from.

	LD6-5412 Base Glass (wt%)	Component	
		DSSF (wt%)	Additives (wt%)
Contribution[a]	100.0	26.7	73.3
Oxide			
SiO_2	55.65		75.91
Na_2O	20.00	74.88	
B_2O_3	5.00		6.82
CaO	4.00	0.01	5.46
Al_2O_3	12.00	12.56	11.82
Minor components*	3.35	12.55	
Total	100.00	100.00	100.00
*Minor components			
Cl	0.36	1.36	
Cr_2O_3	0.04	0.15	
F	0.31	1.15	
I	0.14	0.51	
MoO_3	0.16	0.60	
Fe_2O_3	0.005	0.015	
K_2O	1.51	5.67	
MnO	0.002	0.007	
MgO	0.003	0.008	
P_2O_5	0.19	0.73	
SO_3	0.22	0.83	
SrO	0.12	0.43	
Cs_2O	0.15	0.57	
TcO_2	0.14	0.52	
Total	3.35	12.55	

[a] Contribution (in oxide–equivalent wt%) of component to final glass composition.

Here we present and discuss the results of XANES and XAFS measurements performed on Hanford LLW-based glasses spiked with high concentrations of Cr_2O_3, SO_3, and Cl. The results of supporting Raman and magnetization measurements are also briefly summarized.

EXPERIMENTAL PROCEDURE

Preparation of Glasses

LD6-5412 Base Glass Composition. LD6-5412 base glass slurry was prepared by mixing SiO_2 (56.78 g), H_3BO_3 (9.12 g), $CaCO_3$ (7.15 g), Al_2O_3 (8.66 g), and 71.2

mL of DSSF simulant (described in Table 1) spiked with 0.017 M NH_4ReO_4 (in place of NH_4TcO_4) in a Pt-10%Rh crucible.

Drying and Melting of LD6-5412 Glass. Glass was prepared by first placing the slurry-containing crucible in a drying oven set at 100°C. After 24 hours, the crucible along with the dried DSSF/glass precursor slurry were placed in Deltech DT-31 furnace set at 700°C for 30 minutes. The crucible was then removed and placed in a second Deltech furnace set at 1350°C. The resulting foam that developed during sample heating to the melt temperature subsided after about 30 minutes, assisted by periodic kneading with a Pt rod. The crucible and melt were then partially covered with a Pt lid. After an additional hour at 1350±1°C, the melt was quenched by pouring it onto a stainless steel plate.

L6-5412 Base Glass Composition. L6-5412 base glass batch material was made by intimately mixing dry chemical reagents: SiO_2 (56.78 g), H_3BO_3 (9.12 g), Na_2CO_3 (32.88 g), $CaCO_3$ (7.15 g), Al_2O_3 (12.00 g), Bi_2O_3 (0.014 g), $NaCl$ (0.152 g), Cr_2O_3 (0.036 g), NaF (0.471 g), Fe_2O_3 (0.005 g), K_2CO_3 (0.480 g), MnO (0.007 g), Nd_2O_3 (0.012 g), $NaPO_3$ (1.705 g), Na_2SO_4 (0.569 g), and ZrO_2 (0.005 g) for a total of about 130 g.

L-Series Minor Component Spiked Glass Compositions. L6-5412 batch materials spiked with higher levels of certain minor components (on an oxide-equivalent basis: 0.5, 0.8, and 2.0 wt% Cr_2O_3; 0.4 and 1.0 wt% Cl; and 0.5, 1.0, and 2.0 wt% SO_3) were prepared in the same way as the base glass batch material. However, appropriate amounts of all other components were removed to compensate for the additional amounts of minor components while maintaining a constant proportionality between the other components.

Batch Melting of L-Series Glasses. Using a Pt-10%Rh crucible, each glass was prepared by melting the ≈130 g of batch material in a Deltech DT-31 furnace at 1350±1°C. The melt duration was two hours, during which the crucible was covered with a Pt lid. After the first hour of melting, the molten glass was vigorously stirred using a Pt rod. After the second hour at 1350°C, the glass was quenched by pouring the melt onto a stainless steel plate. Final glass compositions were confirmed using XRF spectroscopy and ICP-AES.

XANES and XAFS Measurements

All XANES and XAFS spectra discussed in this work were obtained at room temperature on beam line X19A at the National Synchrotron Light Source (NSLS). The data were obtained either in transmission or fluorescence mode using samples prepared by thinly distributing ground glasses (-325 mesh) onto cellophane tape. The beam line used a Si (111) double crystal monochromator and standard ionization and fluorescence detectors. The incident and transmission/fluorescence X-ray intensities were recorded as a function of X-ray energy, E, allowing the absorption coefficient, $\mu(E)$, to be determined. X-ray energies were calibrated using appropriate standards. Between three and fifteen scans were taken for each sample, depending on the strength of the signal. Monochromator step increments varied depending on the X-ray energy and type of experiment being performed (see below).

Raman Measurements

Laser Raman spectra of glass chips were obtained at room temperature using a backscattering geometry with a Spex Triplemate Raman spectrometer equipped with a microprobe attachment. The 514.5 nm line of a Spectra Physics Ar^+ laser was used for sample excitation. The spectrometer was calibrated with respect to the characteristic Raman bands in a polycrystalline TiO_2 standard. Data were collected using a CCD detector with an exposure time of 100 seconds and the spectrometer exit slit set at 500 µm. The spectral ranges for the data collected in this study spanned the region between approximately 100 and 1500 cm^{-1} relative to the exciting line. Raw spectra were analyzed and corrected to remove background fluorescence using the Grams 386 spectral analysis program.

Magnetization Measurements

Room temperature DC magnetization responses from the ground glasses (-325 mesh) were measured using a vibrating sample magnetometer. The applied magnetic field strength was monitored using a Hall-effect probe and calibrated with an NMR gauss meter. The induced magnetization was calibrated using a known mass of pure Ni.

RESULTS AND DISCUSSION

The solubility limits of Cr_2O_3, Cl, and SO_3 in the L6-5412 base glass melted at 1350°C were determined to be approximately 0.8, 0.6, and 0.8 wt% respectively.[4,5] The L6-5412 glasses spiked with 0.5 wt% Cr_2O_3 (L6 Cr 0.5), 0.4 wt% Cl (L6 Cl 0.4), and 0.5 wt% SO_3 (L6 S 0.5) that we studied in this work were thus considered to be below their respective solubility limits. The nominal compositions of these glasses are summarized in Table 3 along with that of the L6-5412 base glass. Glasses spiked with higher concentrations of these minor components exhibited precipitation, amorphous phase separation, and, in some cases, macroscopic segregation of secondary phases. This will be discussed in more detail as the results obtained from each minor component are reviewed individually.

Chromium

Increased additions of Cr_2O_3 to the L6-5412 system produced an increase in the melt viscosity[5] as well as a considerable decrease in the chemical durability of the final glass[4] compared to that of the base glass. These trends are strongly dependent on base glass composition and in some cases are opposite in direction from that of the L6-5412 glasses. Clearly, an understanding of the role that chromium plays in affecting these important

Table 3. Nominal oxide-equivalent composition (wt%) of the L6-5412 base glass and the L6-5412 glasses spiked with 0.5 wt% Cr_2O_3 (L6 Cr 0.5), 0.4 wt% Cl (L6 Cl 0.4), and 0.5 wt% SO_3 (L6 S 0.5).

Oxide	Glass			
	L6-5412	L6 Cr 0.5	L6 S 0.5	L6 Cl 0.4
SiO_2	56.78	56.53	56.68	56.65
Na_2O	20.00	19.90	19.96	19.92
B_2O_3	5.00	4.98	4.99	4.98
CaO	4.00	3.98	3.99	3.98
Al_2O_3	12.00	11.94	11.98	11.95
Minor components*	2.22	2.67	2.40	2.52
Total	100.00	100.00	100.00	100.00

*Minor components				
Bi_2O_3	0.014	0.014	0.014	0.014
Cl	0.092	0.092	0.092	**0.400**
Cr_2O_3	0.036	**0.500**	0.036	0.036
F	0.213	0.212	0.213	0.212
Fe_2O_3	0.005	0.005	0.005	0.005
K_2O	0.327	0.325	0.326	0.326
MnO	0.007	0.007	0.007	0.007
Nd_2O_3	0.012	0.012	0.012	0.012
P_2O_5	1.187	1.181	1.185	1.182
SO_3	0.320	0.318	**0.500**	0.319
ZrO_2	0.005	0.005	0.005	0.005
Total	2.217	2.671	2.395	2.518

melt and glass properties is critical in designing new glass systems suitable for handling the problems that will be encountered during Hanford LLW vitrification.

As will be demonstrated below, the principle chromium-bearing species of concern in LLW processing and vitrification are the $Cr^{(III)}$-O and $Cr^{(VI)}$-O complexes. It has been suggested that an increase in the ratio of $[Cr^{(VI)}]$ to $[Cr^{(III)}] + [Cr^{(VI)}]$ in glasses causes a decrease in their chemical durability. The determination of the relative amounts of Cr(III) and Cr(VI) in LLW glasses is thus of obvious importance. The Cr K-edge energy (i.e., the 1s-to-continuum electronic transition) is very sensitive to oxidation state[7] and should enable one to distinguish between $Cr^{(III)}$ and $Cr^{(VI)}$ species occurring in LLW-based glasses. Model compounds for these two oxidation states of chromium include $Cr^{(III)}_2O_3$ and $K_2Cr^{(VI)}O_4$.

With only a few exceptions, $Cr^{(III)}$ complexes are always hexa-coordinate (as in Cr_2O_3) whereas $Cr^{(VI)}$ complexes are four-coordinate tetrahedrons (as in CrO_4^{2-}).[8] Monomeric hexa-coordinate $Cr^{(III)}$-O complexes should occur as regular octahedrons, owing to the fact that there are no Jahn-Teller effects for d^3 transition metal ions nor any significant oxo bond (Cr=O) formation.[8] Aqueous $[Cr(H_2O)_6]^{3+}$ complexes, for example, occur as regular octahedrons. Such symmetric octahedral coordination in $Cr^{(III)}$ complexes prevents d-p orbital mixing, allowing the Cr 3d orbitals to remain nearly pure in character.

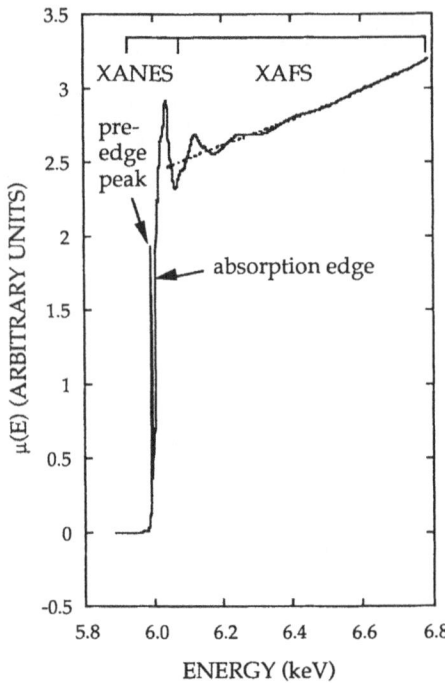

Figure 1. Plot of X-ray absorption coefficient, $\mu(E)$, vs energy, E, for K_2CrO_4. The XAFS and XANES regions as well as the absorption edge (with a corresponding edge energy, E_0), the pre-edge feature, and the smoothly varying background, $\mu_0(E)$ (dashed line) are also indicated.

Polychromium(III) species containing single and double bridging Cr-O-Cr linkages are also known to exist. These complexes possess irregular $Cr^{(III)}$-O octahedrons due to Cr-Cr repulsions and are almost always weakly antiferromagnetically coupled.[8] This type of bonding is observed in Cr_2O_3 which contains irregular octahedrons having several $Cr^{(III)}$-O bond distances and exhibits antiferromagnetic behavior at room temperature.

The tetrahedral coordination obtained by $Cr^{(VI)}$ complexes is realized through strong d-p orbital mixing and results in a significant XANES pre-edge peak due to 1s-to-3d electronic transitions. These transitions are symmetry-forbidden for $Cr^{(III)}$ octahedral complexes. For the chromate anion, CrO_4^{2-}, there exists significant Cr=O character. Chromate anions are thus more acidic than homologous $Mo^{(VI)}$ species, for example, and generally do not form polyanionic species as does the latter.[8] The Cr-O bond distance and number of nearest neighbor information that is obtainable through XAFS measurements should allow us to distinguish between octahedral, i.e., $Cr^{(III)}$, and tetrahedral, i.e., $Cr^{(VI)}$, environments in Hanford LLW-based glasses.

Figure 1 shows a plot of the X-ray absorption coefficient, $\mu(E)$, vs energy, E, for the K_2CrO_4 standard as well as the energy ranges that describe the XANES and XAFS regions. Also indicated in the figure are the 1s-to-continuum absorption edge and the 1s-to-3d pre-edge peak discussed previously. In the following sections we separately discuss the Cr K-edge XANES and XAFS results obtained on selected LLW-based glasses. The nominal concentrations of Cr_2O_3 in the base glasses (0.04 wt%) are too low to have an appreciable effect on glass properties or to obtain reasonable Cr K-edge XANES and XAFS data. In this work, we have studied L-series glasses spiked with 0.5-2.0 wt% Cr_2O_3. As stated above, the L6 Cr 0.5 glass appears to be below the chromium solubility limit for this glass system. Crystals of Cr_2O_3 are clearly visible in the L6 Cr 2.0 glass when observed under an optical microscope. Although phase separation in the L6 Cr 0.8 glass is still being investigated, preliminary transmission electron microscopy results indicate that extremely fine (<20 nm) spherical particles are present. The melt viscosities and chemical durabilities of these same glasses have been previously summarized.[4,5]

Cr K-Edge XANES. Figure 2 shows the Cr K-edge XANES data obtained from the L6-5412 base glass spiked with various amounts of Cr_2O_3 as well as those of the K_2CrO_4 and Cr_2O_3 standards. Data were collected every 0.5 eV in the vicinity around the K-edge. The main edge absorption energies were determined both from the $\mu(E)$ data presented in Figure 2 and, more accurately, from the appropriate zero point of the first derivative of $\mu(E)$ with respect to E. The main K-edge energy for $Cr^{(VI)}$ species (i.e., $CrO_4{}^{2-}$) occurred at 6010.0 ± 0.5 eV compared to 6005.8 ± 0.5 eV for $Cr^{(III)}$ species (i.e., Cr_2O_3). The spectra (and edge energies) of these standard compounds agree with those reviewed by Brown et al.[7] The main Cr edge energies for the L6 Cr 0.5, L6 Cr 0.8, and L6 Cr 2.0 glasses all occurred at 6006 eV, indicating that the majority of the chromium in these glasses is in the +3 oxidation state.

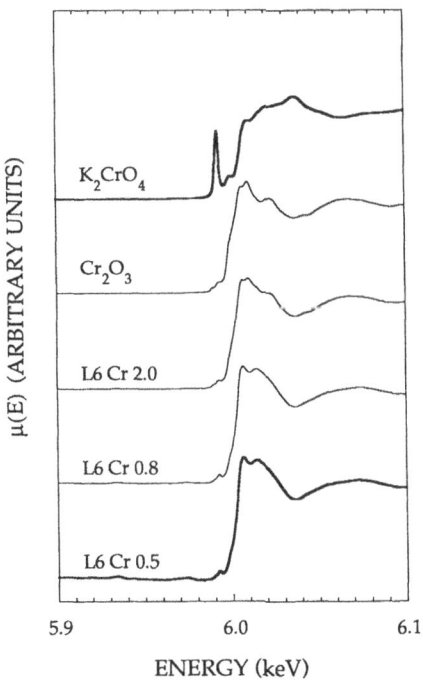

Figure 2. Cr-K-edge XANES data obtained from the K_2CrO_4 and Cr_2O_3 standards and of the L6-5412 base glass spiked with various amounts of Cr_2O_3.

These glasses and the Cr_2O_3 standard all lack a significant pre-edge peak, typical of that which is observed in K_2CrO_4. We thus conclude from these results that chromium ions in the L6 Cr 0.5, L6 Cr 0.8, and L6 Cr 2.0 glasses are primarily (>95%) octahedrally coordinated species in the +3 oxidation state. A more precise estimate of the relative amounts of $Cr^{(III)}$ and $Cr^{(VI)}$ species in these glasses is not possible due to the low intensity of the pre-edge features and their relatively equivalent intensity to that observed in pure Cr_2O_3.

Raman Spectroscopy. In addition to vibrational bands characteristic of network forming and non-bridging oxygen species, the Raman spectrum obtained from the L6-5412 based glass spiked with 0.8 wt% Cr_2O_3 (L6 Cr 0.8) exhibited a very sharp peak at 850 cm^{-1}. The peak is also present in the L6 Cr 0.5 glass, occurring at the same frequency but with a weaker intensity. We have attributed this peak to the presence of isolated $CrO_4{}^{2-}$ in agreement with literature assignments.[9,10] These results, although only qualitative at this point, confirm that a small portion of the chromium in these glasses does exist as $Cr^{(VI)}$. No peak at 850 cm^{-1} was observed in the L6-5412 base glass, most likely due to its very low overall concentration of chromium.

Cr K-Edge XAFS. Standard XAFS data analyses[7,11] were applied to the Cr K-edge data collected from the Cr-spiked LLW-based glasses as well as the Cr_2O_3 and K_2CrO_4 standards. For the work presented here, the oscillations above the absorption edge (see Figure 1), $\chi(E)$, are mainly due to the scattering of the ejected photoelectron back to the chromium central cation off its oxygen neighbors. The $\chi(E)$ data were extracted from $\mu(E)$ by the usual method of normalization and subtraction of the smoothly varying

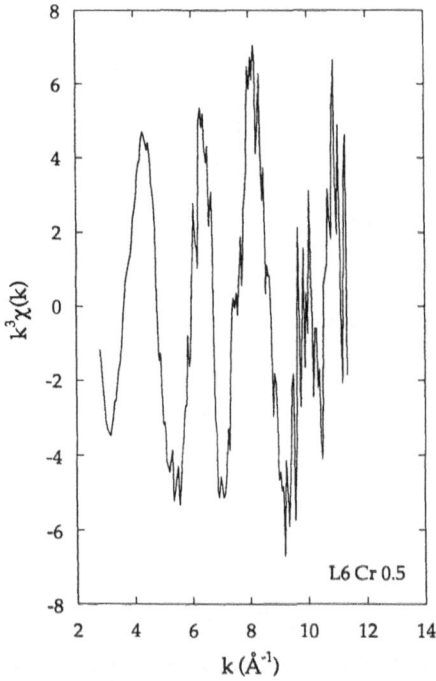

Figure 3. Weighted $k^3\chi(k)$ data obtained from the Cr K-edge L6 Cr 0.5 XAFS.

background, $\mu_0(E)$ (see Figure 1). The data were then converted to $\chi(k)$, where k is the photoelectron wave vector defined as

$$k = [2m (E - E_0) h^{-2}]^{1/2}, \tag{1}$$

m is the electron mass, E_0 is the edge energy, and h is Planck's constant.

Figure 3 shows a single set of k^3-weighted $\chi(k)$ data for the L6 Cr 0.5 glass. Noise was reduced by signal averaging over repeated scans. Weighting of the $\chi(k)$ data by k^n (n=1-3) is required to compensate for the dampening of the XAFS oscillations with increasing k (see Figure 1). For oxygen backscatterers, weighting by k^3 is appropriate. The $k^3\chi(k)$ data above $k\approx 11.5$ Å^{-1} have been discarded due to the occurrence of the Mn K-edge just above this value.

The $k^3\chi(k)$ data obtained from all the analyzed samples and standards were then Fourier transformed to real space over a k-range of 3.6-11.3 Å^{-1}, yielding the radial structure plots (RSPs) indicated in Figure 4. The major feature in each RSP occurs in the range of approximately 1-2 Å and is due to the first nearest neighbor (NN) shell of oxygen anions surrounding the central chromium cation. The radial distances indicated in the figure have not been corrected for phase shifts that occur during scattering of the photoelectron. Although the absolute radial distances associated with the oxygen NN peaks can not be determined from Figure 4, the positions of the

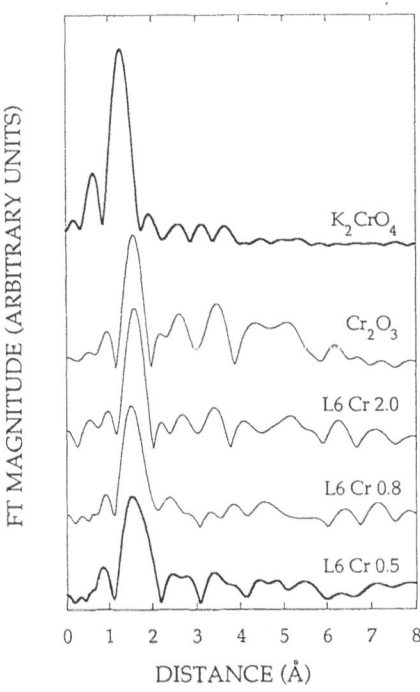

Figure 4. Cr Radial Structure Plots (RSPs) produced by Fourier transforming the $k^3\chi(k)$ data obtained from the K_2CrO_4 and Cr_2O_3 standards and the L6-5412 base glass spiked with various amounts of Cr_2O_3. The radial distances indicated have not been corrected for photoelectron phase shifts.

peaks relative to the standards indicates that the Cr-O bond distances in the LLW-based glasses are most like those found in Cr_2O_3 as opposed to those found in CrO_4^{2-}.

Back Fourier transformation of the RSPs in the radial range that brackets the oxygen first NN shell allows the $k^3\chi(k)$ oscillations associated with chromium-first NN oxygen interactions to be filtered from the other oscillations (and noise) that comprised the original $k^3\chi(k)$ data (e.g., Figure 3). The back Fourier transforms of the Cr-O RSP peaks for the K_2CrO_4 standard and the L6 Cr 0.5 glass are presented in Figure 5.

The Fourier filtered $\chi(k)$ data can be expressed by the general XAFS formula:[7]

$$\chi(k) = [-Nk^{-1}R^{-2} f(k) \exp(-2k^2\sigma^2) \exp(-2R\lambda(k)^{-1})] [\sin(2kR + \phi(k))], \qquad (2)$$

where N is the number of backscatterers (i.e., number of nearest neighbors, NN) around the central atom, R is the average distance between the central atom and the NN shell, k is the photoelectron wave vector, $f(k)$ is the backscattering amplitude, $\phi(k)$ is the total phase shift experienced by the photoelectron during scattering, σ^2 is XAFS Debye-Waller factor which is the sum of both static and vibrational (thermal) disorder effects, and $\lambda(k)$ is the mean free path of the scattered photoelectron (typically <6 Å).

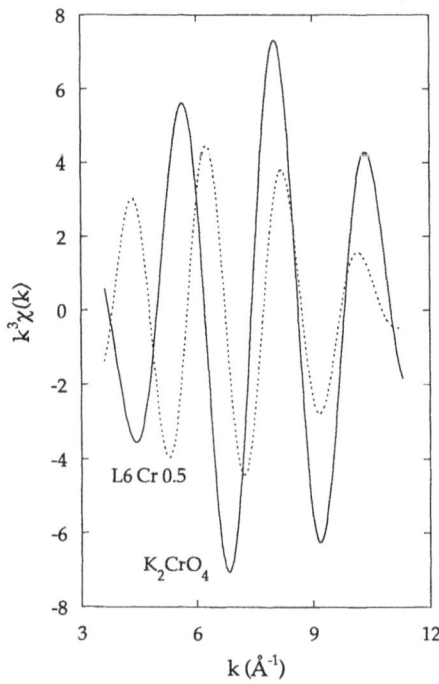

Figure 5. Weighted back Fourier transforms, $k^3\chi(k)$, of the Cr-O RSP peaks for the K_2CrO_4 standard (solid line) and the L6 Cr 0.5 glass (dashed line).

Equation (2) can be summarized as being the product of k-dependent amplitude and phase functions. The differences in the amplitude and phase contributions between two different sets of $\chi(k)$ data, which are clearly discernible in Figure 5, are indicative of differences in the values N, R, and σ^2 between the materials from which the data were obtained. For a reference material with known values of N, R, and σ^2, equation (2) can be used to fit the Fourier filtered first NN $\chi(k)$ data obtained from that material. This allows the determination of $f(k)$ and $\phi(k)$ for a given central atom-backscatterer interaction. For a material with an unknown first NN structural environment that has the same central atom-backscatterer interaction (e.g., both the reference material and the unknown material consist of Cr-O complexes), equation (2) and the reference's $f(k)$ and $\phi(k)$ functions can be used to fit the Fourier filtered first NN $\chi(k)$ data obtained from the material and to make a determination of the values of N, R, and σ^2.

Using this technique and K_2CrO_4 as a reference material (N=4.00, R=1.70 Å, σ^2=0.0100), the Fourier filtered chromium-first NN oxygen $k^3\chi(k)$ data obtained from the L6 Cr 0.5 glass were fit assuming a single coordination environment, which we have labelled O_{mean}, using a non-linear least squares routine. The results of this fitting, which are summarized in Table 4, indicate that the majority of the Cr-O species in the L6 Cr 0.5 glass are hexa-coordinate, presumably as octahedrons. The Cr-O bond distance of 2.06±0.02 Å is close to that observed for Cr_2O_3 (1.99-2.04 Å). These results support those discussed above in that the chromium in the L6 Cr 0.5 glass occurs predominantly as octahedrally coordinated $Cr^{(III)}$-O complexes.

The goodness of fit parameter, χ^2, for this result is relatively high, indicating a less than optimal fit to the data using only a single coordination environment. A much better fit to the data is obtained when two environments ($O_I + O_{II}$) are assumed, as indicated in Table 4 and in Figure 6. Allowing for an even greater number of coordination environments did not produce a significantly better fit to the L6 Cr 0.5 data.

Table 4. Summary of the number of oxygen neighbors, N, their distance, R, from the chromium central cation, and their mean square displacements, σ^2, for the first NN Cr-O shell in the L6 Cr 0.5 glass determined by fitting the $k^3\chi(k)$ data using equation (2) and allowing for either one or two coordination environments. Also indicated are the values of the goodness of fit parameter, χ^2.

Number of environments	Environment label	N	R (Å)	σ^2 (Å^2)	χ^2
1	O_{mean}	6.4±0.6	2.06±0.02	0.0158	0.227
2	O_I	4.8±0.5	2.11±0.02	0.0139	
	O_{II}	1.5±0.2	1.99±0.02	0.0068	
	sum/mean	6.3±0.6	2.07±0.02		0.030

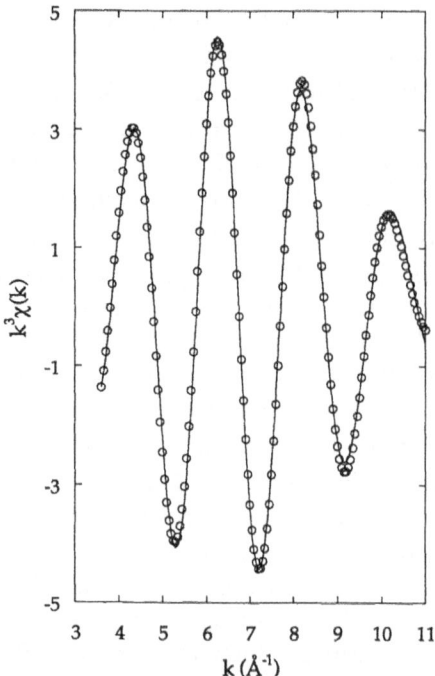

Figure 6. Weighted back Fourier transform, $k^3\chi(k)$, of the Cr-O RSP peak for the L6 Cr 0.5 glass (open circles) and the fit to the data (solid line) using equation (2) and K_2CrO_4 as a reference.

LLW-based glasses containing concentrations of chromium greater than that in the L6 Cr 0.5 glass exhibited phase separation (e.g., Cr_2O_3 crystals were detected). The oxygen NN $k^3\chi(k)$ data would thus be the result of contributions from $Cr^{(III)}$-O complexes contained in both the glassy and crystalline phases.. XAFS data fitting for these glasses is beyond the scope of this article.

Magnetization Measurements. The XANES and XAFS results described above collectively indicate that the chromium in the L6 Cr 0.5 glass occurs primarily as hexa-coordinate $Cr^{(III)}$-O octahedrons. By itself, however, the fact that a model involving two coordination environments (i.e., a distorted octahedron) fits the L6 Cr 0.5 first NN XAFS data better than one involving a single coordination environment (i.e., a regular octahedron) does not completely confirm or deny the notion that the chromium in that glass actually does occur as distorted octahedrons. For example, these subtle differences in the fits to the data may be due to artifacts as a result of the various transformations that the XAFS data undergo.

Magnetization measurements performed on the L6 Cr 0.5 glass indicated that at room temperature, the glass exhibited a low-field susceptibility of 9.46×10^{-5} cgs which saturated at an induced magnetization of 0.28 gauss. For the L6 Cr 0.5 glass, chromium represents $\approx 99\%$ of all the possible magnetic species present. Thus, we can be assured that the observed magnetic behavior in the L6 Cr 0.5 glass is due to the presence of the chromium ions. As we have demonstrated in the above sections, these chromium ions are predominantly in the magnetically active +3 oxidation state.

The susceptibility observed in the L6 Cr 0.5 glass is about two orders of magnitude larger than one would expect based on the assumption that all $Cr^{(III)}$-O complexes were paramagnetic and completely isolated from each other. The extremely large magnetic susceptibility observed for the $Cr^{(III)}$-O complexes in the L6 Cr 0.5 glass indicates that these complexes must be clustering, the resulting super-exchange inducing the magnetic moments of the individual complexes to act collectively.[12] As discussed previously, the Cr-O-Cr bond formation that occurs during clustering causes the formation of distorted $Cr^{(III)}$-O octahedrons. Thus, these magnetization results support the XAFS model involving two coordination environments, i.e., distorted $Cr^{(III)}$-O octahedrons as a result of clustering, used to describe the chromium environment in the L6 Cr 0.5 glass.

Sulfur and Chlorine

Increased additions of SO_3 to the L6-5412 system also produced a decrease in the chemical durability of the final glass[4] compared to that of the base glass. Relatively small increases in the concentration of Cl in L6-5412 batches dramatically increases the melt viscosity.[5] For melts over-saturated with SO_3 or Cl, molten Na_2SO_4 or NaCl salt segregation and subsequent accumulation

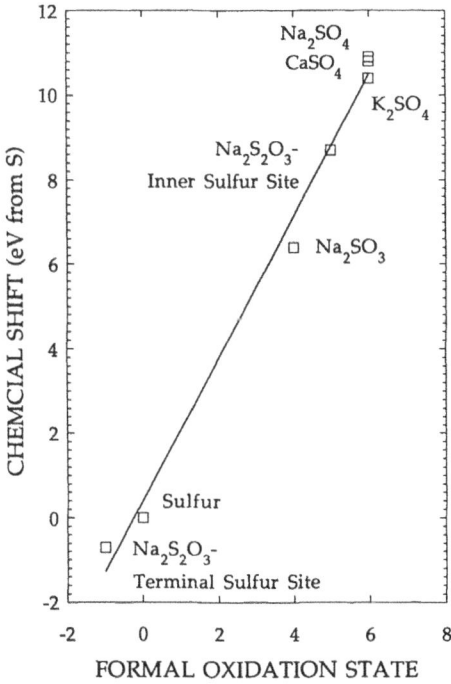

Figure 7. Sulfur K-edge chemical shifts (defined as the edge energy of a compound minus that of elemental sulfur) vs formal oxidation state for various inorganic sulfur-oxygen standard compounds.

at the melt surface occurred.[5] Other troublesome components, such as certain radionuclides, may become entrained during molten salt segregation to melt surfaces, potentially causing an increase in their volatilization rates.

Again, an understanding of how these minor components affect the melt characteristics, chemical durability of the final glasses, and other properties is important. One of the best methods of studying the chemical environment of sulfur species is XANES. The use of XANES in studying the environment local to Cl in LLW glasses also shows promise, although availability of suitable standards appropriate for Cl-containing glasses is limited.

Sulfur K-edge XANES measurements were performed on a series of inorganic sulfur-oxygen standard compounds and various simulated Hanford waste glasses. Data were collected every 0.2 eV in the vicinity around the K-edge. The edge energies were determined from the appropriate zero point of the first derivative of $\mu(E)$ with respect to E. For the standard compounds, Figure 7 shows a plot of their S K-edge chemical shift, which is defined as the edge energy of the compound minus that of elemental sulfur, vs their formal oxidation state. Note the sensitivity of the chemical shift to oxidation state, which is typical of sulfur-oxygen species.[13]

The S chemical shifts obtained from the LD6-5412 and the L6-5412 base glass spiked with various amounts of SO3 (e.g., L6 S 0.5) as well as a host of other SO3-containing LLW and HLW glass compositions produced under a

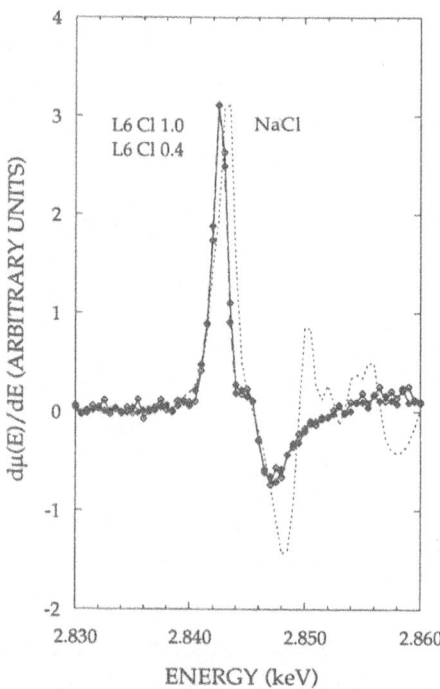

Figure 8. Normalized first derivative XANES plots of the Cl edges for the NaCl standard (dashed line) and the L6-5412 base glass spiked with 0.4 and 1.0 wt% Cl (data points).

variety of melt conditions (oxidizing, reducing, etc.) were all identical at 10.7±0.2 eV. This indicates that regardless of the glass composition or preparation method the sulfur in these glasses occurs in the +6 oxidation state, most likely as isolated tetrahedral oxoanions[14], SO_4^{2-}, associated with the alkali or alkaline earth cations in the glass.[4]

Raman spectra obtained from L6-5412 and LD6-5412 base glass compositions as well as L6-5412 glass spiked with various amounts of SO_3 all exhibited a strong vibrational band at 981-990 cm^{-1}. This band corresponds to that of isolated SO_4^{2-} groups, such as those found in Na_2SO_4, in agreement with literature assignments.[15]

The chemical environment of Cl in LLW glasses and other glass compositions is most uncertain. It has been suggested that Cl can substitute for bridging oxygen in glass[16], forming, for example, Si-Cl-Si linkages. Substitution of chlorine for non-bridging oxygen is another possibility. Isolated alkali or alkaline earth chloride species may also exist.

Figure 8 shows normalized first derivative XANES plots of the Cl edges for the L6-5412 base glass nominally spiked with 0.4 and 1.0 wt% Cl as well as for NaCl. XRF analysis on the L6 Cl 1.0 sample indicated that it contained only 0.6 wt% Cl (i.e., the final glass contained only its solubility limit of Cl, the excess formed surface-segregated NaCl as previously discussed). The normalized XANES spectra of the two glasses are identical, suggesting that the environment of the Cl in these glasses is the same.

Huggins and Huffman[17] have performed Cl K-edge XANES measurements on standard compounds relevant for the speciation of chlorine in coals. In their work, chemical shifts were determined by the position of the major maximum in the first-derivative XANES spectra of the compounds with respect to that of NaCl. They determined that the chemical shifts of all the alkali and alkaline earth chlorides relevant to the glass compositions studied here (i.e., NaCl, KCl, and $CaCl_2$) occurred at +0.4 to +2.2 eV, while those of $Ca(OCl)_2$ and $KClO_3$ occurred at -2.6 eV and +5.8 eV respectively. Based on an identical method of edge energy determination, the L6 Cl 0.5 and 1.0 glasses exhibited chemical shifts of -1.0±0.2 eV from that of the NaCl standard (see Figure 8), indicating that the majority environment of the chlorine in these glasses is not representative of those of alkali and alkaline earth chlorides and oxychlorides.

In contrast, a variety of solid organic compounds containing covalent C-Cl bonds exhibited chemical shifts in the range -1.9 to -0.5 eV.[17] One might expect that compounds possessing analogous Si-Cl bonds would have chemical shifts occurring in the same direction (i.e., < 0.0) as those of former, but perhaps of slightly different magnitude. Based on these results, we believe that the majority of soluble chlorine in the L6-5412-based glasses occurs as non-bridging Si-Cl groups.

SUMMARY

We have investigated the chemical and structural environments of chromium, sulfur and chlorine in simulated Hanford Site low-level radioactive waste (LLW) glasses using X-ray Absorption Near Edge Structure

(XANES) and X-ray Absorption Fine Structure (XAFS) spectroscopies. Complementary Raman spectroscopy and magnetization measurements were performed on some of these glasses as well.

XANES results indicated that octahedral $Cr^{(III)}$-O complexes were the predominant chromium species that occurred in all the Cr-spiked L6-5412 glasses investigated. Raman studies on these same glasses confirmed the presence of a small concentration of isolated tetrahedral $Cr^{(VI)}O_4{}^{2-}$ species.

Fitting of the $Cr^{(III)}$-O XAFS data obtained from the L6-5412 glass spiked with Cr_2O_3 below the solubility limit of the glass indicated that the $Cr^{(III)}$-O octahedrons were basically hexa-coordinate ($\approx$6 O at 2.06 Å). However, the best fit to the XAFS data was obtained when two coordination environments ($\approx$4 O at 2.11 Å and $\approx$2 O at 1.99 Å) were allowed. This distorted environment is indicative of $Cr^{(III)}$-O octahedron clustering, and was confirmed by the high magnetic susceptibility observed for this glass.

Sulfur XANES chemical shift values of a variety of LLW glasses indicated that they ubiquitously contained $S^{(VI)}O_4{}^{2-}$ species. Raman studies on these same glasses confirmed the presence of isolated $SO_4{}^{2-}$ groups, which we believe are associated with alkali (i.e., Na^+) cations.

The chemical environment of Cl in LLW glasses is most uncertain, although the XANES results obtained in this work suggest that a non-bridging Si-Cl or related species might be the predominant form.

ACKNOWLEDGMENT

This work was supported by the U.S. Department of Energy under contract DE-AC06-76RLO 1830. The authors would like to thank the experimental assistance of our coworkers at the Pacific Northwest Laboratory: J. Coleman, J. Liu, D. McCready, R. Sanders, D. Smith, and M. Schweiger. H. Li is grateful to Associated Western Universities, Inc. for his postdoctoral appointment at the Pacific Northwest Laboratory. We also acknowledge the use of beam line X19A at the National Synchrotron Light Source (operated by the U.S. Department of Energy) and the experimental assistance of F. Lu and Y. Ma.

REFERENCES

1. Pacific Northwest Laboratory is operated for the U.S. Department of Energy by the Battelle Memorial Institute under contract DE-AC06-76RLO 1830.
2. M.S. Gerber, "The Hanford Site: An Anthology of Early Histories," Westinghouse Hanford Company document WHC-MR-0435 prepared for the U.S. Department of Energy Office of Environmental Restoration and Waste Management, Richland, WA (1993).
3. Westinghouse Hanford Company Report WHC-SD-WM-RD-044, p. A-3.
4. H. Li, J.G. Darab, P.A. Smith, X. Feng, and D.K. Peeler, Chemical durability of low-level simulated nuclear waste glasses with high-concentrations of minor components, in "INMM 36th Annual Proceedings", Institute for Nuclear Materials Management, Northbrook, IL (in press, 1995).
5. H. Li, J.G. Darab, P.A. Smith, M.J. Schweiger, D.E. Smith, and P.R. Hrma, Effect of minor components on vitrification of low-level simulated nuclear waste glasses, ibid ref. 4.
6. P.A. Smith, J.D. Vienna, and P. Hrma, The effects of batch reactions on laboratory scale waste vitrification," *J. Mater. Res.* 8, 2137 (1995).

7. G.E. Brown, Jr., G. Calas, G.A. Waychunas, and J. Petiau, X-ray absorption spectroscopy and its applications in mineralogy and geochemistry, in "Spectroscopic Methods in Mineralogy and Geology," Reviews in Mineralogy, Vol. 18, F.C. Hawthorne, ed., Mineralogical Society of America, Washington, D.C., pp. 431-512 (1988).

8. F.A. Cotton and G. Wilkinson, "Advanced Inorganic Chemistry," 4th ed., John Wiley & Sons, New York, pp. 719-736 (1980).

9. S.A. Brawer and W.B. White, Raman spectroscopic study of hexavalent chromium in some silicate and borate glasses, *Mat. Res. Bull.* **12**, 281 (1977).

10. W.P. Griffith, Advances in the Raman and infrared spectroscopy of minerals, in "Spectroscopy of Inorganic-Based Materials," R.J.H. Clark and R.E. Hester, eds., John Wiley & Sons, New York, pp. 119-186 (1987).

11. P.A. Lee, P.H. Citrin, P. Eisenberger and B.M. Kincaid, Extended x-ray absorption fine structure - its strengths and limitations as a structural tool, *Rev. Mod. Phys.* **53**, 759 (1981).

12. L. Néel, *Compt. rend.* **252**, 4075 (1961); **253**, 9 (1961); **253**, 203 (1961).

13. A. Vairavamurthy, B. Manowitz, W. Zhou, and Y. Jeon, Determination of hydrogen sulfide oxidation products by sulfur K-edge X-ray absorption near-edge structure spectroscopy, in "Environmental Geochemistry of Sulfide Oxidation," ACS Symposium Series, Vol. 550, C.N. Alpers and D.W. Blowes, eds., American Chemical Society, Washington, DC, pp. 412-430 (1994).

14. F.A. Cotton and G. Wilkinson, ibid ref. 7, pp. 502-541.

15. K. Nakamoto, "Infrared Spectra of Inorganic and Coordination Compounds," 2nd ed., John Wiley & Sons, New York (1970).

16. K.H. Sun and A. Silverman, Lewis acid-base theory applied to glass, *J. Am. Ceram. Soc.* **28**, 8 (1945).

17 F.E. Huggins and G.P. Huffman, Chlorine in coal: an XAFS spectroscopic investigation, *Fuel* **74**, 556 (1995).

INDEX

Goethite, 176
G. Seaborg Institute for Transactinium Science, 169

Hanford tanks, 238
Helicity, 188
Helium, 129
HFIR-ORNL, 48
Humic acid, 215
Hydrated sodium aluminate, 83

IBM, 107
Imaging
  phosphors, 22
  plates, 22
Industrial protein crystallography, 8
Initiator factor 3, 29
Interface, 107
Iron-carbonyl, 25

Kesterson reservoir, 174
Kilo-Dalton, 1

Lanthanides, 229
Lawrence Berkeley National Laboratory, 119
Liquid crystal polymers, 57
Low level waste glass, 237

MacEXAFS, 160
Macrocyclic Polyether, 149
Macromolecular crystallography, 21
Magnetic materials, 107
Magnetometry, 187
Methylamine, 119
MEXAFS, 195
MgO supported molybdenum catalyst, 65
Molecular beam scattering, 119
Monofilament drawing, 60
Monolayer detection in SXF, 112
MOS devices, 110
Multilayers,
  copper-cobalt, 113
  heteromagnetic, 187
  magnetic, 112
Multi-wavelength Anomalous -Dispersion Phasing (MAD), 6, 24
Myoglobin, 26

Na $AlO_2 \cdot 5/4H_2O$, 83
National Synchrotron Light Source (NSLS), 2, 21
  U4b, 187
  U7a, 58
  X3b1, 84
  X14a, 57–58
  X19a, 160
  X23a2, 58
  X23b, 150
Naval Research Laboratory (NRL), 149, 187
Nd, 231
Near-threshold ionization, 129

Neutron diffraction, 48
NEXAFS, 56
NMR, 149, 230, 240
Northwestern University, 57
Np, 180

Occupied states, 212
Organic aluminum fluorides, 42
Orthophosphates, 230
Oxynitrides, 110
Ozone, 119

Pacific Northwest National Laboratory, 237
Participator electron, 211
Parylene, 190
Perclene, 39
Phosphates, 229
Photoabsorption, 187
Photodissociation, 119
Photo-double-ionization, 129
Photoelectron spectroscopy, 107
Photoionization spectroscopy, 119
Photo-oxidation, 157
Photosynthetic oxygen-evolving manganese-oxygen cluster, 141
Photosystem II, 141
Phytates, 229
Phytic acid, 229
Poly(cis-benzoxazole) PBO, 60
Polymers, 207; see also Deformation of polymers, Liquid crystal polymers
  surfaces, 70
Polystyrene, 59, 207
Poly(trans-benzothiazole) PBZT, 60
Polyurethane, 59
Powder diffraction, 37, 83, 90
Protein crystallography, 1, 21
Protein data bank–Brookhaven National Laboratory, 2
Protein structure table, 5
Pu, 237
  metallurgy, 181
  in nitric acid, 180

Quantum well states, 113

Radionuclides, 176
Rare earth transition metal, 187
Receptors, 2
Resonant photoemission, 207
Resonant Raman spectroscopy, 137
Resonant valence band map, 210
Ribosomal helper protein, 29
Rietveld analysis, 83
ROMO II-HASY Lab, 218
Rydberg series, 133

Sandia National Laboratory, 57
Satellite, 214
Savanna River Ecology Laboratory, 149
Se bioremediation, 181

The manufacturer's authorised representative in the EU is Springer
Nature Customer Service Centre GmbH, Europaplatz 3, 69115 Heidelberg,
Germany. If you have any concerns regarding our products, please
contact ProductSafety@springernature.com

Printed and bound by CPI Group (UK) Ltd, Croydon, CR0 4YY
24/04/2026
02096348-0019